Sour-Gas Processing and Sulfur Recovery

Sour-Gas Processing and Sulfur Recovery

PPC Books Division
The Petroleum Publishing Company
Tulsa, Oklahoma

P. O. Box 1260 (1421 South Sheridan Road)
Tulsa, Oklahoma 74101

Library of Congress Catalog Card Number: 78-71327
International Standard Book Number: 0-87814-101-4
Printed in U.S.A.

1 2 3 4 5 83 82 81 80 79

TABLE OF CONTENTS

Part 1 Sour-Gas Treating

Part 2 Sulfur Recovery

Part 3 Tail-Gas Cleanup

Part 1
Sour-Gas Treating

Guidelines set for handling sour gas

B. Gene Goar
Thomas O. Arrington
Goar, Arrington & Associates Inc.
Tyler, Tex.

THE ever-increasing demand for more energy in the United States is generating stronger interest in processing and conditioning of sour natural gas for pipeline sales. Use of fossil fuels in the United States doubles about every 20 years. Tremendous quantities of sour natural gas will be produced in the future.

The large reserves of sour gas in the southwest are located in the Permian basin of West Texas and the Smackover trend running through East and Northeast Texas, Louisiana, Mississippi, Alabama, and western Florida. Huge amounts of sour gas will be produced and processed from the Smackover trend in the future.

Future sour-natural-gas processing facilities will necessarily become more complex and more expensive due to existing and forthcoming federal and state air-pollution regulations and the increased concern for protection of the general public. The processing of sour natural gas can be a complex problem, involving technical, economic, and environmental considerations.

A sour-gas plant must meet the sales-gas specifications, show an attractive payout, provide for a safe operation, and comply with all pertinent air-pollution regulations.

Project evaluation. When a producing company evaluates a "grass-roots" sour-natural-gas processing-plant project, many process considerations and economic factors must be reviewed to decide upon the optimum overall scheme. A detailed evaluation must be performed to establish the best combination of processing steps for gas cleaning, gas treating or sweetening, LP-gas recovery required for dew-point control or liquid sales, gas dehydration, sulfur recovery, and, if required, tail-gas cleanup (TGCU). The major factors affecting the outcome of the evaluation are listed in Table 1.

Table 1

Evaluation factors

SOUR-GAS PLANT

- Daily volume of sour gas to be processed.
- Estimated life of reservoir at daily flow volume.
- Amount of sour components (H_2S and CO_2) to be removed.
- Amount of sulfur to be recovered.
- Presence and amount of inerts and other contaminants such as N_2, O_2, COS, CS_2, and mercaptans.
- LP-gas liquids content of gas (gpm of C_2+).
- Sales-gas specifications to be met.
- Air-pollution regulations which will dictate if sulfur recovery and/or TGCU is required.
- Availability of a close market for plant products.
- Price and total revenue obtained for plant products, i.e., sweet sales gas, LP-gas liquids, and sulfur.
- Investment required to construct and operate plant, and to construct gas pipelines.

Once drilling activity in a sour-gas field nears completion, the producer must decide upon the best approach to processing the sour gas. One question is whether it is more economical to lay a sour-gas line to the nearest gas-processing plant, which handles similar gas and may have excess capacity, and pay a processing cost (¢/Mcf) to have the gas conditioned or to lease equipment (skid-mounted plant) for on-site processing. Another question is whether it is most feasible to build a grass-roots plant to be owned and operated by the producer or should the plant be owned by one company or group of companies who contribute to the overall investment and operating cost, and share in the overall revenue.

The answers to these questions depend upon many of the major factors cited in Table 1.

If the decision is made to build a grass-roots plant, the producer should employ an engineering company to study the project and prepare their recommendations for the optimum gas-processing scheme, environmental impact, plant revenues to be realized, and development of a definitive estimate for construction and operation of the plant. An economic analysis and justification must then be prepared which should show an attractive payout and return on investment (ROI) for the project.[1 2] (References to be listed at the end of Part 2.)

Table 2

Sales-gas specs

SWEET AND DRY SALES GAS

- Maximum hydrogen sulfide content = 0.25 grains/100 scf.
- Maximum mercaptan content = 0.20 grains/100 scf.
- Maximum total sulfur content = 1-5 grains/100 scf.
- Maximum carbon dioxide content = 1-3% by volume.
- Maximum water content = 4-7 lb/MMscf.
- Minimum gross heating value = 950 BTU/scf.
- Maximum hydrocarbon dewpoint at 800 psig = +15° F. (often waived in U.S., but not in Canada).
- Maximum delivery temperature = 120° F.
- Minimum delivery pressure = 700 psig.
- Gas commercially free from sand, dust, gums, and free liquids.
- Gas shall be as free of oxygen as possible, but not to exceed 0.4% by volume (contract may permit no oxygen).

Typical sweet and dry sales-gas specifications set by intrastate and interstate utility-company contracts are listed in Table 2. One of the most-important aspects of the overall project is the collection of accurate and meaningful data on the inlet sour-gas composition, presence of contaminants, and quantity of gas to be treated. Numerous questions should be asked as listed in Table 3.

The importance of obtaining an accurate sour-gas analysis and satisfactory answers to the preceding points cannot be overemphasized. The whole plant design and economics depend upon it.

The presence of even small amounts of impurities such as COS, CS_2, and

Table 3

Needed data

FOR INLET GAS STREAM

- Were wells flowed for awhile, before sampling, to insure that water was blown from well bores and flow lines and a representative sample obtained?
- Was the flow potential measured for only one or two wells, or were several wells tested to get an average value?
- How and when was the gas sampled for analysis? Were the samples taken in sample bombs by water displacement, evacuation, or how?
- Was the acid-gas (H_2S and CO_2) content of the gas determined at the wellhead or in the laboratory? It should be determined at the wellhead.
- Are O_2, COS, CS_2, mercaptans, or other trace impurities present in the gas? If so, to what extent? (These normally require a chromatographic analysis, preferably performed at the wellhead.)
- Will corrosion inhibitors be used for down-hole corrosion control or collection-line protection? If so, what types and how much of each?
- What quantity of free hydrocarbon liquids will be produced with the sour gas (bbl/MMscf)?

mercaptans can have a large bearing on the selection of the type of treating process to be used. Therefore, the gas must be diligently tested for the presence of these and other impurities, possibly O_2, Hg, cyanides, and others.

Gas cleaning. Many of the operational problems associated with a sour-gas plant arise from solids carried into the plant along with the sour gas. Most sour-gas gathering lines are fabricated from carbon-steel pipe. Since the gas usually contains H_2S, CO_2, and water (and sometimes O_2), formation of iron sulfide, iron oxide, and other corrosion by-products may be quite high initially.

Iron sulfide particles formed in gas-collection lines can exist in sizes as small as 0.1 μ (typical particle size of tobacco smoke) to as large as about 80-100 μ.

Provisions should be made for an adequate inlet-gas separator large enough and efficient enough to remove essentially all solids and free liquids. In many plants, most of the iron sulfide found in the amine system enters with the sour gas. Efficient removal of this iron sulfide from the gas can prevent many operational and maintenance problems.

Of course, removal of free liquids is essential to satisfactory operation of the plant. If iron sulfide and/or liquids are known to be a problem, a special highly efficient separator should be used, such as a "filter-sep" type, centrifugal type, or wash-liquid type.

Gas treating

Today, the field of gas treating is so broad and the available processes are so numerous that selection of an optimum process is a complex problem. Gas composition, available gas pressure, quantity and type of sour components to be removed, product gas specifications, and subsequent processing of the acid gas stream are all important factors to be considered.[3 4 5] Several different types and categories of processes are available such as:

- Chemical solvent processes—alkanolamines (MEA, DEA, DGA, MDEA, Adip), and alkaline salt solutions (hot pot, Catacarb, Benfield).
- Physical solvent processes—Sulfinol, Selexol, and Fluor solvent.
- Direct conversion processes—Stretford, Thylox, Takahax, and Ferrox.
- Dry-bed processes—iron sponge and molecular sieves.

Chemical solvent processes

All of the chemical solvent processes involve actual reaction of the sour components with the treating chemicals. They are summarized and discussed in the literature.[3-9] The highlights of the major processes considered to be of commercial importance in the U.S. and Canada will be reviewed briefly.

Alkanolamines. Of the alkanolamines, the monoethanolamine (MEA) process has been by far the most widely used process for gas treating. Advantages of the MEA process are high reactivity, low solvent cost, good chemical stability, ease of reclamation, low hydrocarbon content of the acid gas produced, and lower plant investment when compared with many other processes.

Disadvantages of MEA are irreversible degradation by COS, CS_2, and O_2 in the gas, higher vaporization losses than any other amine-type process, ineffectiveness for removing mercaptans, nonselectivity for H_2S in the presence of CO_2, and higher utility costs than some other processes.

Unless COS, CS_2, or O_2 are present, MEA may be the best choice for treating a sour-gas stream containing low-to-medium acid-gas concentrations, with low-to-medium acid-gas partial pressures.[3] Some very general guidelines might be a sour gas containing from 4 grains H_2S/100 scf to 15 mol % total acid gas, with acid-gas partial pressures up to 100 psia.

The diethanolamine process (DEA) is the second most widely used gas-treating process.[10] DEA's advantages with regard to MEA are its resistance to degradation from COS and CS_2 and lower vaporization losses.

Its disadvantages are lower reactivity, higher solvent circulation rates (conventional system), and higher solvent cost. DEA is often used to treat refinery sour-gas streams because of the presence of COS and CS_2.

In recent years, DEA has gained much wider acceptance in the United States for sour-natural-gas treating. This is due to successful experience gained with "Hi-Loaded" DEA treating in western Canada.

In the Hi-Loaded version of the DEA process, lower circulation rates and much lower utilities are experienced than with MEA.

The diglycolamine (DGA) process[11] competes with the MEA and DEA processes, but has not gained as wide acceptance as DEA for high-pressure sour-gas treating. In some applications, it presents certain advantages over MEA and DEA.

Fluor Corp. recently announced a new improvement in the DGA process, which they call the "Fluor Improved Economine (DGA) Process" or the "Improved DGA Process."

Fluor indicates that extremely high acid-gas loadings, low solution circulation rates, and many process economies are attainable with their new Improved DGA process.[12 13]

The methydiethanolamine (MDEA) process is discussed in some detail in the literature.[14 16] The major advantage which MDEA can offer over other amine processes is its selectivity for H_2S in the presence of CO_2. Where this is a desirable situation, MDEA can be used to good advantage.

If the gas is contacted at a sufficiently high pressure (800-1,000 psig), pipeline-quality gas (¼ grain H_2S/100 scf) can possibly be obtained under certain conditions. At the same time, from 40% to 60% of the CO_2 present may be allowed to flow through the contactor, unreacted.

In cases where a high CO_2/H_2S ratio is present, MDEA can be used to improve the quality of the acid-gas feed stream (higher H_2S content) to the Claus sulfur-recovery plant. But of course, the higher CO_2 content of the treated residue gas must be tolerated.

The Shell Adip process uses DIPA (diisopropanolamine) as the solvent.[17] It has been used most widely in Europe, but appears to be gaining popularity in the U.S. and Canada.

It can compete with DEA and DGA for certain applications. It exhibits H_2S-selectivity abilities also.

Alkaline salt solutions. These solutions include the hot potassium carbonate, or "hot pot" process. It uses an aqueous solution of 25-35 wt % K_2CO_3.[3 4]

It is most applicable to large gas streams containing an appreciable amount of CO_2. If H_2S is present, with little or no CO_2, hot pot will not effectively sweeten the gas. Hot pot will remove some COS and CS_2 by hydrolysis of these compounds.

If H_2S is to be removed to meet normal pipeline specifications, a special design or a two-stage system may have to be used. Normally, a minimum contact pressure of about 300 psig is required for hot pot.

Catacarb is a licensed version of the hot pot process which contains activator chemicals and corrosion inhibitors to increase the reactivity and loading of the solution. The Catacarb process has been applied most widely to CO_2 removal from hydrogen plant and ammonia plants, both in the U.S. and overseas.

The Benfield process is also a catalyzed and licensed version of the hot pot process, containing activators to improve the treating capabilities of the solution. It has been applied most widely to hydrogen, ammonia, and SNG plants overseas.

The Giammarco-Vetrocoke (G-V) process is also a catalyzed version of the hot pot process, using arsenic-containing activators to improve reactivity of the solution.[4 5] Many plants have been built worldwide; but only one plant has been built in the U.S. to treat sour natural gas. The poisonous nature of the arsenic-bearing solution has been the process' greatest drawback.

Physical solvent processes

The physical-solvent processes all use organic solvents and accomplish acid-gas removal mainly by physical absorption, rather than chemical reaction, which is directly proportional to the acid-gas partial pressure in the sour-gas stream.

These processes are most applicable to high-pressure gas streams containing appreciable quantities of sour components. The high-acid-gas loadings realized account for low solvent-circulation rates, reduction in equipment size and cost, and low utilities which are characteristic of these processes.

In general, the physical-solvent processes have two major disadvantages: (1) the solvent's affinity for absorbing heavy hydrocarbons (it can cause dark sulfur production in the Claus plant), and (2) the expensive nature of the solvents. All of these processes are licensed by the developer and require payment of a royalty fee for their use.

The Shell Sulfinol process is unique in this class of processes, in that it uses a mixture of solvents (Sulfolane and DIPA) which allows it to behave as both a chemical and physical solvent process.[17 18]

Sulfinol has good affinity for sour components at low-to-medium partial pressures and extremely high affinity for sour components at high partial pressures.

It should normally be applied to gas streams containing mostly light hydrocarbons. Sulfinol appears to have its greatest advantage when the H_2S/CO_2 ratio is 1:1 or greater, and when the acid-gas partial pressure exceeds 110 psia.[18]

The Allied Selexol process uses as a solvent dimethyl ether of polyethylene glycol (Dmpeg). Typically, no water is used with the solvent. This process is applicable for bulk removal of CO_2 and H_2S, primarily CO_2. Most applications have been on streams containing large amounts of CO_2 (18-43 mol %) and less than 1 mol % H_2S.

Selexol is not particularly attractive for a condition of low-acid-gas partial pressure.[5] Selexol has not been widely used or accepted for treating natural gas containing appreciable H_2S.

Fluor solvent, Lurgi Purisol, Lurgi Rectisol, GAF M-Pyrol, and Estasolvan[3 5 9] are all physical solvent processes. But none of them has gained very wide acceptance in natural-gas treating in the U.S. or Canada.

Direct-conversion processes

The direct conversion processes have been used most widely in Europe for removal of H_2S and recovery of sulfur from manufactured gas, coal gas, and coke-oven gases.

The processes are usually best suited for sour gases containing 1 to 1,000 grains of H_2S/100 scf, and with a maximum sulfur production of 10-15 long tons/day (lt/d).[4 7 19] These processes do not remove any CO_2 from the feed gas.

Most of the direct-conversion processes are based on a scheme, using oxidation-reduction reactions, involving absorption of H_2S in an alkaline solution containing oxygen carriers. The H_2S is oxidized to elemental sulfur.

The oxygen is furnished by air which is normally added to the bottom of the regenerator or oxidizer vessel. The air also acts as a flotation agent for the sulfur, which is collected at the regenerated solution interface as froth. The sulfur is recovered by filtration and melting.

The leading processes in this class of treating systems are Stretford, Thylox, Takahax, and Ferrox. Only Stretford has been applied to natural-gas treating in the U.S. or Canada in recent years. Only a few Stretford plants have been built for this purpose. Considerable operational problems have been encountered.

Dry-bed processes

Two types of dry-bed processes are considered to be of commercial importance: iron sponge and molecular sieves. Of these, the iron sponge is used most widely. Molecular sieves have a special area of application.

The iron sponge or iron oxide process is one of the oldest treating processes still in use today.[3-5] It uses wood shavings impregnated with ferric oxide. Partial regeneration with air is possible, but eventually the bed becomes plugged with sulfur and must be replaced.

The process is usually limited to treating gas streams of small-to-medium volumes, containing 1,000 grains of H_2S/100 scf or less, and containing up to 8-10 lt/d of sulfur.[3 4] These limitations are due to economics of bed replacement.

Advantages of this process are essentially complete H_2S removal, ease of operation, simplicity, and low cost of installation.

LP-gas liquids recovery

Some sales-gas contracts require that the hydrocarbon dew point not exceed 15° F. at 800-psig pressure. However, at present this specification is usually not enforced very widely. The producer usually is more concerned about liquids recovery for revenue purposes than to meet dew-point controls.

Four basic schemes may be used to recover liquids from natural gas: (1) conventional refrigeration (normally using propane), (2) cryogenic refrigeration (using a turboexpander), (3) lean oil absorption (ambient and refrigerated), and (4) dry-bed adsorption. The type of process selected depends upon the quantity of gas and its liquid content (gal/mcf of C_2+).

Conventional refrigeration

If the gas stream is small-to-medium in size (10-50 MMscfd) and contains appreciable liquids (3-16 gal/Mcf of C_2+), propane refrigeration might be selected as the optimum scheme. If the market situation is favorable for "moderately deep" ethane recovery (70-75% of C_2 and 93% of C_3+), the refrigeration system would probably be operated at −45° F. to chill the gas stream to $-40°$ F.

If ethane recovery is not desired, and only 80%+ recovery of propane and heavier components is desired, the refrigerant system might be operated at 5-10° F. to chill the gas stream to 15°-20° F.

If ethane recovery is desired, the normal scheme for small-to-medium-size plants is to demethanize the liquids and sell them as "ethane and heavier" mix via a liquids pipeline. In some larger plants, the liquids are first demethanized and de-ethanized and later fractionated into specification propane, butane, and natural-gasoline products, and sold as such.

In plants designed for only propane and heavier recovery, the liquids may be de-ethanized and sold as a mix, or fractionated into specification products. The quantity of liquids produced and the market situation will dictate the optimum scheme for processing of the liquids recovered from the gas.

Often, trace amounts of H_2S are removed from the gas with the liquids, even though the gas has been treated (sweetened) upstream. If so, molecular sieves, solid caustic potash (KOH), or possibly iron sponge treatment will remove the H_2S to permit the liquids to pass the specification "copper-strip" corrosion test.[20 21] A recent paper describes the interactions of sulfur compounds on the copper-strip corrosion test.[22]

Cryogenic refrigeration

The use of turboexpanders to achieve cryogenic processing of natural gas is of fairly recent vintage, and is gaining popularity due to the trend toward greater liquids recovery.[23-25]

If the gas is available at significantly higher pressure than required for the gas pipeline, "free" pressure drop is available and can be used to good advantage in a turboexpander. Even if high pressure is not available, use of this cryogenic scheme followed by recompression is advantageous in many applications.

If the gas stream is of medium-to-large size (30-250 MMscfd) and contains a medium amount of liquids (3-10 gal/Mcf of C_2+), cryogenic processing of the gas might be selected.

If deep ethane recovery is desired (90-95% of the C_2+), the turboexpander process will probably be operated at $-150°$ F. to $-225°$ F., depending upon the gas composition. The normal scheme is to demethanize the liquids and sell the ethane-plus mix via a liquids pipeline.

For both conventional and cryogenic refrigeration processes, the degree of liquids recovery may be limited by the minimum gas gross-heating-value specification defined in the sales-gas contract.

If the natural-gas stream contains appreciable nitrogen, which would lower the sales-gas heating value below the minimum contract requirement, cryogenic processing of the gas for upgrading of the BTU content should be evaluated.

In this method, essentially all hydrocarbons (C_1+) are liquefied and separated from the nitrogen. The turboexpander outlet typically operates at $-300°$ F. or colder.

The separated hydrocarbon liquids are heat-exchanged with the incoming gas stream. The methane is revaporized, and essentially all of the ethane and heavier fractions are available for liquid sales.

The deep-cryogenic N_2 rejection scheme for BTU upgrading should be considered if the gas-stream volume is 10-15 MMscfd or larger, depending upon the N_2 content and gas composition. Typically, smaller gas flows will not justify the expensive cost of such plants. Special circumstances can alter these guidelines.

Lean-oil absorption

The lean-oil absorption (LOA) process may be applied to medium-to-large gas streams (50-500 MMscfd) of low-to-medium liquids content (0.5-5.0 gal/Mcf of C_3+) for moderate-to-high liquids recovery (say 70-99% of C_3+).

The ambient LOA scheme would typically recover the lower percentage of liquids indicated here. The refrigerated LOA process would chill the incoming gas and lean oil to 20° F. to $-50°$ F., depending upon the gas composition and degree of liquids recovery desired. Typically, the refrigerated LOA process would recover 50-75% of the C_2 and 99% or more of the C_3+ fractions.

The LOA process is much more complicated (and normally more expensive) than a turboexpander process for a typical application in the medium-plant-size range. Many producers were reluctant to accept the mechanical stability of a turboexpander when these devices first came on the market (10-12 years ago).

With the advancement of this technology and demonstration of turboexpander machines in many gas plants built, industry is accepting the process more readily. Due to the popularity, simplicity of construction and operation, and lower capital investment, the turboexpander cryogenic process is preferred over the refrigerated LOA process in many applications. Certain other applications may still fit the refrigerated LOA process best.

Dry-bed adsorption

The dry-bed adsorption (DBA) process is used mostly for dew-point control on small-to-medium gas streams (1-10 MMscfd) which contain a low liquids content (0.5-1.0 gal/Mcf of C_3+ or less). An added advantage of this scheme is that gas dehydration is normally accomplished simultaneously.

Liquid recoveries of 10-15% of the C_4's and 50-90% of C_5+ fractions are typical. If the gas is very sour, it must be sweetened before processing in a DBA unit, since H_2S tends to poison the desiccant.

The DBA process functions best at an optimum contact pressure of about 600 psig or higher. The process is a semicontinuous (cyclic batch) scheme, normally employing three or more DBA vessels (with beds) cycling between (1) processing, (2) regeneration, and (3) cooling.

A stabilized condensate product (12-lb Rvp or 26-lb Rvp liquid) is typically produced. The desiccant normally used in DBA units is some form of silica gel (Mobil Sorbead or Davison silica gels are common).

The conclusion (Part 2) will appear next week. END

The authors . . .

B.G. Goar

T. O. Arrington

B. Gene Goar is a consulting chemical engineer with Goar, Arrington & Associates Inc., Tyler, Tex. He specializes in gas processing, gas and liquid treating, gas dehydration, sulfur recovery, and abatement of air pollution from SO_2 emissions. Goar has authored numerous technical articles and contributed to books on those subjects. He was previously associated with Howe-Baker Engineers Inc., Stearns-Roger Corp., Pan American Petroleum Corp., and Struthers Scientific & International Corp. Goar is a graduate of the University of Texas in chemical engineering.

Thomas O. Arrington is also a consulting chemical engineer with Goar, Arrington & Associates Inc., Tyler, Tex. His 14 years' experience have been gained with Goar, Arrington & Associates Inc., Howe-Baker Engineers Inc., Fluor Corp., and E. I. du Pont de Nemours & Co. He is most heavily oriented toward process design and engineering and specializes in crude and heavy-ends processing, refinery hydrotreating processes, light-ends recovery, gas treating, and all types of distillation and fractionation. Arrington is a graduate of Louisiana Tech University and a registered professional engineer in Texas and Louisiana.

More guides offered for handling sour gas

B. GENE GOAR
THOMAS O. ARRINGTON
Goar, Arrington & Associates Inc.
Tyler, Tex.

CONTINUING the options and steps in treating sour natural gas discussed in part 1 of this two-part article (OGJ, June 26, p. 160) dehydration of the sales gas is an essential part. Basically, three types of gas dehydration are used in natural-gas plants. These are ethylene glycol (EG) injection for conventional refrigeration schemes, triethylene glycol (TEG) contacting for residue-gas drying, and dry-bed-adsorption drying, (DBAD) for residue (sales) gas streams or for cryogenic processing pretreatment.

Normally, sour gas is treated (sweetened), processed for liquids recovery, and then dehydrated, if necessary. Sometimes only gas treating is used and then followed by dehydration.

The presence of H_2S and CO_2 in a natural-gas stream tends to increase the temperature at which hydrates form. If long collection lines in cold climates are encountered, glycol (TEG) dehydration at the well sites or methanol injection may be required to prevent hydrate formation in the sour-gas gathering system. If hydrates form, eventual line plugging will result.

EG injection. If a conventional refrigeration process is used for liquids recovery in the plant, EG is normally injected into the inlet gas stream at several points in the process to prevent hydrate formation in the plant. Of course, dehydration to some degree is obtained also with use of the EG.

Wet EG is collected, regenerated, and recycled back to the process. Water dew points of the gas obtained with EG injection depend upon the refrigeration temperature used. But usually the gas will meet sales-gas specifications for water content, if the contact pressure is sufficiently high.

TEG contacting. If no liquids recovery is planned, TEG is normally used in a gas-contacting tower to accomplish dehydration after gas sweetning. Circulation rates may vary from 2.5 to 4.0 gal of TEG/lb of H_2O removed, typically.

Dew-point depressions of 70°-110° F. may be achieved, depending upon the number of contact trays (normally 6-8) and the degree of TEG regeneration (normally 99-99.8% lean TEG). Sales-gas water contents of 4-7 lb/MMscf can normally be achieved readily with TEG dehydration.

Dry-bed-adsorption drying. If cryogenic (turboexpander) processing is planned, then methanol injection or a dry-bed pretreatment unit is required. The preferred method is to use a molecular-sieve dry-bed adsorption unit ahead of a cryogenic plant, which normally yields a water dew point of −150° F. or lower.

Also, the molecular-sieve unit will remove trace amounts of CO_2 from the gas. Neither H_2O or CO_2 can be tolerated to any significant amount in a cryogenic plant (formation of solids and eventual plugging occur).

A dry-bed unit, containing alumina or silica gel as a desiccant, may be used to dry residue (sales) gas. If appreciable LP-gas liquids are present in the gas, they may be adsorbed and carbonized (fouling bed) during the high-temperature regeneration cycle.

Normally, dry-bed-dehydration units should be applied to lean residue-gas streams wihch have been sweetened and processed for liquids removal. Such a dehydrator will yield water dew points of −100° F. or lower. The typical sales-gas specification of 4-7 lb of H_2O/MMscf can be achieved easily with such a unit.

Table 4

Claus SRU

DESIGN CONSIDERATIONS, OPERATING CONTROL VARIABLES

Design considerations[19 26 27 28 29]
- Composition of acid gas feed.
- Combustion of acid gas.
- For a once-through scheme, retention time of combustion gases at elevated temperatures.
- Catalytic-converter feed-gas temperature.
- Optimum reheat scheme(s).
- Space velocity in converters.
- Sulfur condensing temperatures.
- Coalescing and separation of entrained sulfur from condenser effluent gases.
- Turndown of plant throughput.

Operating control variables[19 27 30]
- Control of H_2S/SO_2 ratio at optimum value of 2:1.
- Catalytic converter feed-gas temperatures (Inlet temperature must be kept high enough to prevent sulfur condensation in catalyst bed).
- Catalyst maintenance.
- Sulfur condenser outlet temperatures.
- Elimination of hydrocarbons and other impurities in feed gas.
- Proper operation of instrumentation at severe turndown conditions.

Sulfur recovery

If the acid-gas stream removed from a sour-gas feed contains an equivalent of more than 3-5 lt/d of sulfur, and if the plant is to be located in Texas, normally the Texas Air Control Board (TACB) would require installation of a sulfur-recovery unit. Several key points should be considered in the selection of a Claus sulfur-recovery unit (SRU) process scheme.

In the sulfur-recovery-unit process selection, a choice must be made between the once-through or the split-stream Claus process.[19] If the acid-gas feed contains 30-40 mol % H_2S or greater, the once-through scheme should be selected, since it gives the highest overall sulfur recovery and permits maximum heat recovery at a high temperature.

In this scheme, all of the acid gas is fed to the reaction furnace, along with enough air to burn one-third of the H_2S to SO_2 and all hydrocarbons completely.

Normally, 150-300-psig steam is generated in the waste-heat boiler. In most cases, the high-pressure steam is used in a steam turbine (with 40-psig exhaust) to drive the combustion-air blower for the unit. Often the amount of steam generated is adequate to drive the air blower.

The low-pressure steam from the turbine exhaust is flowed to the amine-stripper reboilers, where it usually supplies 40-50% of the reboiler steam requirement. This steam generation/utilization system makes for a very thermally efficient overall plant operation.

The major design considerations and operating control variables for a Claus SRU are listed in Table 4.

Lean acid-gas streams (with high impurities such as CO_2, H_2O, and N_2) present many design and operating problems.[26] Consideration should be given to enriching the H_2S content of such a stream, instead of constructing an SRU that will be an operational nightmare.

Technology is available to industry for enriching a lean-acid-gas stream containing 10-20% H_2S to an upgraded acid-gas stream containing 50-70% H_2S.[26]

Tail-gas cleanup

If a sulfur-recovery unit is to be installed in Texas, and if its size is greater than 50-100 lt/d, a TGCU unit may be required by the TACB.

Several viable TGCU processes are available to industry. The six leading processes are: the (1) Shell SCOT, (2) Parson Beavon, (3) Wellman-Lord, (4) SNPA Sulfreen, (5) Amoco CBA, and (6) IFP process.[31 32]

The SCOT, Beavon, and Wellman-Lord processes can achieve 250 ppmv or less SO_2 in the stack-gas effluent. The Sulfreen, CBA, and IFP processes cannot achieve such a low level of SO_2 emissions, with 1,500-2,000-ppmv SO_2 in the stack gas being more representative of these processes.

If the tail gas from the SRU contains less than 20-40% CO_2, the SCOT process might be selected on the basis of low investment, low utilities, and wide commercialization. The SCOT is based on high-temperature catalytic hydrogenation of the sulfur compounds in the SRU tail-gas stream, followed by cooling and absorption of the H_2S in a selective amine solvent, with eventual H_2S recycle back to the Claus plant.

If an extremely high CO_2 content is present in the SRU tail gas, probably the Beavon or Wellman-Lord process would be selected. The Beavon is more suitable for smaller, as well as larger plants. It is based on high-temperature catalytic hydrogenation of the SRU tail gas, followed by cooling and removal of the H_2S by the Stretford process.

The Wellman-Lord (Davy Powergas) process is better suited for larger SRU plants requiring more-stringent SO_2 removal. It is based on incineration of the SRU tail gas, followed by cooling and absorption of the SO_2 in a solution of sodium sulfite. Regeneration of the solution frees SO_2 for recycle back to the Claus plant.

The SNPA Sulfreen and Amoco CBA processes are similar in nature. Both employ the concept of an extended Claus reaction on a cool catalyst bed. The overall sulfur recovery obtained may be only 98-99%, which is lower than that for the SCOT, Beavon, or Wellman-Lord processes (which is normally 99.5-99.9%). Typically, the Sulfreen and CBA processes have been applied to huge SRU plants (500-1,000-lt/d size) in Canada and France.

The IFP process is based on the extended Claus reaction in a liquid, catalyzed solvent system. Its overall recovery is not equivalent to that of the SCOT, Beavon, or Wellman-Lord processes. IFP's process is losing popularity in the U.S. and Canada because of its inability to meet expected future state and EPA regulations.

Environmental impact

Since the processing of sour natural

gas necessarily produces hydrogen sulfide and/or sulfur dioxide as waste emissions, these air pollutants must be dealt with in light of current and possible future federal and state regulations.

In Texas, regulations of three governmental bodies must be considered: Railroad Commission of Texas, Texas Air Control Board, and U.S. Environmental Protection Agency. The applicable regulations of each of these governmental bodies will be discussed.[33-42]

Railroad Commission of Texas. Due the extreme toxicity of hydrogen sulfide, the Railroad Commission of Texas amended, effective Sept. 1, 1976, Rule 36 to provide additional protection to the general public.

Hydrogen sulfide is an extremely toxic gas. An exposure of only 250 ppmv/hr is considered to be a hazardous limit (concentration that may cause death), and 600 ppmv is considered to be a lethal concentration (concentration that will cause death with short-term exposure).

The threshold limit (concentration at which it is believed that all workers may be repeatedly exposed day after day without adverse effects) has been set at only 10 ppmv by the American Conference of Governmental Industrial Hygienists. Hydrogen cyanide is only slightly more toxic than hydrogen sulfide, and hydrogen sulfide is much more toxic than carbon monoxide.

Basically, Rule 36 applies to gas-processing facilities in which the H_2S concentration is greater than 100 ppmv and the "100-ppmv radius of exposure" is greater than 50 ft. Depending upon whether or not the "100-ppmv radius of exposure" penetrates a public area or the "500-ppmv radius of exposure" penetrates a public road, various degrees of compliance are required. The ultimate compliance requirements are to provide:

- Adequate markers and signs.
- Adequate facility security (protection from public access).
- For new construction or modifications, facilities that satisfy the requirements of NACE Standard MR-01-75 and API RP-14E, Sections 1.7 (c), 2.1 (c), and 4.7.
- Adequate safety and control equipment.
- A comprehensive contingency plan.
- Adequate personnel training.

NACE Standard MR-01-75, entitled "Materials for Valves for Resistance to Sulfide Stress Corrosion Cracking in Production and Pipeline Service," specifies the materials of construction for valves and valve components that are acceptable in hydrogen sulfide service.[34]

The principle of sulfide stress-corrosion cracking is a widely known, extremely complex phenomenon. It is generally felt that wet environments below 150° F. in which the H_2S partial pressure is greater than 0.001 atm may cause certain carbon steels to be susceptible to stress-corrosion cracking.

Generally, carbon steels with a Rockwell C hardness of 22 or less are not nearly as susceptible to stress cracking as those with a Rockwell C hardness greater than 22.

NACE Standard MR-01-75 addresses itself to valves and valve components only. It specifies handling, manufacturing, and heat-treating procedures, as well as types of carbon steels, etc., acceptable for valve components in sour service. In addition, the standard specifies other materials of construction that are satisfactory for use in valves in sour gas service.

At present, a NACE committee is in the process of preparing a broader standard for materials suitable for use in sour-gas service. This standard is to be much more general and will not be limited to valves and valve components only. It is expected that the new standard will be issued in the near future.

Texas Air Control Board. In addition to having to meet the requirements of the Railroad Commission of Texas, gas-processing facilities must meet the emission requirements of the Texas Air Control Board (TACB).

The Texas Clean Air Act, as administered by the Texas Air Control Board, currently provides for emission standards for new gas-processing facilities on a case-by-case basis, consistent with ambient-air-quality standards and best-available control technology considering economics. Emission limits depend upon location and proximity to highly industrialized or populated areas.

Each new facility will require a "permit to construct" and a "permit to operate" from the Texas Air Control Board. In general, approximate guidelines for emission limits for sour-gas field-processing facilities by the State of Texas are as follows:

For less than 3-5 lt/d; process to use gas treating only, probably no SRU, flaring or incineration of tail gas; 0% expected sulfur recovery.

For greater than 3-5 lt/d, but less than 30 lt/d; process to use gas treating plus probably two-stage SRU, incineration of SRU tail gas; 93-95% expected sulfur recovery.

For greater than 30 lt/d, but less than 50 lt/d; process to use gas treating plus probably three-stage SRU, incineration of SRU tail gas; 94-96.5% expected sulfur recovery.

For greater than 50 lt/d; process to use gas treating, plus two or three-stage SRU plus probably TGCU, incineration of TGCU tail gas; 99.5+% expected sulfur recovery.

The foregoing guidelines are the authors' interpretations and do not reflect any official rules of the TACB.

Federal Environmental Protection Agency. Up until the present, there have been essentially no comprehensive standards governing the emissions from field sour-gas treating and sulfur-recovery facilities. At present, any new gas-treating facility moving into an area could decrease the ambient air quality no greater than that allowed by the Significant Air Quality Deterioration Standards published in the Federal Register of Dec. 5, 1974.

In the past, air-pollution regulations for these field facilities have basically been established by state regulatory bodies such as the Texas Air Control Board, as provided for by the Federal Clean Air Act. However, it now appears that the issuance of stringent federal standards for field units is just around the corner.

Emission standards for sulfur-recovery facilities located in refineries were "proposed" in the Federal Register "40 CFR Part 60," issued Oct. 4, 1976. These standards limit the concentration of SO_2 in the stack gases discharged to the atmosphere to 250 ppmv at zero percent oxygen and on a dry basis.[40] On Mar. 22, 1978, the EPA announced that the regulations of 40 CFR Part 60 were now in effect.[43]

In January 1977, the EPA issued the "Standards Support and Environmental Impact Statement" for emission regulations governing field gas-processing plants.[42] It is expected that the proposed regulations will be very similar to the regulations now in effect for refinery-located sulfur-recovery facilities, i.e., 40 CFR Part 60.

It is believed that the "New Source Performance Standards" for natural-gas field-processing plants, to be proposed in November 1978, by the EPA, will basically set down the following approximate requirements:

- Plant fuel gas H_2S concentration will be limted to 10 grains/100 scf (160 ppmv).
- The concentration of sulfur dioxide (SO_2) in tail or stack gases discharged to the atmosphere will be limited to 250 ppmv on a zero percent oxygen and dry basis. To correct to 0% oxygen, multiply the actual SO_2

concentration in the stack gas by the factor:

$$F_{O2} = 20.9/(20.9 - \%\ O_2) \qquad (1)$$

To correct to a dry basis, multiply the actual SO_2 concentration in the stack gas by the factor:

$$F_{H2O} = 100/(100 - \%\ H_2O) \qquad (2)$$

The % O_2 and % H_2O values used in Equations 1 and 2 are actual volumetric percentages as measured in the stack gas.

- For sulfur-recovery processes (such as Stretford) that do not include incineration, the total concentration of H_2S, COS, and CS_2 (calculated as SO_2) in the tail gas discharged to the atmosphere will be limited to 300 ppmv. In addition, the H_2S concentration will be limited to 10 ppmv. These values are on a 0% oxygen and dry basis.
- Current thinking is to exempt sulfur-recovery facilities of less than 2-lt/d capacity from the proposed standards.
- Also, current thinking is to exclude small sulfur-recovery units added to gas-sweetening plants that existed before the promulgation of these standards.

Typically, the concentration of sulfurous compounds in tail-gas streams from sulfur-recovery units, without tail-gas cleanup, are on the order of 10,000-15,000 ppmv. Therefore, application of the above emission limits would dictate that new field sour-gas-treating facilities include gas treating, plus a two or three-stage Claus sulfur-recovery unit, plus a tail-gas clean up unit.

Of course, SRU's of less than 2 lt/d would be exempt from this requirement.

However on the bright side, some current thinking within EPA is to relax the proposed SRU regulations for field-located SRU's having a feed "H_2S/CO_2 ratio" of less than 1:1, and/or a capacity below 20 lt/d.

References

1. Berlie, E. M., "Estimating Capital and Operating Costs," Gas Processing/Canada, September-October 1972, pp. 28-34.
2. Estep, J. W., and Plum, E. W., "Economics of Sour Gas Industry," Paper presented at 19th Canadian Chemical Engineering Conference & 3rd Symposium on Catalysis, Edmonton, Alta., Oct. 19-22, 1969.
3. Maddox, Dr. R. N., Gas and Liquid Sweetening, Second Edition, Campbell Petroleum Series, Norman, Okla., 1974.
4. Kohl, A. L., and Riesenfeld, F. C., Gas Purification, Second Edition, Gulf Publishing Co., Houston, 1974.
5. Goar, B. G., "Today's Gas-Treating Processes—1," OGJ, July 12, 1971, pp. 75-79.
6. Blake, R. J., "How Acid-Gas Treating Processes Compare", OGJ, Jan. 9, 1967, pp. 105-108.
7. Riesenfeld, F. C., and Blohm, C. L., "Acid Gas Removal Processes Compared," Hydrocarbon Processing & Petroleum Refiner, April 1962, pp. 123-127.
8. Fitzgerald, H. J., and Richardson, J. A., "How Gas Composition Affects Treating Process Selection," Hydrocarbon Processing, July 1966, pp. 125-129.
9. Swaim, Jr., C. D., "Gas Sweetening Processes of the 1960's," Hydrocarbon Processing, March 1970, pp. 127-130.
10. Smith, R. F., and Younger, A. H., "Tips on DEA Treating," Hydrocarbon Processing, July 1972, pp. 98-100.
11. Dingman, J. C., and Moore, T. F., "Gas Sweetening with Diglycolamine," Proceedings of the 1968 Gas Conditioning Conference, University of Oklahoma, Norman, Okla., 1968.
12. Bucklin, R. W., "Natural Gas and Related Operations—1975," Paper presented at Seminar on Petrochemical Industries, Baghdad, Iraq, October 1975.
13. Bucklin, R. W., "Removal of Hydrogen Sulfide from Natural Gas by DGA," Paper presented at 57th Annual GPA Convention, Mar. 20-22, 1978, New Orleans.
14. Miller, F. E., and Kohl, A. L., "Selective Absorption of Hydrogen Sulfide," OGJ, Vol. 51, No. 51, April 1953, pp. 175-183.
15. Pearce, R. L., and Brownlie, T. J., "Selective Hydrogen Sulfide Removal," Proceedings of 1976 Gas Conditioning Conference, University of Oklahoma, Norman, Okla., 1976.
16. Vidaurri, F. C., and Ferguson, R. C., "MDEA Used in Ethane Purification," Proceedings of 1977 Gas Conditioning Conference, University of Oklahoma, Norman, Okla., 1977.
17. Klein, J. P., "Development in Sulfinol and Adip Processes Increase Uses," Oil and Gas International, September 1970, pp. 109-111.
18. Goar, B. G., "Sulfinol Process Has Several Key Advantages," OGJ, June 30, 1969, pp. 117-120.
19. Goar, B. G., "Today's Sulfur Recovery Processes," Hydrocarbon Processing, September 1968, pp. 248-251.
20. Perry, C. R., "Several Treating Options Open For Ethane—Recovery Plants," OGJ, May 23, 1976, pp. 76-79.
21. Cummings, W. P., "Special Considerations For Product Treating," Paper presented at Rocky Mountain Regional GPA Meeting, Sept. 25, 1975, Denver.
22. Pyburn, C. M., Cahill, F. P., and Lennox, R. K., "The Effect of Sulfur Compound Interactions on the Copper Corrosion Test in Propane," Paper presented at 57th Annual GPA Convention, Mar. 20-22, 1978, New Orleans.
23. Randall, B., "Cryogenic Processing of Natural Gas," Proceedings of 1973 Gas Conditioning Conference, University of Oklahoma, Norman, Okla., 1973.
24. Ewan, D. N., et al., "Why Cryogenic Processing (Investigating the Feasibility of a Cryogenic Turbo-Expander Plant)," Paper presented at 54th GPA Annual Convention, Mar. 10-12, 1975, Houston.
25. Maddox, R. N., and Bretz, K. E., "Turbo-Expander Applications in Natural Gas Processing," Journal of Petroleum Technology, May 1976, pp. 611-613.
26. Goar, B. G., "Impure Feeds Cause Claus Plant Problems," Hydrocarbon Processing, July 1974, pp. 129-132.
27. Goar, B. G., "Sulfur Recovery From Natural Gas Involves Big Investment," OGJ, July 14, 1975, pp. 78-85.
28. Palm, J. W., "Watch These Trends in Sulfur Plant Design," Hydrocarbon Processing, March 1972, pp. 105-108.
29. Goar, B. G., "Tighter Control of Claus Plants Needed by TGCU System," OGJ, Aug. 22, 1977, pp. 134-137.
30. Norman, W. S., "There Are Ways to Smoother Operation of Sulfur Plants," OGJ, Nov. 15, 1976, pp. 55-60.
31. Goar, B. G., "Claus Tail-Gas Cleanup, Part 1 of 2 Parts—Cost, Air Regulations Affect Process Choice," OGJ, Aug. 25, 1975, pp. 96-112.
32. Goar, B. G., "Claus Tail-Gas Cleanup, Part 2 of 2 Parts—Dry Bed Processes Can Cover A Wide Application Range," OGJ, Aug. 25, 1975, pp. 96-103.
33. Railroad Commission of Texas, Order No. 20-65,354, "Special Order Amending Rule 36 of the General Conservation Rules of Statewide Application, State of Texas, Having Reference to Oil and Gas Operation in Hydrogen Sulfide Areas," Mar. 15, 1976.
34. National Association of Corrosion Engineers, NACE Standard MR-01-75, "Material for Valves for Resistance to Sulfide Stress Cracking in Production and Pipeline Service," Approved March 1975.
35. Private Communications with Rearick-Hanson Associates, April 1975.
36. Private Communications with Railroad Commission of Texas, September 1976.
37. Private Communications with the Texas Air Control Board, 1976.
38. Saulmon, W. D., "Tougher Air-Quality Standards Face Sulfur-Recovery Plants," OGJ, May 9, 1977, pp. 53-57.
39. U.S. Environmental Protection Agency, "Air Quality Implementation Plans, Prevention of Significant Air Quality Deterioration," Federal Register, Vol. 39, No. 235, Dec. 5, 1974, pp. 42,510-42,517.
40. U.S. Environmental Protection Agency, "Standards of Performance for New Stationary Sources, Petroleum Refinery Sulfur Recovery Plants," Federal Register, 40 CFR Part 60, Vol. 41, No. 193, Oct. 4, 1976, pp. 43,866-43,874.
41. Private Communications with Environmental Protection Agency, May 1977.
42. U.S. Environmental Protection Agency, "Standards Support and Environmental Impact Statement, An Investigation of the Best Systems of Emission Reduction for Sulfur Compounds from Crude Oil and Natural Gas Field Processing Plants," Research Triangle Park, N.C., January 1977.
43. "Sulfur-emission Limits Seen No Problem for U.S. Refiners," OGJ, Apr. 17, 1978, p. 58.

About the panel . . .

THE Gas Processors Association Program Committee for the 1978 57th Annual Convention recognized that in the ensuing months the Environmental Protection Agency would be setting sulfur dioxide regulations both for gas processors and for refiners. Regulations were put into effect for refiners on Mar. 22 this year (OGJ, Apr. 17, p. 58). Similar regulations are expected for gas processors. The program committee assembled some leading authorities on hydrogen sulfide removal methods for a panel session at the annual GPA convention to assist gas processors in being prepared to meet the forthcoming regulations.

In this issue, the first installment of a five-part series kicks off the formal presentations made by the panelists. Moderator for the panel was Laurance S. Reid, Laurance S. Reid Associates, Norman, Okla. The panel comprised Charles R. Perry, Perry Gas Companies, Odessa, Tex.; R. W. Bucklin, Fluor Engineers & Constructors Inc., Irvine, Calif.; R. L. Pearce, Dow Chemical U.S.A., Freeport, Tex.; L. V. Kunkel, L. V. Kunkel Engineering, Naperville, Ill.; and B. Gene Goar, Goar, Arrington & Associates Inc., Tyler, Tex.

GPA explores H_2S removal methods

THE AMINE-sweetening process has been widely accepted in removing CO_2 and H_2S from natural-gas streams. Monoethanol amine has been used for many years. It was first commercially applied as the Girbitol process in 1930. Lately, Diethanol amine has come into favor for gas treating. Both processes use essentially identical equipment.

These opening comments were made by Charles R. Perry, Perry Gas Companies Inc., Odessa, Tex., as the lead panelist at a session on H_2S removal methods for natural-gas streams at the recent Gas Processors Association Convention in New Orleans. Laurance S. Reid, Laurance Reid Associates, Norman, Okla., moderated the session.

The panel went into H_2S removal with MEA, DEA, MDEA, and DGA, as well as into sulfur recovery by the Claus process and into Claus tail-gas cleanup.

Fig. 1

Amine treater flow diagram

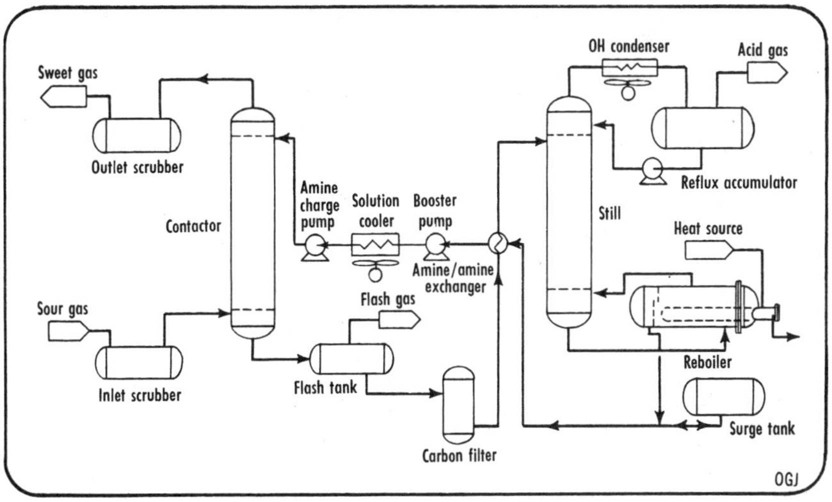

H_2S removal with MEA and DEA

Perry continued his presentation with a description of the amine-sweetening process and indicated differences in design calculations between MEA and DEA sweetening essentially as follows:

Process description. Referring to Fig. 1, natural gas containing CO_2 and H_2S is contacted in countercurrent gas liquid absorption process in a trayed or packed tower to provide intimate contact for a chemical reaction between H_2S-CO_2 (acid gas) and amine. Good design practice dictates a scrubber on the inlet gas to remove entrained liquids including distillate and water from the gas before it enters the contactor. Also, a scrubber on the outlet gas to recover any amine solution carried over from the contactor should be provided.

Rich amine from the bottom of the contactor is flashed at reduced pressure to remove entrained gases, including part of the acid gas, and then heated in a rich lean amine exchanger. The solution is then fed to the stripper tower where the solution is regenerated and denuded of acid gas by steam-stripping. Acid gases are concentrated in the overhead accumulator and disposed of by burning in a flare, reboiler, or other incineration device.

In cases where sulfur exceeds 8-10 tons/day, the acid gases may be processed for sulfur recovery. Lean amine solution from the bottom of the reboiler is exchanged with rich amine in the solution exchangers, then pumped in multistage pumps back to the contactor to complete the process loop.

Inasmuch as a clean solution is a key to the success of a treating system, good filtration is essential. Activated carbon filters have been found to provide the best and most-economical filtration. Also, MEA solutions may be reclaimed in a side-stream reclaimer. DEA solutions cannot be reclaimed due to the high boiling point of DEA. Perry's experience indicates reclaiming is not necessary for DEA solutions.

Design calculations. The input data needed for design calculations are the minimum, normal, and maximum gas flow, pressure, temperature and composition; specifications for the treated gas; and the utilities available.

Circulation of amine solution is determined by the acid-gas content of the inlet gas, strength of the solu-

Table 1

Regeneration vessel sizes, in.

Solution circl. rate, gpm	Still diameter	Surge tank Diam.	Surge tank Length	Reflux accum. Diam.	Reflux accum. Length	Flash tank Diam.	Flash tank Length	Carbon filter Diam.	Carbon filter Length
10	16	24	72	16	36	24	72	16	84
25	24	42	96	24	48	42	96	24	84
50	30	48	144	30	96	48	144	36	96
100	42	60	192	42	96	60	192	48	96
200	60	84	288	60	96	84	288	60	96
300	72	84	384	72	96	84	384	72	96
400	84	96	384	84	96	96	384	84	96

Table 2

Expense estimate

YEARLY AMINE-UNIT OPERATING EXPENSES

Basic data

MMcfd, average ______ Capital investment ______
MMcfd, max. design ______ Manpower required* ______
Gpm design circulation ______

Expenses:

1. Direct labor: ____men at $______/year .... = $______
2. Fringes, FICA, WC insurance, etc., at 30% of direct labor .... = ______
3. Supervision (use 50% of direct labor if no better estimate available) = ______
4. Utilities, ____hp connected × 6,535 × $______/kw-hr .... = ______
5. Chemicals and supplies, ($5,000 + 100 × ______gpm) .... = ______
6. Repairs, 3% to 5% of capital investment .... = ______
7. Overhead allocation (use 100% of direct labor if no better estimates available) .... = ______
8. Depreciation (use 10% of capital investment if accelerated depreciation not used) .... = ______
9. Interest on investment capital (suggest 10% of capital cost) .... = ______
10. Insurance and taxes (3% of capital cost) .... = ______

Total cost per year .... = ______
Cost/day = cost/year ÷ 365 .... $______
Cost/Mcf = cost/day ÷ average Mcfd throughput .... = ______

*For 100 gpm or less, use 2 men, for 100 gpm to 400 gpm, use 4 men, and for 400 gpm to 1,000 gpm, use 6 men.

tion, and the type of amine to be used. Fig. 2 provides a quick determination of circulations required as a function of gas volume, acid-gas content, and type of amine.

Amine circulation can be calculated by these formulas:

For MEA: gpm = $41.0 \times QXZ$.
For DEA: gpm = $45.0 \times QXZ$ (conventional loading)
gpm = $32.0 \times QXZ$ (proprietary high-load processes)

where:

Q = gas to be processed, MMscfd
X = acid-gas content, volume %
= mol % CO_2 + (grains H_2S/632)
Z = amine concentration, wt %

Formulas are based on mol loadings of 0.33 mol/mol for MEA, 0.5 mol/mol for conventional DEA, and 0.7 mol/mol for proprietary high-load DEA.

Pressure and gas volume are major influences on the contactor sizing. Referring to Fig. 3, the size of the contactor may be determined as a function of pressure and gas flow.

Table 1 provides data on standard sizes of the regeneration vessels, including the still, surge tank, reflux accumulator, flash skimmer tank, and carbon filter.

Estimates of heat loads and surface areas to be required are shown here as a function of circulation rates for amine solutions:

	Duty-BTU/hr	Area-sq ft
Reboiler (direct fired)	72,000×gpm	11.30×gpm
Solution-exchangers	45,000×gpm	11.25×gpm
Solution coolers (air cooled)	15,000×gpm	10.20×gpm
Reflux condenser (air cooled)	30,000×gpm	5.20×gpm

Power requirement for the process is largely for pumping of the amine solution. The following tabulation gives approximate power requirements as a function of amine solution circulation:

hp = gpm×ΔP/1,713 (efficiency of pump)

For:

Solution pumps, hp = gpm×psig×0.00065
Booster pumps, hp = gpm×0.06
Reflux pumps, hp = gpm×0.06
Aerial cooler, hp = gpm×0.36

Capital investment for plants to use either the MEA or DEA process can be broken into two segments. First, the contactor investment is a function of pressure and size and is presented in Fig. 4.

Also in Fig. 4 is a cost for the regenerator unit as a function of amine circulation.

In these curves, it is assumed that the regenerator equipment will be skid-mounted and assembled in a shop rather than a field construction. The cost figures do not include transportation or foundations and piping required at the location. Add about 40% to these costs for "turnkey" installed costs.

Operating expenses for the amine unit will be largely operating labor and power. Typical operating expenses for a treating unit are presented in Table 2.

H_2S removal by DGA

Panelist R. W. Bucklin, Fluor Engineers & Constructors, Irvine, Calif., discussed H_2S removal from natural gas by DGA (diethylene glycolamine). Bucklin focused his comments on new developments in DGA:

DGA developments. The original laboratory investigation of DGA[1] was done in Fluor's research laboratory as early as 1950. The promising potential of DGA as a competitive treating solution was recognized and a patent[2] for its use was issued in 1955. Subsequent developments in morphaline manufacture made DGA commercially available in early 1964.

In 1965, the first commercial Econamine[3] plant was put on stream by El Paso Natural Gas Co. at Monument, N.M.[4] Subsequently, over 30 plants using DGA in conventional amine-plant configurations have been put in operation to process over 2 billion scfd of natural gas. These plants were retrofits of plants using other amines as well as grass-root plants.

Recent innovations in DGA plant design have been developed by Fluor in response to specific client needs. These designs are known as Improved Econamine or improved DGA. Fluor has done extensive testing in Saudi Arabia to expand its DGA data base and to assure the reliability of these innovations.

Over 4 billion scfd of gas will be processed by improved DGA plants now under construction. A patent covering improved DGA technology has been applied for.

Important DGA characteristics. DGA has several important physical and chemical characteristics. Aqueous solutions are comparable with ethylene glycol in freezing point, viscosity, and dew-point depression of natural gas. It is thermally stable at temperatures in excess of 400° F.

It is a primary amine reacting with acid gases in much the same way as monoethanolamine. It has a relatively low vapor pressure, so chemical losses in the treated gas are not significant.

DGA can be used in a wide range of aqueous solution strengths, ranging from 5 to 90 weight % DGA. It is generally noncorrosive to carbon steel, and has little tendency to foam even when the solution has a high solids content. DGA is suitable for use in the Arctic as well as hot desert areas barren of cooling water.

When used in higher concentrations. DGA can remove 4-15 scf of acid gas/gal. In some situations, loadings approaching stoichiometric can be obtained with successful treating.

DGA reacts with carbonyl sulfide and carbon dioxide to form urea-type degradation products. These degradation products are thermally reversible at elevated temperatures and regenerator pressures. Vacuum reclaiming is not required.

DGA can produce treated gases to satisfy transmission pipeline H_2S requirements, LNG plant CO_2 requirements, and stringent COS specifications. DGA is also effective in removing mercaptans.[5 6] The COS and mercaptan removal capabilities can introduce significant cost savings by eliminating requirements for downstream liquid product treating units.

Disadvantages of DGA. DGA does have disadvantages. It is more expensive than most of the commonly used amines. It is a good extraction agent for aromatics. Therefore, special design provisions are sometimes needed to assure production of bright sulfur from an associated Claus unit.

Overall savings in capital and operating costs of Improved Econamine are not significantly impacted by the higher amine cost. The special design provisions required by certain sulfur plants are relatively minor cost items.

DGA supply. DGA supply has been the concern of many potential users. There are now three plants in operation producing DGA, two in the United States and one in Europe. The requirements of all known present and future users are about two-thirds of the current production capacity.

Fig. 2

Amine circulation required

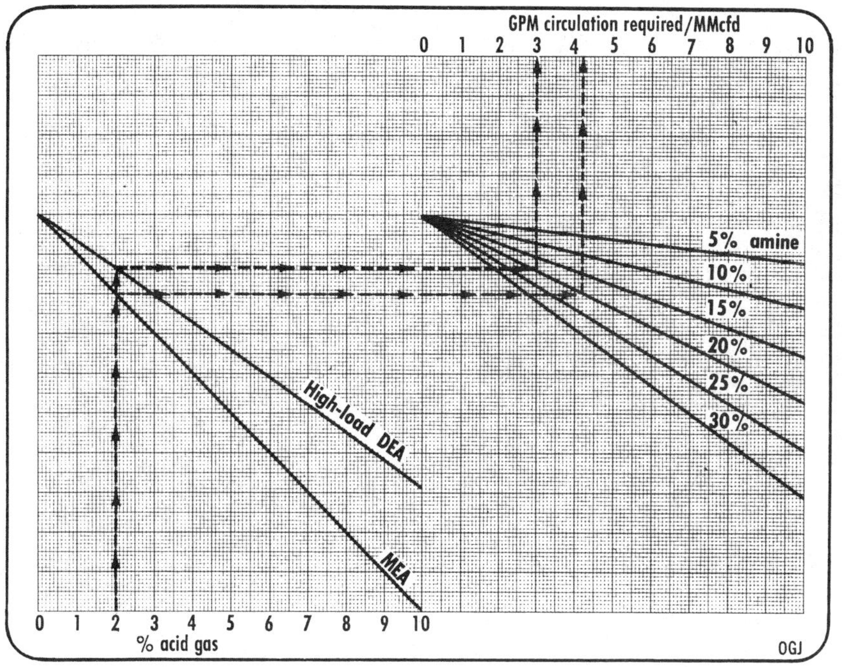

Analytical developments. A necessary component of Fluor's gas-treating development program has been the development of an accurate, portable, on-line sulfur analyzer. It was found that the treating performance could not be accurately determined from samples transported to remote laboratories due to the instability of some sulfur compounds as well as the tendency of sample containers to adsorb significant amounts of sulfur compounds.

Fluor's test kit can determine the contaminants in a gas for design purposes as well as treated gas purity and overall plant performance.

Filters and chemical losses. Experience has shown that filter operations can be high man-hour consumers as well as a major potential for chemical losses. Several DGA plants are operated without filters. Some DGA plants with leaf-type precoat filters have had chemical losses almost proportional to the number of filter changes. Good reclaimer operation can significantly reduce filtering requirements.

In one plant, the DGA solution had become contaminated with soluble ferric ions before start-up. The soluble iron acted as a chelating agent with DGA and entrapped iron sulfide particles. The result was a waxy-looking material that would plug a precoat filter almost instantaneously.

It was necessary to reclaim the solution and concentrate the iron in the reclaimer where it could be eliminated from the system. If not removed by either filtration or reclaiming, the wax-like chelate product would probably plug an exchanger in the cooler areas of the plant.

Process selection. The factors governing a process selection are capital cost, operating costs, reliability, flexibility, availability, and ease of operation. Improved Econamine is an excellent competitor in all of these basic areas.

A comparison of circulation rates and overall energy requirements is a good guide to probable comparative capital costs for grass-roots plants. Here's a comparison of these parameters for conventional MEA, conventional DGA, and Improved DGA.

Assume sour-gas feed rate of 100 MMscfd of 900 psig, 90° F. gas containing 5 mol % H_2S and 5 mol % CO_2. The treated gas specification is less than a quarter grain of H_2S/100 scf and less than 1,000-ppm volume CO_2.

	Conventional MEA[7]	Conventional DGA[7]	Improved DGA
Weight %	18	60	65
Circulation rate, gpm	2,460	1,405	845
Reboiler duty, MMBTU/hr	177	127	51
Pickup, scf/gal	2.82	4.94	8.22

Fig. 3

Contactor capacity

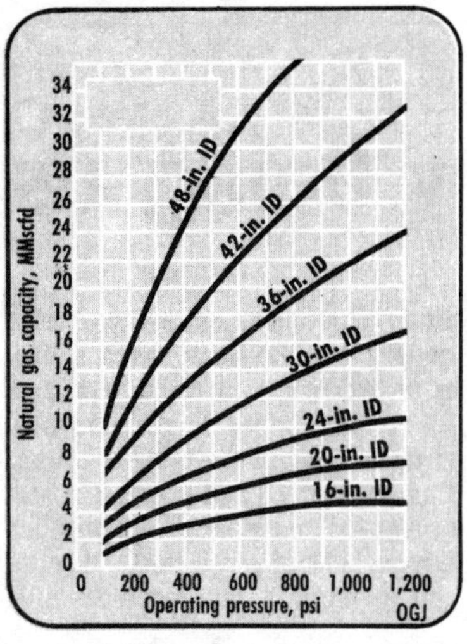

Fig. 4

Amine unit costs vs. capacity

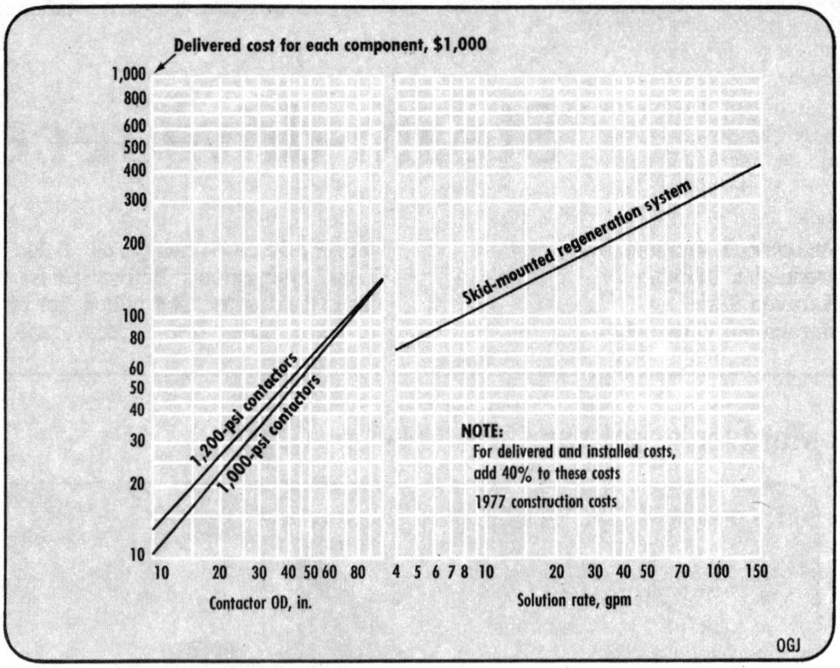

The conventional plants are typical industrial designs conforming to arbitrary design rules. The reader can assign his own values for comparison if he has information to give him a better degree of confidence.

DGA has some flexibility aspects not enjoyed by any other amine. The low freezing point of aqueous solutions makes it particularly suitable for very cold climates. At the other end of the scale, its treating capability at elevated temperatures provides a unique advantage in excessively warm climates.

DGA is suitable for air cooling with ambient temperatures in excess of 120° F.

This feature can have additional environmental advantages since problems of water supply and waste-water disposal can be avoided. Air-cooled plants are also somewhat simpler to operate and maintain than plants requiring cooling water.

The low corrosivity of DGA with carbon steel has given Econamine plants a favorable record of maintenance and plant availability. Conservative designs using alloy steels at historically critical amine-plant locations have been used in most of the new plants. These plants have experienced virtually no corrosion. Operating experiences in retrofit plants indicate that DGA is less corrosive than MEA.[8]

The wide range of flexibility for process and ambient conditions makes Econamine plants relatively simple to operate. Operating manpower requirements are no different from other amine plants.

The value of the energy required to operate the process is the most-significant operating cost of an amine plant. Today, operating labor and maintenance costs are relatively minor especially when differential costs among competing processes are considered.

Improved Econamine should be considered by energy-conscious operators who can take advantage of reduced fuel bills and thus enjoy more treated gas available for sale.

In summary, new developments in DGA technology are well oriented to both capital-cost reduction and energy conservation. Improved DGA does represent better designs due to the increase in confidence gained from operating experience and plant tests. Improved DGA must be considered to assure the choice of the best option in a process selection.

Literature cited

Perry-Bibliography

Fails, J. C., and Harris, W. D., "Practical Way to Sweeten Natural Gas." OGJ, July 11, 1960.

Goar, B. G., "Today's Gas Treating Processes," Proceedings of the 1971 Gas Conditioning Conference, University of Oklahoma, Norman, Okla.

Kohl, A. L., and Riesenfeld, F. C., Gas Purification, McGraw-Hill Book Co. Inc., New York City, 1960.

Maddox, R. N., and Burns, M. D., "MEA Process to be Considered First," OGJ, Aug. 21, 1967.

Perry, C. R., "Basic Design and Cost Data on MEA Treating Units," Portable Treaters Inc., Odessa, Tex.

Perry, C. R., "Design and Operating Amine Units for Trouble Free Unattended Operation," Proceedings of the 1969 Gas Conditioning Conference, University of Oklahoma, Norman, Okla.

Rosen, W. D., "Here's a Quick Design Method for Amine Sweetening Plants," OGJ, Mar. 18, 1968.

Shell, A. D., "Consider Basic Problems in Improving MEA Filtration," OGJ, Feb. 12, 1968.

Swaim, C. D., "Gas Sweetening Processes of the 1960's," Hydrocarbon Processing, Vol. 49, No. 3, Mar. 1970.

Zapffe, F., "Gas Sweetening," Proceedings of the 1965 Gas Conditioning Conference, University of Oklahoma, Norman, Okla.

Bucklin-References

1. DGA is a colloquial name for "Diglycolamine," a registered trademark of Jefferson Chemical Co., for 2-2 aminoethoxy ethanol. It is also known as diethylene glycolamine and Beta, Beta'- hydroxy amino ethyl ether.
2. U. S. Patent 2,712,978.
3. Econamine is a registered trademark of Fluor Corp. for treating processes using DGA.
4. Holder, H. L., "Diglycolamine — Promising New Acid-Gas Remover," OGJ, May 2, 1966, pp. 83-86.
5. Harbison, J. L., and Dingman, J. C., "Mercaptan Removal Experiences in DGA Sweetening of Low Pressure Gas," Gas Conditioning Conference Proceedings 1972, pp. H-1 to H-4.
6. Williams, W. W., "Treatment of Gas Plant Liquids with Diglycolamine Agent," Gas Conditioning Conference, University of Oklahoma, 1973.
7. Adapted by permission from Dingman, J. C., and Moore, T. F., "Compare DGA and MEA Sweetening Methods," Hydrocarbon Processing, Vol. 47, pp. 138-140, 1968.
8. Mason, S. R., and Griffith, T. E., "MEA-DGA Switch Saves Steam," Gas Conditioning Conference, Norman, Okla., 1969.

GPA panelists continue probe into H_2S removal, sulfur recovery

PRESENT and potential stack-emission standards for gas-processing plants handling sour gas are forcing gas processors to look closely at every available process for gas sweetening, sulfur recovery, and sulfur-plant tail-gas cleanup. A group of authorities explored these topics at a panel session at the recent Gas Processors Association Convention in New Orleans.

Part 1 of this series covering formal presentations focused on gas sweetening with MEA and DEA. This second part goes into gas sweetening with the MDEA process. Part 3 goes into improvements in sulfur recovery with the Claus process.

H_2S removal with MDEA

Hydrogen sulfide removal with methyl diethanolamine (MDEA) was explored by R. L. Pearce, Dow Chemical U.S.A., Freeport, Tex. His presentation was essentially as follows:

The MDEA process has been studied by several investigators and has developed into a commercial process. MDEA does not possess all the desirable characteristics of an ideal solvent, but it does have a good overall balance of favorable properties.

The low energy requirements, coupled with favorable selectivity and capacity, makes the process worthy of serious evaluation when searching for a versatile acid-gas-removal process.

Development. The first serious effort to use methyl diethanolamine (MDEA) as an H_2S removal solvent was by Fluor Corp. in the late 1940's and early 1950's.[1, 2] Fluor did both pilot-plant and commercial-scale testing. Dow Chemical Co. has investigated MDEA in pilot studies[3] and commercial applications. Graff Engineering Corp. has reported the use of MDEA for H_2S removal from fuel gas and Claus tail-gas streams.[4]

Phillips Petroleum Co. discussed both pilot and commercial experience in a recent publication.[5] In addition, Badische Anilin und Soda-Fabrik A.G. (BASF)[6] and Shell Chemical[7] have patented processes using MDEA for H_2S removal.

It is estimated that eight to 10 MDEA units are operating in the U.S. The upgrading of the H_2S content of Claus sulfur-plant feed, recovery of H_2S from sulfur-plant tail gas, and treating of natural gas for H_2S removal to pipeline specifications are the known areas of application.

All of these plants use MDEA's selectivity for H_2S in the presence of CO_2 to produce a regenerated acid gas much richer in H_2S than produced with conventional acid-gas-removal processes.[3, 4, 5] Since CO_2 absorption is minimized and MDEA has lower heats of reaction than other amine-type solvents, much lower utility requirements are possible.[3]

Process description. The use of MDEA for the removal of H_2S will result in a flow sheet similar to that found in conventional amine treating units. In fact, a number of MDEA units have been retrofits of plants using other amines.

A typical flow sheet is shown in Fig. 5. Sour-gas feed enters the bottom of the absorber while an aqueous solution of MDEA (usually 30-50% by wt) is pumped to the top of the absorber.

This lean solution absorbs most of the H_2S and some of the CO_2 as it passes countercurrent to the gas. The purified gas exits at the absorber top. The extent of selectivity obtained in the absorber depends upon the H_2S requirements in the treated gas and the desired H_2S level in the offgas from the regenerator.

Rich solution containing absorbed

Fig. 5

MDEA process

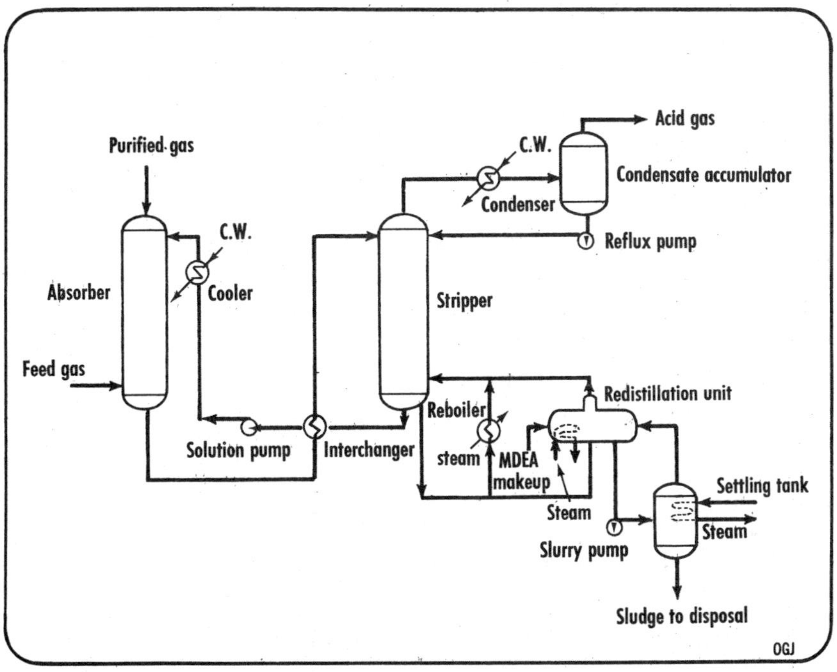

Fig. 6

Acid-gas loading

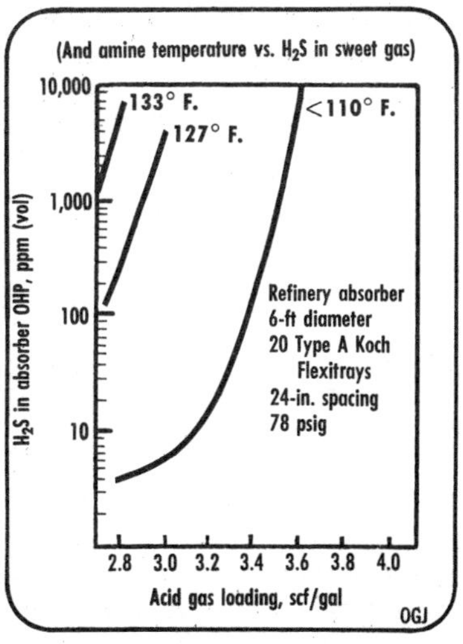

acid gas is heated by interchange with the hot lean solution and flows to the stripper for MDEA regeneration. The absorption reactions are reversed with heat supplied by stripping steam generated in the reboiler.

Removal of the acid gases is facilitated by the stripping steam which is then condensed, separated from the acid gas, and returned to the stripper column as reflux. The acid gas continues on to sulfur recovery.

The hot lean solution is pumped from the stripper bottom and cooled by interchange with the cool rich solution. Additional heat is removed in a water-cooled (or air) exchanger before being fed to the absorber.

Process chemistry. The H_2S, organic acids, and a portion of the CO_2 are chemically absorbed from the gas by the 20-50% by wt of aqueous solution of MDEA.

$$R_2NCH_3 + H_2S \rightleftarrows R_2NCH_3^+ \cdot HS$$

$$R_2NCH_3 + CO_2 \rightleftarrows R_2NCH_3^+ \cdot HCO_3$$

The acid gases are absorbed in the absorber; and when heat is applied in the stripper, these reactions are reversed freeing the acid gas. Reactions between MDEA and organic acids are not reversible.

In the event contamination by organic acids is significant, a side stream of the lean MDEA may be sent to a reclaimer, where a portion of the degraded solution can be recovered by high-temperature distillation ranging from 250°-350° F.

Many primary, secondary, and tertiary amines form water-soluble salts with hydrogen sulfide and carbon dioxide. The primary and secondary amines form carbamates as intermediate compounds similar in nature to those formed by ammonia.

The carbamate is formed instantaneously. Thus it provides a transport mechanism to assist in the reaction rate of CO_2 and amines.

The inability of tertiary amines to react with carbon dioxide to form carbamates provides a much slower reaction rate than that of hydrogen sulfide, which may react directly with either primary, secondary, or tertiary amines.

The rate of reaction of carbon dioxide with various tertiary amines is slower in all cases than with primary and secondary amines. Thus, much of the research regarding selective H_2S solvents revolved about tertiary amines.

H_2S removal efficiency. MDEA is normally regarded as a selective H_2S removal solvent with only moderate low-level removal capabilities. This concept has come about due to the highest level of importance attached to selectivity or CO_2 rejection as contrasted to H_2S removal efficiency. With proper design, MDEA can make pipeline specification gas and still provide good H_2S selectivity.

The data of Miller and Kohl[2] show pipeline-quality gas in one case where selectivity was under study (Table 3). Vidaurri and Ferguson[5] obtained pipeline-quality gas in a few pilot-plant runs (Table 4) and in several refinery tests (Fig. 6). Table 5 summarizes the test data from Tables 3 and 4. A licensee of Dow Chemical has a conversion unit producing less than 4-ppm H_2S gas (Table 6).

When pipeline H_2S-specification gas is desired, then the process can be designed to produce an acceptable product in most cases.

Table 3

Pilot-plant results with aqueous MDEA solution

Run No.	Absorber pressure, psia	Gas/liquid rates, scf feed gas/gal sol'n.	Wt % MDEA in solution	Feed gas Vol % CO_2	Feed gas Vol % H_2S	Treated gas Vol % CO_2	Treated gas ppmv H_2S	Selectivity‡	Acid gas absorbed, mol gas/mol MDEA, (scf/gal)
1	57	17.9	35	4.1	0.42	1.72	*17.6	1.72	0.00(0.5)
2	58	17.1	35	23.9	0.82	14.6	*32.0	2.56	0.17(1.6)
3	65	17.6	37	35.5	0.34	25.5	†0.8	3.55	0.23(2.3)
§4	65	17.1	35	48.1	0.41	41.3	*16.0	7.05	0.23(2.1)
5	60	22.4	32	25.4	5.0	19.6	†9.3	4.38	0.30(2.5)
6	215	19.6	30	7.5	3.2	2.15	*28.8	1.40	0.32(2.5)
7	595	87.5	35	8.0	0.70	3.47	*4.0	1.77	0.51(4.8)
¶8	760	24.3	11	11.1	2.3	1.61	*7.2	1.17	0.96(2.8)
¶9	760	47.2	13	14.1	2.5	8.8	*320	2.63	0.97(3.3)

*Determined by continuous I_2 titration. Includes compounds other than H_2S, up to the equivalent of 4-ppmv H_2S. †Determined by cadmium sulfide precipitation. ‡Defined in Table 6. §One-half absorber length used. ¶Three-fourths absorber length used.

Table 4

Test results from 2-in.-diameter contactor

Solvent	No. trays	Contactor press., psig	Solvent loading, scf acid gas/gal,	CO_2 Feed, %	CO_2 % passed	H_2S Feed, %	H_2S Prod, ppm	Selectivity
20% MDEA ...	16	50	3.0	2.1	72	8.8	6	3.56
20% DEA	16	50	2.7	3.0	14	8.7	10	1.16
65% DGA	16	50	3.3	3.2	0	10.9	0	0.31
15% MDEA ...	8	250	2.7	24.0	68	2.0	6	3.12
15% MDEA ...	8	250	3.1	21.5	75	1.9	70	3.99
20% MDEA ...	16	65	2.4	2.6	64	8.0	8	2.77
20% MDEA ...	16	65	2.8	2.0	77	8.0	12	4.35
20% MDEA ...	16	65	3.2	2.1	77	8.0	25	4.35
20% MDEA ...	16	65	3.4	3.1	67	9.0	97	3.03
20% MDEA* ..	16	65	2.8	2.0	77	8.0	12	4.35
20% MDEA† ..	16	65	2.8	1.9	72	8.0	54	3.57

*Absorber OH temp. = 85° F. †Absorber OH temp. = 110° F.

Table 5

Test tables 3 & 4 summary

(Results with aqueous MDEA solution)

Source	Selectivity range	Loading mol/mol	CO_2/H_2S
Miller et al.[2]	1.17-7.05	0.06-0.97	2/1-100/1
Vidaurri et al.[5]	0.31-4.35	0.15-0.77	1/4- 12/1
Pearce et al.[3]	2- 17	0.05- 1.0	2/1-100/1

Table 6

H_2S removal—commercial plant

H_2S partial pressure, psi	MDEA, %	Loading, mole/mole	H_2S in treated gas, ppm
0.3	35.0	0.7	< 4
0.4	32.0	1.0	<10

Fig. 7

Contact-time effect

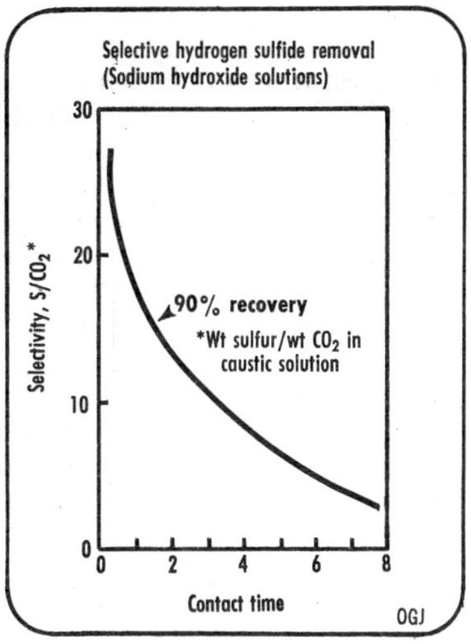

Fig. 8

Acid-gas absorption

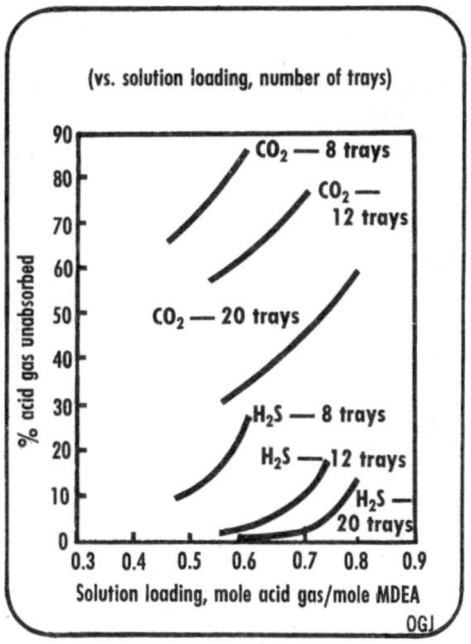

Fig. 9

MDEA selectivity

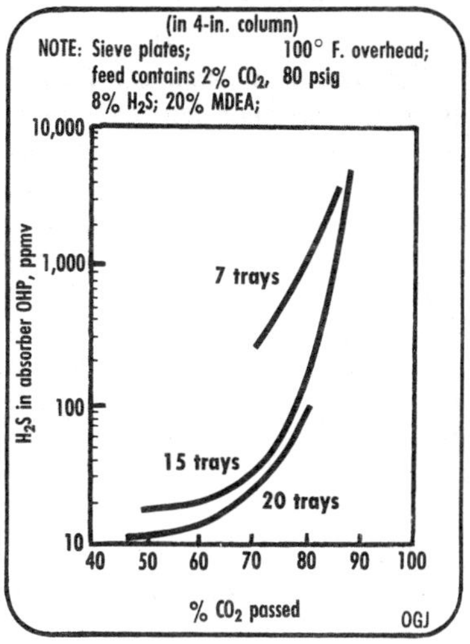

MDEA selectivity. An important attribute of MDEA is the ability to selectively remove H_2S from gases containing both H_2S and CO_2. The selectivity depends upon several variables. Since the approach is generally one of nonequilibrium design, the selectivity correlations may vary among investigators.

Contact time and contact stages are variables that affect selectivity. They are the most readily controlled. Thus much of the research has concentrated on these characteristics.

Selectivity is affected by contact time to such an extent that differential reaction rates between H_2S and CO_2 constitute a significant part of the selectivity. In fact, some selectivity can be built into even monoethanolamine or caustic, providing short contact times are used.

The concept of building selectivity into a process by control of contact time can best be illustrated by the Dow process which uses a caustic solution for the selective removal of hydrogen sulfide from gases containing carbon dioxide.

Examination of Fig. 7 will show the effect of contact time on selectivity. A 20-fold increase in the H_2S to CO_2 weight ratio is obtained in the liquid absorbent.

The process is based solely upon the faster reaction rate of hydrogen sulfide with alkaline solutions over that of carbon dioxide, thus relying upon intimate contact at very short residence times.

The process is nonregenerative, but the sodium hydrosulfide solution from the absorption process is suitable for use in Kraft pulp production.

The data of Frazier and Kohl[1] show the effect of contact stages on selectivity in Fig. 8. Vidaurri and Ferguson[5] show selectivity and treated gas purity as a function of contact stages in Fig. 9.

The definition of H_2S selectivity will generally vary from one investigator to another. Therefore, for the sake of continuity the same method of expressing selectivity has been used in this article for all investigators' work.

Selectivity is defined in Table 7, which also shows selectivity for MEA and DEA as well as heats of reaction values for various amines. Pearce and Brownlie[3] have correlated acid-gas partial pressure and contact stages to selectivity as shown in Fig. 10.

Regeneration. Operation of the regenerator at solution concentrations ranging from 14 to 20 wt % MDEA, solution loadings from 0.43 to 0.75 mole acid gas/mole MDEA, and CO_2/H_2S mole ratios from 0.03 to 0.19 produced data on the H_2S content remaining in the lean MDEA as a function of the reflux rate.

Fig. 11 shows the relationship between residual H_2S in lean solution and the ratio of reflux to acid gas in the regenerator.[2] Very effective stripping of the MDEA solution can be provided at a reflux ratio of about 1 mole of water refluxed to the regenerator per mole of acid gas stripped. In these commercial scale tests, a reflux ratio of 1 represented a total requirement of about 2 lb of reboiler steam/lb of acid gas.[2]

The effect of stripper reflux is also shown by Vidaurri and Ferguson[5] in Fig. 12, at somewhat higher values. However, these data are reported at 99° F. and 129° F. absorber contact temperatures.

In general, 1 mole of stripping steam per mole of acid gas is quite adequate to produce a lean solution capable of achieving 10-ppm H_2S or less in most cases.

Energy requirements. The total energy requirement for MDEA gas removal is considerably less than for other amines used in gas processing as shown in Table 7. The lower heats of reaction (Table 7) coupled with only partial removal of CO_2 and relatively low reflux requirement make the MDEA process one of the lowest consumers of energy of the available processes.

A comparison of MDEA vs. MEA is shown in Table 8[3] covering two different levels of H_2S removal on the same gas. The MDEA is a 50% solution compared with a 30% MEA — both inhibited and both loaded to maximum permissible levels.

Corrosion. A 20-week commercial-scale test with about 20 wt % MDEA solution showed only minor corrosion in the reboiler and on the lean solution side of the rich-lean solution heat exchanger.[2] Commercial experience in the 15-20 wt % MDEA range has shown corrosion to be slight or less than MEA under like circumstances.[5]

Maximum energy savings will result from operation at high MDEA concentrations (about 50 wt %) and maximum acid-gas loadings.[7] Static corrosion studies under these conditions show corrosion rates in the order of 26 mils/year.[3]

Commercial-plant operation in the 15-30 wt % MDEA range with moderate-to-high loadings indicate the use of carbon-steel construction.

Product availability. The product is commercially available from two or more suppliers. The current price for bulk quantities is about 90¢/lb and reflects the present limited market. The

Fig. 10

Selectivity of MDEA

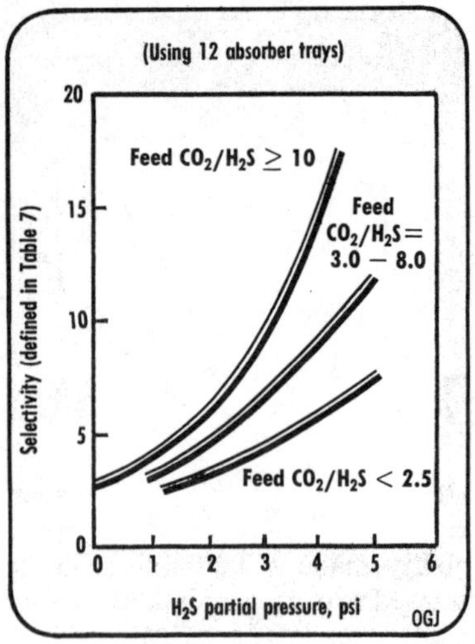

Fig. 11

Regen. performance

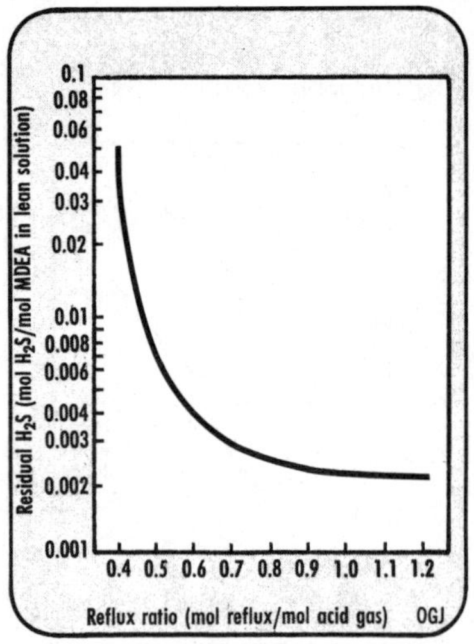

Fig. 12

Stripper reflux

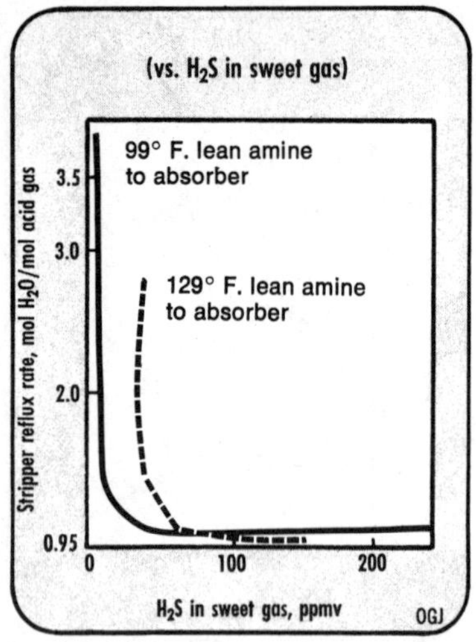

price of MDEA is a disadvantage when compared with 35¢-38¢/lb for MEA and DEA. However, the overall economics are decidedly in favor of MDEA if complete CO_2 removal is not required.

Physical properties. The physical properties of MDEA, MEA, and DEA are listed in Table 9 for comparison. Note that MDEA has a −6° F. freezing point. Small amounts of water sharply reduce the freezing point, as is common with the other amines.

The vapor pressure is sufficiently low to allow absorption of acid gases at low pressure without undue loss of solvent. The thermal and oxidative stability of MDEA coupled with the 447° F. boiling point suggests atmospheric reclamation. The heat of vaporization contributes to low energy requirements in a side-stream reclaimer where MDEA is taken as an overhead product.

The moderate viscosity coupled with the other physical properties points out the similarity between the three amines. The hydrocarbon solubility of MDEA is comparable to DEA.

Part 3 will appear in the Aug. 7 issue.

Table 7

Selectivity factors

(Acid gas selectivity, capacity and heats of reaction for various amines)

Compound	Selectivity*	Capacity, mol gas/mol amine		Heat of reaction, BTU/lb	
		H_2S	CO_2	H_2S	CO_2
MDEA	3.85	0.10	0.12	450	577
MEA	0.89	0.07	0.50	650	820
DEA	2.27	0.09	0.32	493	650

*Selectivity = (mol % H_2S)$_{feed\ gas}$ − (mol % H_2S)$_{treated\ gas}$/(mol % H_2S)$_{feed\ gas}$ = (mol % CO_2)$_{feed\ gas}$ − (mol % CO_2)$_{treated\ gas}$/(mol % CO_2)$_{feed\ gas}$

Table 8

Operation of MDEA compared with MEA*

*Base conditions: 6,925 mol/hr, feed gas at a temp. = 105° F. and pressure = 75 psia. 43 moles/hr feed H_2S = 0.62 vol %. 696 moles/hr feed CO_2 = 10.05 vol %.

	Case 1		Case 2	
	MDEA	MEA	MDEA	MEA
Solution circulation, gpm	545	742	255	732
Reboiler duty, BTU/hr	28	46	13	45
Required H_2S in treated gas, ppm	<50	<1,000	....	....
Delivered H_2S in treated gas, ppm	<50	<5	<1,000	<10

Table 9

Amine physical properties

Property	MDEA	MEA	DEA
Molecular weight	119.17	61.09	105.14
Density, lb/gal	8.67 at 68° F.	8.47 at 68° F.	9.09 at 86° F.
Boiling point, °F.			
at 14.7 psia	447.1	338.7	*515.1
at 0.97 psia	327	212	369
at 0.19 psia	262	156	302
Vapor pressure, psia			
at 68° F.	<0.0002	0.004	<0.0002
Freezing point, °F.	−5.8	50.5	82.4
Flash point, °F. (Cleveland open cup)	260	200	280
Solubility in water, % by wt at 68° F.	Complete	Complete	96.4
Viscosity, cp	101 at 68° F.	24.1 at 68° F.	380 at 86° F.
Heat of vaporization, BTU/lb	223 at 14.7 psia	355 at 14.7 psia	288 at 0.44 psia
Approximate cost, $/lb	0.90	0.38	0.39

*DEA decomposes at this temperature.

References

1. Frazier, H. H., and Kohl, A. L., "Selective Absorption of Hydrogen Sulfide from Gas Stream," Industrial and Engineering Chemistry, 42 (11, pp., 2,288-2,292) November 1950.

2. Miller, F. E., and Kohl, A. L., "Here's A New Solution for Selective Absorption of Hydrogen Sulfide," OGJ, 51, (51), pp. 175-183, Apr. 27, 1953.

3. Pearce, R. L., and Brownlie, T. J., "Selective Hydrogen Sulfide Removal," Gas Conditioning Conference, Norman, Okla., Mar. 8-10, 1976.

4. Crow, J. H., and Baumann, J. C., "Versatile Process Uses Selective Absorption," Hydrocarbon Processing, pp. 131-132, October 1974.

5. Vidaurri, F. C., and Ferguson, R. G., "MDEA Use in Ethane Purification," Gas Conditioning Conference, Norman, Okla., Mar. 7-9, 1977.

6. Bartholome, E., U.S. Patent 3,622,267; November 1971.

7. Sykes, J. A., Jr., U.S. Patent 3,965,244, June 1976.

Guidelines can help choose proper process for gas-treating plants

R. N. TENNYSON
R. P. SCHAAF
Fluor Engineers & Constructors Inc.
Los Angeles

CHOICE of solvent for gas-treating plants is important because solvent-circulation rate usually determines equipment sizes and overall cost of the unit. Many processes have been developed (Table 1). But their effectiveness in removing impurities is limited by the solvents used in treating feed gas.

Solvent selection should be based on composition, temperature, and pressure of feed gas, and required restraints on product gas. These parameters will determine whether a chemical or physical solvent is more economical.

Chemical solvents are characterized by relatively high heats of solution and their capacity to absorb acid gases without a high sensitivity to pressure.

Since they usually consist of an aqueous solution of a reagent, concentration is important. If a unit design is based on a certain loading, a 30 wt % solution will require almost twice the circulation rate as that of a 60 wt % solution and with considerably higher costs.

Fig. 1

Gas treating for CO_2 removal

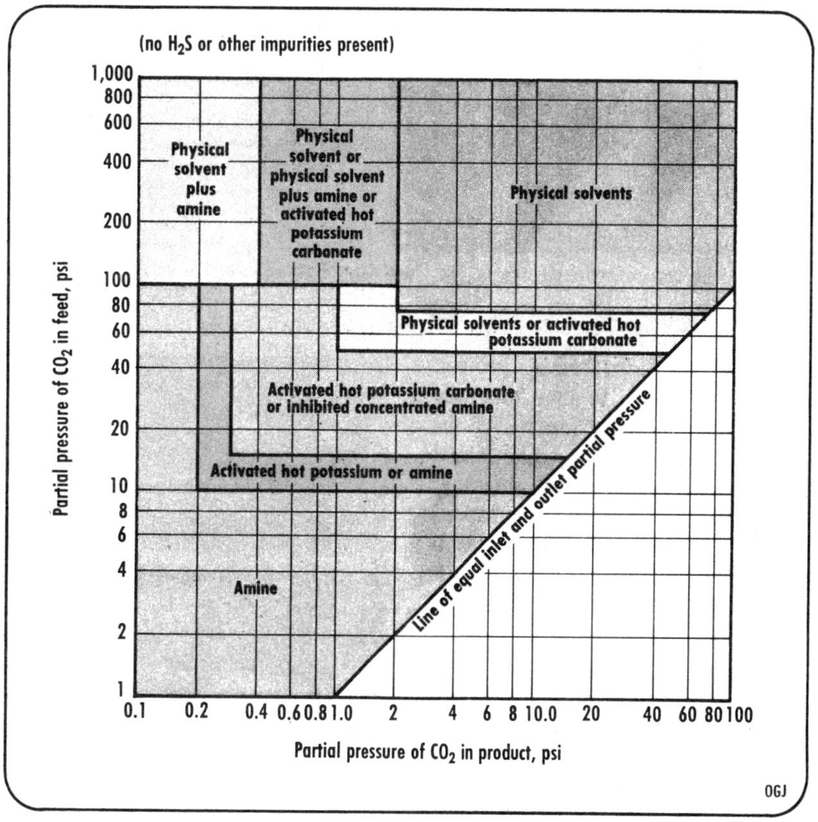

Substantial amounts of heat are needed to regenerate the rich solution. Heat is needed for: (1) dissolution of acid gases from the solvent, (2) raising temperature of solvent to regenerator bottom condition (usually the boiling point of the solution), and (3) generating stripping vapors which ultimately pass overhead from the regenerator with the acid gases.

Physical solvents have low heats of solution and can absorb acid gases in proportion to their partial pressures. At low or moderate pressures (atmospheric to 200 psi), little absorption capacity occurs unless acid-gas concentration in feed gas is high. However, as partial pressure of acid gases increases, absorptive capacity of physical solvents also increases.

As an example, with acid-gas partial pressures at 75 psig, capacity of physical solvents in commercial use far surpasses that of chemical solvents. Therefore, required circulation rate of solvent is considerably less in these systems.

An even-greater advantage can be seen with high partial pressures of acid gases. In a few cases, moderate pressure absorption with physical solvents has been economical in low-temperature operations (−70° F.).

Although dissolved impurities can be released in a physical-solvent system by reducing pressure, some applications use heat to regenerate solvent. Amount of heat required in these applications, however, is much less than in chemical-solvent systems.

Absorption pressure also has a profound effect on attainable purity

Table 1

Available gas treating processes*

Chemical solvents
Amines—Monoethanolamine, diethanolamine, triethanolamine, diglycolamine, diisopropylamine
Activated hot potassium carbonate—Benfield, Catacarb, Lurgi, Vetrocoke
Others—Alkazid, ammonia, copper liquor, Scott, Stretford, tripotassium phosphate
Lesser known—Ferrox, Konox, Manchester, Seaboard, Sulfox, Thylox, sodium phenolate

Physical solvents
Fluor solvent, Purisol, Rectisol, Selexol, Sulfinol**, water

Solid bed
Activated carbon, iron oxide, molecular sieves, zinc oxide

Other
Distillation

*This is not a complete list as new processes are being constantly introduced. **A mixed solvent—included in physical solvents for convenience.

Fig. 2

H_2S removal with no CO_2 or other impurities

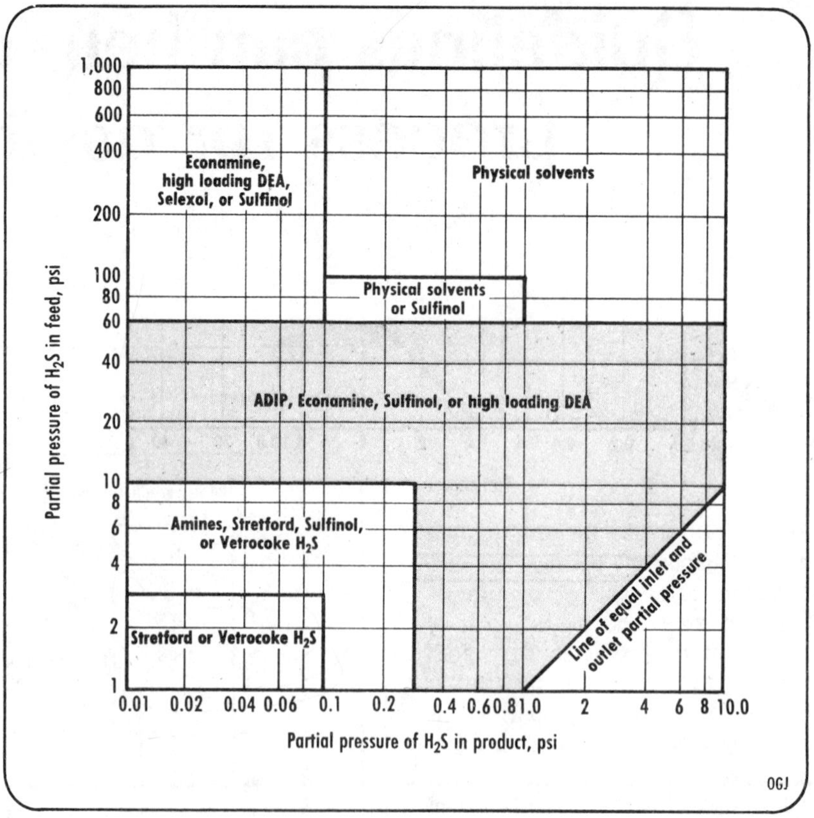

when physical solvents are used. At higher pressures, a more-complete removal of acid gas is attainable. In units designed to remove a single acid gas component from the feed gas by flashing only (no stripping) in solvent regeneration, attainable purity in product gas can be approximated by calculating:

$$C_p = 150\ P_f/P_a$$

where:

C_p = vol % acid gas in product gas
P_f = final flash pressure of lean solvent, psia
P_a = pressure at top of absorber, psia

Importance of partial pressures. Economics of acid-gas removal are influenced by partial pressure of acid gas in the feed and the corresponding value in product gas.

If partial pressure of acid gas in the feed is quite low, all physical solvents can be eliminated. If partial pressure of acid gas in the product is quite low, physical solvents can be eliminated also, unless total pressure is high.

When it has been decided whether a chemical or physical solvent, or combination of both in series, is to be used in the system, differences in properties in the selected group can eliminate other processes before comparing in greater detail is needed. For example, if physical solvents are eliminated in the choice, required product purity can help to eliminate some of the chemical processes. Amines usually reduce acid-gas concentrations to a few ppm economically, but hot potassium carbonate solutions may be economical only to 1,000-ppm levels of treating.

Side reactions between solvent and a minor constituent in the feed can also rule out one of more processes.

Chemical solvents

Amines. Although developed in the 1920's, monoethanolamine (MEA) is still widely used for removing CO_2 in feed gas when H_2S is not present. However, corrosion has been a major difficulty with MEA systems and only dilute aqueous solutions (15-20 wt %) have been used for treating. Newly developed corrosion inhibitors are now allowing concentrations of over 30 wt % MEA with resulting decreases in circulation rates.

MEA can reduce concentrations of both CO_2 and H_2S to low ppm levels, but a relatively high quantity of heat is needed for regeneration. If COS is present in the feed, MEA should not be used due to the resulting irreversible reaction and ultimate degradation of solvent.

Diethanolamine (DEA) is similar to MEA but it can be used when COS is present in feed gas. Technology has been developed for using DEA to treat natural gas with exceptionally high acid - gas loadings to meet pipeline specifications (¼ grain/100 scf, or about 4 ppm).

Although the higher molecular weight of DEA is a disadvantage, its use is economical because it can accommodate two or more times the loading of MEA.

Although diglycolamine has been known for some years as a good solvent for acid gases, it has just recently received greater attention for reducing gas-treating costs. In some installations, it has replaced MEA, but an even greater number of new plants has been put into operation. Just last summer, Fluor announced the Improved Econamine Process which further reduces treating costs.

Other amines currently in commercial use include triethanolamine (TEA) and diisopropanolamine (DIPA, ADIP).

Potassium carbonate. Although car-

Fig. 3

Simultaneous removal of H_2S and CO_2

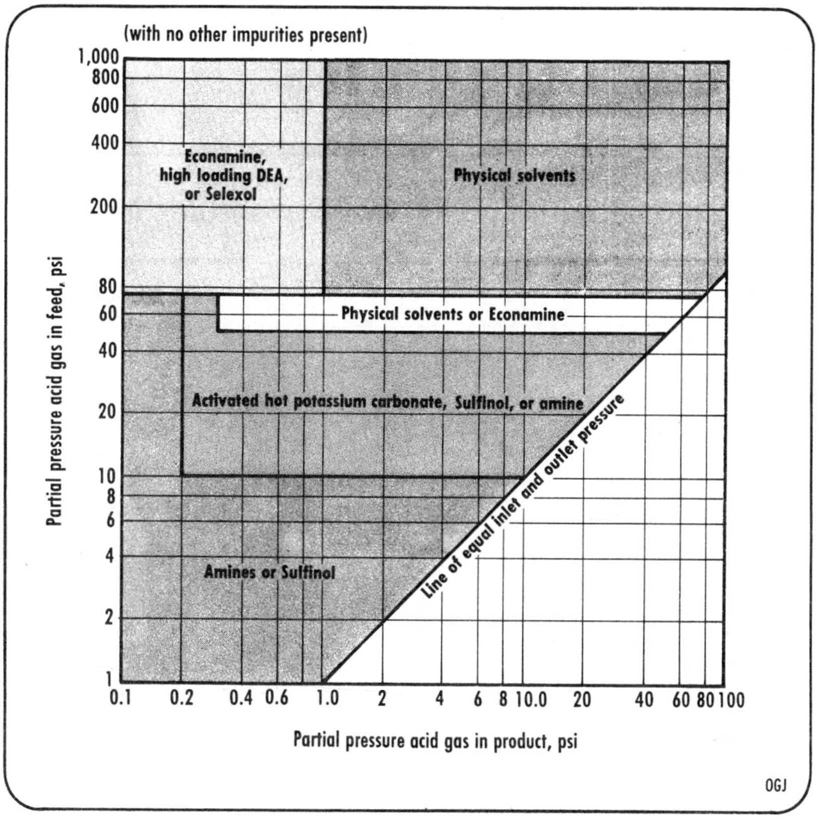

bonates react reversibly with weak acidic compounds such as CO_2, they have not been widely used before the 1960's because absorption rates were quite low, circulation rates were high, and large absorbers were needed in the process.

Several developments have led to widespread commercial application of activated hot potassium carbonate for CO_2 removal particularly in ammonia, hydrogen, and town-gas plants. Since these processes operate at high temperatures, an advantage is gained in that gas to be treated is frequently available at high temperature and must be reheated after the CO_2 removal. With these processes, the gas need not be cooled as far or reheated as much as would be needed in other CO_2 removal systems that operate at more-conventional temperatures. In addition, relatively low heat is required for regenerating hot potassium carbonate systems.

Physical solvents

Originally, water was used as a physical solvent for acid-gas removal, but its use has been discontinued due to its extremely high circulation-rate requirements.

Solvents now in use are: n-methyl pyrrolidone (Lurgi Purisol), dimethyl ether of polyethylene glycol (Allied Selexol), propylene carbonate (Fluor solvent), methanol (Lurgi Rectisol), and sulfolane. All of these solvents absorb both CO_2 and H_2S, but differences in solubilities of these two components in the various solvents causes some to be preferred for CO_2 removal, others for H_2S removal, and still others for selective removal of both.

For acid-gas removal, sulfolane is not normally used by itself, but is combined in a mixture with an amine. Shell's Sulfinol process uses a mixture of sulfolane and (usually) diisopropylamine.

Process selection

CO_2 removal, no H_2S. This treating difficulty is usually encountered in ammonia synthesis - gas purification and in natural-gas treating. Feed-gas compositions normally range from 3-4% CO_2 to as high as 65%. Absorber operating pressures are from a few inches of water to over 1,000 psig. Thus, inlet CO_2 partial pressures can vary from 1 psi to 650 psi or higher. Product-gas specifications are usually from 0.1% to 3% CO_2.

For treating gas with a low partial pressure of CO_2 (10-15 psi), amine is the preferred process. Above this range and below inlet CO_2 partial pressure of 75-100 psi, activated hot potassium carbonate solutions can be considered. Above this range, physical solvents are advantageous. These statements assume the process is capable of producing specified CO_2 content in the product gas.

Amine will reduce CO_2 content of a gas stream to about 0.005-psi partial pressure (15 ppm at 300-psi total pressure). Activated hot potassium carbonate processes can reduce CO_2 partial pressure in a product stream to 0.2-0.25 psi, and physical solvents produce gases with CO_2 partial pressures of 1-3 psi. Fig. 1 shows range of treating for various processes.

For treating gas containing high inlet partial pressures of CO_2 to produce a gas with low partial pressure, it is often economical to combine a physical solvent with an amine plant.

Physical-solvent processes do have a disadvantage which can eliminate them in favor of activated hot potassium carbonate processes. These solvents can remove some of the C_3+ hydrocarbons from the feed gas and exhaust them with waste CO_2.

H_2S removal with no CO_2. H_2S removal is usually necessary in coke-oven gas, refinery gas, or natural-gas treating. In these streams, feed - gas H_2S composition can range from 1-2 grains/100 scf to as high as 35%. Absorber operating pressures range from a few inches of water to over 1,000 psig. Product gas specifications require 0.1 to 1.0 grains/100 scf.

Selection of an H_2S removal process must consider final disposition of waste H_2S. Normally, it is converted to elemental sulfur, so Stretford and Vetrocoke processes are suitable. Amine and physical-solvent processes usually exhaust H_2S in concentrations suitable for feed to catalytic sulfur-recovery plants.

Stretford and Vetrocoke processes are preferred for feed gases containing low concentrations of H_2S. Amine processes become competitive for H_2S removal with partial pressures of H_2S near 10 psi and feed gas at about 3 psi.

Adding a sulfur-recovery plant to an

amine plant increases capital-investment requirements so that competitive range of Stretford or Vetrocoke process is increased. Although the breakpoint is not well established, these processes should be compared with an amine plant combined with a Claus-type sulfur-recovery plant.

With larger plants, amine-Claus sulfur-recovery combinations are favored. Above 60-psi H_2S inlet partial pressure, physical solvent processes should be considered. Normally, conventional physical-solvent processes cannot reduce H_2S concentration in product gas to meet ¼ grain/100 scf specifications so it may be necessary to use an amine "cleanup" after them. However, the Selexol process is an exception. Fig. 2 compares each process and assumes that each, except Stretford and Vetrocoke processes, is followed by a sulfur-recovery unit. These data also assume no other impurities are present.

Simultaneous CO_2 and H_2S removal. Removal of both CO_2 and H_2S from a gas stream is the most-common treating problem met, and most of the processes previously mentioned can perform this. Amine, activated hot potassium carbonates, and physical solvents can remove both H_2S and CO_2 in the same absorber and exhaust the two components together in about the same ratio in which they were present in the feed. In some designs, configurations can be changed to remove these components completely. If H_2S concentrations in the exhaust is 15% or higher, this gas can be used as feed to a catalytic sulfur plant.

Amine or Sulfinol systems should be closely considered for lower feed partial-pressure ranges. In middle-pressure ranges, activated hot potassium carbonate systems can be included but the effect of the ratio of CO_2 to H_2S must be considered in the preliminary process selection. If CO_2-to-H_2S ratio is high, activated potassium carbonate systems are competitive; if it is low, Sulfinol process is a better choice. However, high-concentration amines, DEA or Econamine, should be included in considerations also.

As shown in Fig. 3, boundary areas between processes depend to some extent on the ratio of CO_2 to H_2S in the feed gas. This is true for the physical-solvent lower-use boundary as well as for those previously discussed. Presence of C_3+ hydrocarbons will raise partial pressure of acid gases at which physical solvents are competitive.

Fig. 4

Selective H_2S removal in presence of CO_2

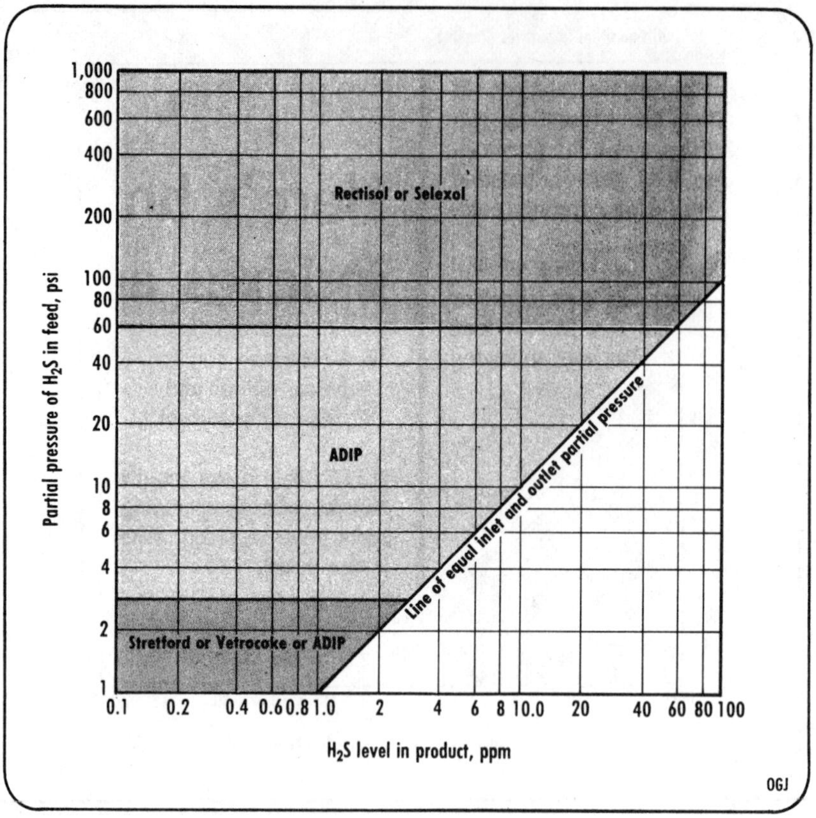

Selective H_2S removal. Frequently in natural-gas treating, it is desirable to remove H_2S to low residual levels without removing CO_2. In other applications, it may be desirable to remove H_2S from CO_2. In both cases, a selective process for H_2S is an economical choice.

Among the large number of acid-gas-removal processes available, only a few are sufficiently selective for this purpose. Although only five processes are shown in Fig. 4, others are available which are economically sound under specific circumstances.

At low partial pressure, Stretford or Vetrocoke H_2S processes can reduce H_2S content of gases to low levels. Each of these processes has a disadvantage, however. Stretford process is limited in quantity of CO_2 it can tolerate in the feed gas and Vetrocoke H_2S contains arsenic which many operators do not want to use.

In the medium-pressure range, ADIP is usually more competitive. Higher-pressure areas usually use selective physical solvents such as Rectisol and Selexol.

Other impurities. Process schemes must be revised due to certain impurities in feed gas, such as NH_3, HCN, COS, CS_2, SO_2, SO_3, mercaptans, and hydrocarbons or tars.

If MEA is used, HCN, COS, SO_2, and SO_3 can react with it and partially deactivate the solution. During regeneration, a portion of the COS and CS_2 are freed from the MEA and return the amine portion to its active state. MEA combined with HCN, SO_2, and SO_3, along with a portion of that still combined with COS and CS_2, can be reclaimed to its active state by treating it with sodium carbonate or caustic soda in the reclaimer. Thus, permanent MEA loss from these impurities will be caused by only a portion of the COS and CS_2 present in the gas to be treated.

If the sour gas contains appreciable

amounts of COS and CS_2, then DEA would be an economical choice since it is not affected by these impurities.

COS and CS_2 are not harmful to activated hot potassium carbonate processes either. These impurities are partially hydrolyzed to CO_2 and H_2S and absorbed as such. Nonhydrolyzed portions pass through with the product stream. HCN is almost completely absorbed and is removed from the solution along with CO_2 and H_2S.

SO_2 and SO_3 react irreversibly with hot potassium carbonate solutions. If these impurities are present in significant quantities, costs for prewashing them from feed gas vs. possible solution loss should be carefully compared.

As example of the effect of impurities on process selection, consider removal of CO_2 and H_2S from a feed gas at 100 psig with the following composition.

CO_2 = 30%
H_2S = 1.5%
COS = 0.1%
HCN = 0.1%
CO, H_2, and light hydrocarbons = 68.3%

Using data shown in Fig. 3, considering only concentrations of CO_2 and H_2S, this treating could be handled most economically by using either an activated hot potassium carbonate process or an amine process. Presence of the large quantity of COS makes chemical consumption excessive if MEA is used and would dictate use of DEA. Reclaiming would be a necessity if either DEA or MEA were selected since the inlet gas contains HCN.

Since neither COS or HCN contribute to solution degradation and reclaiming is not needed, the presence of these and other impurities (including hydrocarbons) favors the selection of the hot potassium carbonate process and it would probably be selected after a detailed investigation of the available processes. END

Improve economics of acid-gas treatment

Fig. 1

Simple flow chemical-solvent process

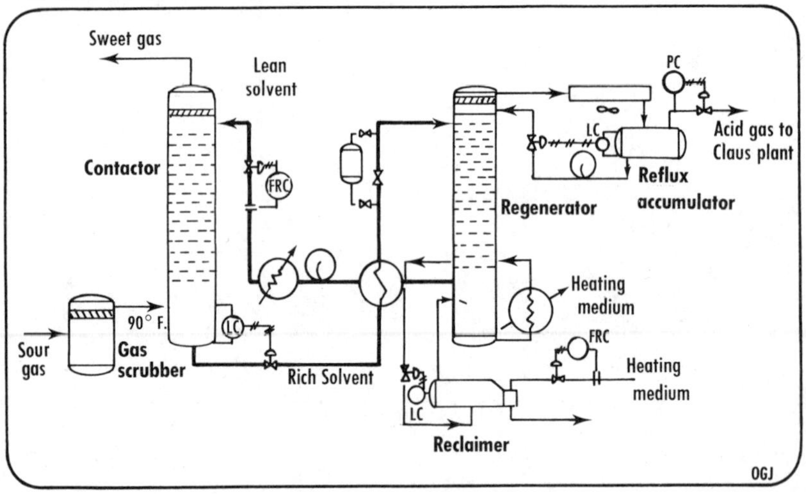

Table 1

Comparison of acid-gas treatment solutions

Function	MEA	DEA	K_2CO_3	Sulfinol	DGA
Reconcentrator solution conc. wt %	15	25	30	80	50
Reconcentrator rich solution loading, mol AG/mol solu.	0.4	0.4	(3)	(3)	0.3
Reconcentrator stripping rate, lb/steam/gal*	1.2	1.2	0.6	0.8	1.35
Spec. sweet gas without add. treatment†	Yes	Yes	No	Yes	Yes
Interaction with hydrocarbons (C_3+)	No	No	No	Yes	Slight (0.01 mol%)
Removal of RSH	Part.	Part.	No	Yes	Yes§
Removal of COS	Yes	Yes	Yes	Yes	No
Nonregen products due to RSH	No	No	No	No	No
Nonregen products due to COS	Yes	No	No	No	No
Reclaimer required	Yes	No	No	Yes	Yes
Royalty payment	No	No	No	Yes	No‖
Chemical cost per gal of solution, $¶	0.21	0.28	0.44	2.00	1.41
Solution loading, ft³ AG/gal	3.2	3.05	‡5.0	‡5	4.7
Stripping rate, lb/steam/ft³ AG	0.37	0.39	0.12	0.16	0.29

*Varies with H_2S/CO_2 ratio. †Pipeline specification—typical at ¼ grain/100 cu ft ≈ 6 ppm and 0.01 vol % CO_2. ‡Varies considerably with AG partial pressure in sour gas. §Reported 100 grains inlet RSH down to 0.2 grains at 120-psi pressure with 5% H_2S, 6% CO_2 inlet. ¶Prices are relative. Check your local chemical dealer for up-to-date prices. ‖Patent rights retired 1972.

ROBERT S. SMITH
Worley Engineering Inc.
Houston

THE recently increased price of energy and associated operating expenses has put pressure on operators to improve economies of acid-gas treatment plants.

Acid-gas treatment plants lend themselves to improved economics on two fronts: (1) change acid-gas treatment chemical, and (2) improve energy efficiency of process.

A change in acid-gas treatment chemical could reduce circulation rate by increasing acid-gas loading, reduce "stripping rate" in the reconcentrator, or improve removal of mercaptan and carbonyl sulfide contaminants. Table 1 compares five commonly used gas-treatment chemicals for "energy-related" properties.

The acid-treatment process may be changed to dramatically reduce energy costs. Figs. 1, 2, and 3 illustrate the process capital equipment complexity increasing in efforts to decrease energy costs.

Simple-flow process. The simple flow process (Fig. 1) is generally used for MEA, DEA, DGA processes for treating gas streams of between 5-100 MMscfd.

The economies of this process may be improved by changing to a solvent of higher acid-gas loading properties.

Split-flow process. Fig. 2 illustrates this more-complex process. An energy-conservation measure has been introduced with this process by additional heat-recovery equipment. Increased energy efficiency may be examined without capital improvement by a solvent change that would allow higher acid-gas loading.

Power-recovery process. Capital expenditure is required to convert to the power-recovery solvent process

(Fig. 3). Utility costs for fuel and electricity are reduced measurably by this process. The cost of capital equipment to recover this energy loss becomes important.

Economics of converting to this process must be examined on an individual basis. Generally, an economic justification can be made for conversion of high-pressure plants with throughput greater than 50 MMscfd.

Example problem

A plant operating at 100 MMscfd at 1,000 psi with 6% H_2S and 4% CO_2 wishes to examine the economics of: (a) converting from MEA to DGA; and (b) converting from a simple flow to a power-recovery system.

Step 1: Determine new solution circulation rate, compare with old rate.[1]

1. $$\text{Gpm}_{MEA} = 20.6 \times \frac{(\text{MMscfd})(\text{AGMF})(\text{MW of solvent})}{(\text{AG mole/mole solvent})(\text{wt \% solvent})}$$
$= 20.6\ (100)\ (.10)\ (61.1)\ /\ (0.4)\ (15)$
$= 2{,}097$

Where

AGMF = acid-gas mole fraction.

2. $\text{Gpm}_{DGA} =$
$20.6\ (100)\ (10)\ (105.1)\ /\ (0.3)\ (50)$
$= 1{,}443$

Step 2: Determine energy savings from MEA to new solvent.

$\text{bhp} = 0.0005833 \times \text{gpm} \times \Delta P_{psi}\ /\ \text{efficiency}$
$= 1{,}528_{MEA},\ 1{,}052_{DGA}$

$\text{kw/hr} = 300\ \text{kw-hr}\ @\ \$0.030/\text{kw-hr}$
$= \$75{,}600/\text{year}$

This will justify change-out from a larger to a 1,100-hp driver/pump. The motor starter would not need be changed. Some salvage credit could be assigned to the larger motor driver.

Fig. 2

Split flow chemical-solvent process

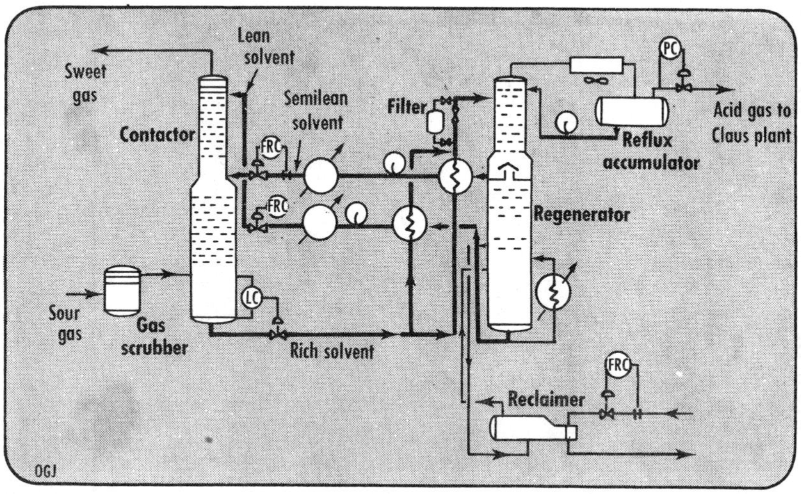

Fig. 3

Power recovery chemical-solvent process

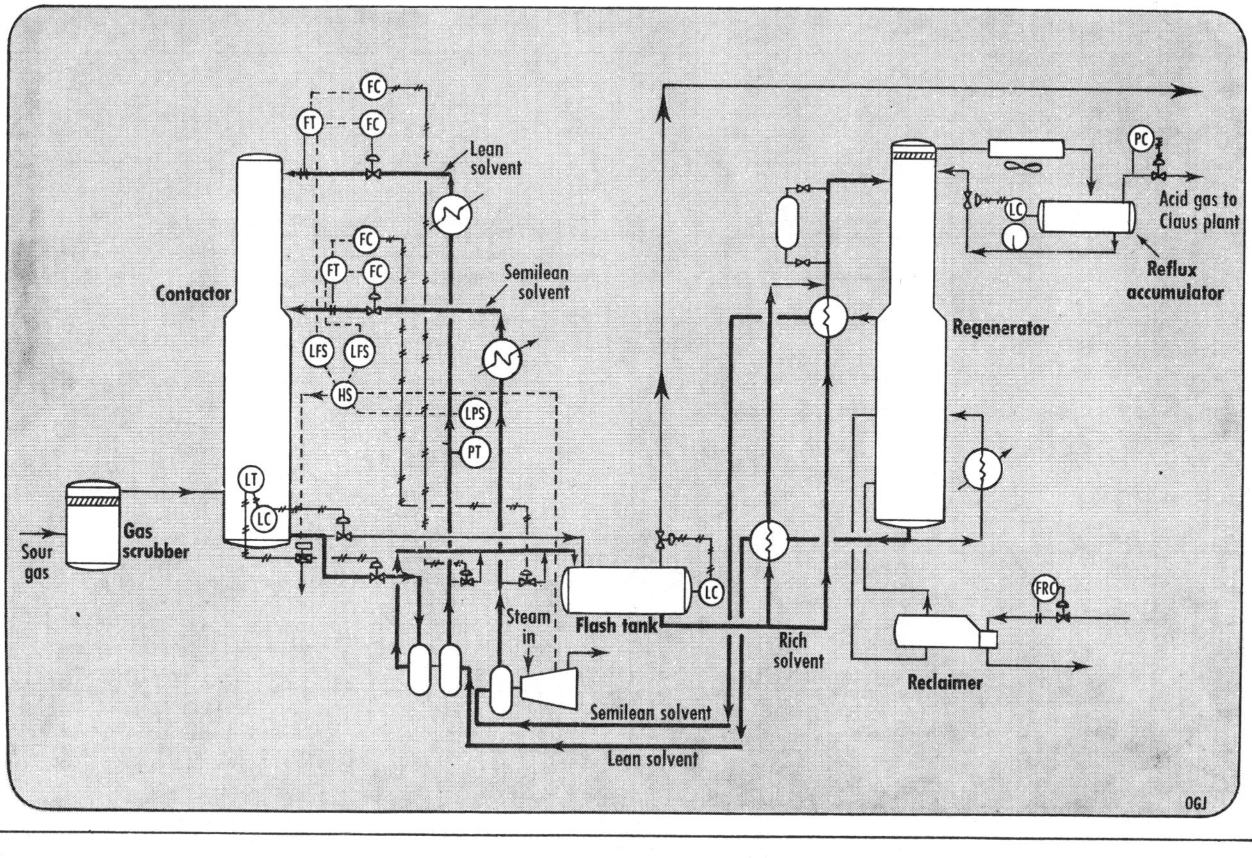

Step. 3: Determine capital and operating cost of power recovery vs. simple process.

Although several power-recovery configurations are possible, a conservative control, start-up, and recovery option is pursued as shown on Fig. 3.[2]

A power recovery of 60% of main solvent booster pump is examined. Fig. 4 illustrates the total operating costs of a solvent pump at various utility charge rates. At $0.030/kw-hr, our incremental power savings at 60% power recovery would be $124,000/year.

Capital cost for conversion is estimated at $180,000. If a capitalized operating cost is used, escalating at 7%/year for 20 years, the following simple comparison can be made:

Capital cost = $180,000
Capitalized operating cost saving
= 11.65 × $124,000
= $1,444,000

Fig. 4

Total operating cost

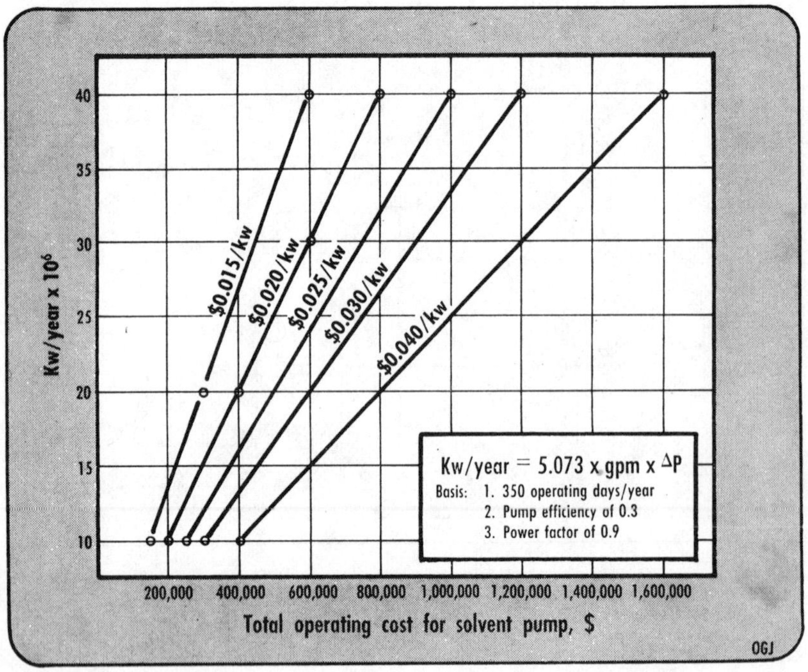

Although the main thrust of energy saving will be modifications to equipment and type of solvent, additional capital savings may result in a new plant design.

One area of savings is tower design. The gas-treatment plant has a gas contactor and a solvent reconcentrator. Generally, the gas-contactor sizing is controlled by the inlet gas rate.

However, the reconcentrator sizing is a direct function of solvent circulation rate. Figs. 5 and 6 illustrate the size of solvent reconcentrators as a function of solvent rate with other parameters held constant.

A decrease in solvent circulation rate by change-out to a higher acid-gas loading solvent may make a dramatic difference in capital cost of the solvent reconcentrator by reducing the tower diameter and resulting tray diameter.

References

1. Campbell, John M., "Gas Conditioning and Processing," Second Edition.
2. Jenett, Eric, "Hydraulic Power Recovery Systems," Chem. Engr., June 17, 1968.

Figs. 5, 6

Solvent reconcentrator

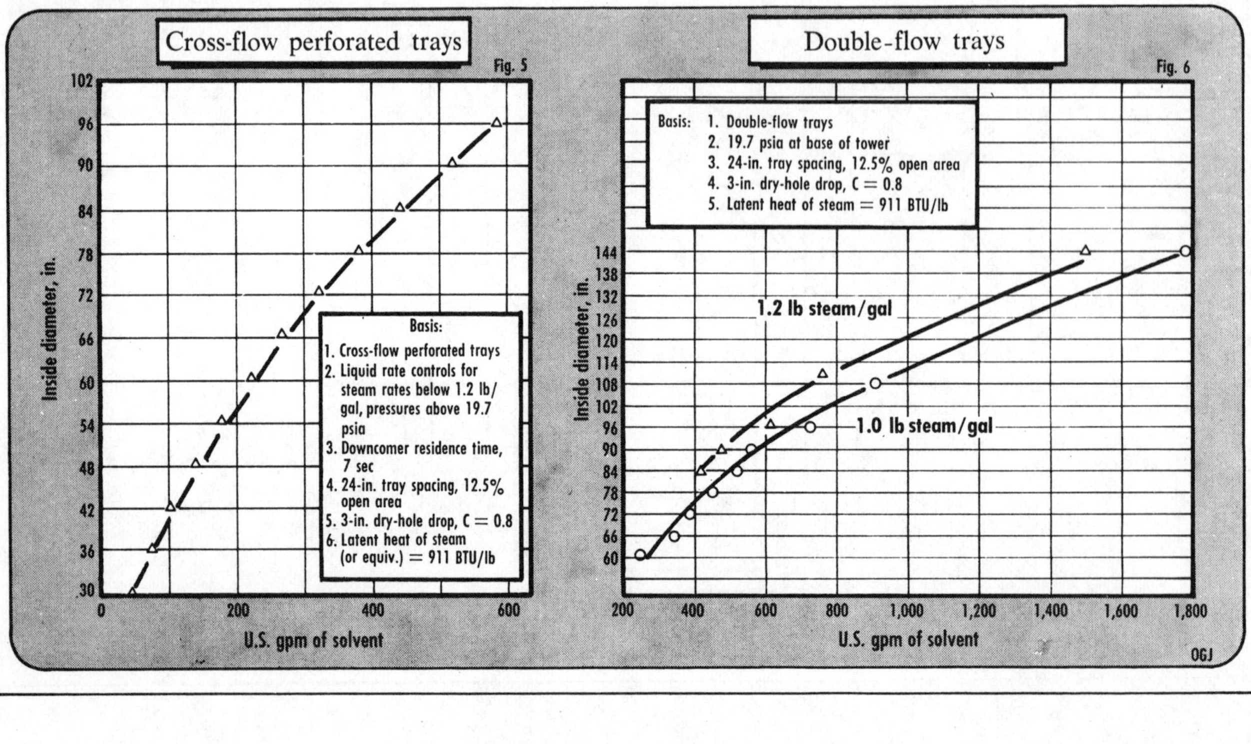

Gas-process-plant conversion cuts energy use, emissions

D. K. JUDD, Northern Natural Gas Co., Midland, Tex.

NORTHERN Natural Gas Co. has recently and successfully completed a major in-house plant revision, conversion of its Mitchell plant near Fort Stockton, Tex., from DEA solvent to a physical solvent, Allied Chemical's Selexol.

The revision has reduced fuel-gas consumption to near zero. An ecological benefit was also gained as the plant overall emission level is now practically nil.

Plant evolution. Northern Natural Gas Co.'s Mitchell plant was built in the mid-1950s in conjunction with development of Puckett field. Located 35 miles Southeast of Fort Stockton, Tex., the plant processes a gas from the Ellenburger formation that contains 28 mol% CO_2.

The plant was built in three stages. Stearns-Roger engineered the first installation; Dresser Engineering did the last two. The original plant was a small MEA treating unit which came on stream in 1954. It was designed to process 35 MMscfd of inlet gas, producing 25 MMscfd of residue gas with less than 1 mol% CO_2.

The plant used steam turbines as prime movers, and turbine exhaust steam provided regeneration heat. Therefore, a steam-boiler installation accompanied the plant. Since the plant location was remote and the electrical requirements small, electricity was generated on site.

In March 1957, plant capacity was increased to 145 MMscfd inlet (105 MMscfd residue) gas. This was done by addition of six identical, large absorbers and four identical regeneration units. Additional steam-boiler and cooling-tower capacity was installed to meet utility requirements.

In May 1958, plant capacity was again increased, to 201 MMscfd inlet gas. Three additional absorbers, two complete regeneration trains, and two more steam boilers were commissioned at this time. The Mitchell plant was now capable of delivering 145 MMscfd of residue gas to the pipeline.

The plant additions of 1957 and 1958 consisted of identical process modules. Each process module consisted of an absorber unit and two complete regeneration trains. The multiple-unit concept reduced investment costs while providing a high degree of flexibility.

In the 1950s, the Mitchell plant was one of the largest MEA plants in existence for removing high concentrations of CO_2 from natural gas. At full capacity, the plant circulated 14,000 gpm of 20% MEA, consumed 960,000

Adapted from paper, "Northern Natural Gas Converts Mitchell Plant to Selexol," presented at 28th Annual Gas Conditioning Conference, University of Oklahoma, Norman, March 6-8, 1978.

lb/hr of steam, vented 56 MMscfd of CO_2 and 45 MMscfd inlet gas was burned as fuel. The plant equipment contained 475,000 gal of MEA solution, and there were literally acres of aerial coolers within the plant.

In 1974, the Mitchell plant treating solution was changed from MEA to DEA, and boiler-feed water economizers were added. The DEA process requires less heat input than the MEA process for regeneration and because of its less corrosive nature, the concentration can be higher.

Circulation rate was dropped to about 10,000 gpm and the fuel-gas requirement dropped to about 25 MMscfd.

Part of the reduction in solution rate and steam was due to slightly lower throughput as the plant residue gas volume had declined to about 130 MMscfd. Because of the decline, the original Stearns-Roger unit was shut down in 1973.

In 1975, the DEA operation was modified to high concentration, high loading. The concentration was increased from 25 to 30 wt % DEA. Circulation rates were dropped to about 8,500 gpm, and fuel-gas consumption dropped to 20 MMscfd.

At the same time, the Mitchell operation was modified to high load DEA, exploratory design studies were begun for conversion to a physical solvent. Northern had one existing Selexol plant operating in the Fort Stockton area, the Oates plant. Operating experience with the Oates plant had been favorable after overcoming initial operating problems.

The Mitchell plant feed gas, containing more CO_2 (28 mol% vs. 18 mol%) and less H_2S (1 grain/100 scf vs. 8 grain/100 scf) is a better feed gas for this particular process.

The design studies for conversion of the Mitchell plant to Selexol solvent proceeded favorably. The project was officially authorized in late 1976 and construction completed in January 1978. Plant start-up took place during late January and February 1978.

Design considerations. The composition of the Ellenburger gas is given in Table 1. This is not expected to change significantly over the remaining years of the field life.

The gas availability is expected to decline steadily in the future. The projected rates through 1990 are:

Year	Ellenburger gas flow, MMscfd
1978	146
1979	132
1980	121
1981	110
1982	101
1983	92
1984	84
1985	75
1986	69
1987	63
1988	52
1989	48
1990	44

The preceding rates reflect changing compression loads as well as declining field rates, hence do not follow a smooth curve.

In the Mitchell DEA operation, about 10% of the incoming gas was burned as fuel to generate steam. Steam turbines powered the process pumps, and the turbine exhaust was used to provide reboiler heat. Raw inlet gas was burned directly in the boilers, eliminating purification costs for the fuel gas. Inspection of the above flow rates through 1990 reveals that about 35 Bcf of raw gas would be burned as fuel over this period.

A physical-solvent plant would allow the boilers to be shut down. To provide an alternate source of power, a very basic decision was made to purchase electricity for the process needs. This is not the cheapest option relative to operating costs, but is economic and is in the correct direction for future energy usage. It permits coal to be used to generate electricity while putting natural gas into the pipeline for heating homes. It provides the equivalent of 25 Bcf of new gas reserves.

Table 2 compares the material balances for the three basic operational phases, MEA, HL-DEA, and Selexol solvents. The comparison is made at an assumed constant inlet gas rate of 200 MMscfd. This is reasonable for past operations with MEA and DEA but is entirely fictional for Selexol.

The inlet rate has now declined to about 150 MMscfd, but comparison at constant rates give a more lucid picture as to the advantages of the physical solvent process.

Selexol process at Mitchell. The Selexol process technology is licensed from Allied Chemical Corp. The Selexol solvent is the dimethyl ether of polyethylene glycol. It is nontoxic, noncorrosive, and has a very low heat of absorption. Selexol has a low vapor pressure which minimizes solvent losses to the residue gas. Selexol will dehydrate the gas.

The Selexol process differs from the DEA process in that no chemical bond is formed with the acid gases. Selexol is a physical solvent and works best when the acid-gas concentration and

Table 1

Composition of Ellenburger gas to Mitchell plant

Component	Mol percent
N_2	0.45
CO_2	27.85
H_2S	0.0013*
C_1	70.35
C_2	0.83
C_3	0.22
iC_4	0.05
nC_4	0.08
iC_5	0.03
nC_5	0.03
C_6	0.05
C_7+	0.06
	100.00

*0.8 grain/100 scf.

operating pressure are high and the driving force is great.

Low process temperatures allow the Selexol to hold more acid gas—this can be visualized from the absorption factor, $A=L/VK$. Pressure reduction alone will regenerate the solvent in most cases. Auto refrigeration from the flashes in Northern Natural's case, reinforced by a small mechanical refrigeration unit, provides the process chilling. An inlet heat exchanger recovers the maximum practical amount of cooling from the residue gas.

Rich Selexol from the absorber section flows to a sump tank whose purpose is to allow separation of the entrained methane which is returned to the absorber as recycle. The solvent is regenerated via a series of three flashes at successively lower pressures.

The first, or high-pressure, flash releases a gas containing about 50% methane which is recycled to the plant inlet by a centrifugal compressor. The gas from the other two flashes is rejected and delivered to the adjacent Sacroc compressor plant. Sacroc further compresses the CO_2-rich gas and pipelines it north to the Snyder, Tex., area for injection into an oil-producing formation as a means of secondary recovery.

The rich Selexol is let down through two stages of hydraulic turbines and the energy recovered is used to pump the lean solvent back to the contactors. It is necessary to supplement the turbine power with an electrical motor to pump the entire solvent stream.

Process flow. Fig. 1 illustrates the process flow for the Mitchell Selexol plant. Table 3 gives the design material balance. Gas from the field enters the plant inlet scrubber at

Table 2

Overall plant material balances

Component	MEA solvent Inlet	MEA solvent Fuel	MEA solvent Residue	HL-DEA solvent Inlet	HL-DEA solvent Fuel	HL-DEA solvent Residue	Selexol Inlet	Selexol Fuel	Selexol Residue
	Mol/Hr								
N_2	98.8	14.1	84.7	98.8	10.3	88.5	98.8	..	98.8
CO_2	6,115.5	875.6	62.2	6,115.5	636.8	71.3	6,115.5	..	550.0
H_2S	0.3	......	......	0.3	......	......	0.3	..	......
C_1	15,448.0	2,211.8	13,236.2	15,448.0	1,608.7	13,839.3	15,448.0	0.9	14,907.3
C_2	182.3	26.1	156.2	182.3	19.0	163.3	182.3	..	156.2
C_3	48.3	6.9	41.4	48.3	5.0	43.5	48.3	..	0.5
iC_4	11.0	1.6	9.4	11.0	1.0	9.9	11.0	..	0.2
nC_4	17.6	2.5	15.1	17.6	1.8	15.8	17.6	..	0.4
iC_5	6.6	0.9	5.7	6.6	0.7	5.9	6.6	..	0.3
nC_5	6.6	0.9	5.7	6.6	0.7	5.9	6.6	..	0.4
C_6+	24.2	3.5	20.7	24.2	2.5	21.7	24.2	..	2.2
	21,959.2	**3,143.9**	**13,637.3**	**21,959.2**	**2,286.5**	**14,265.1**	**21,959.2**	**0.9**	**15,716.3**
MMscfd	200.0	28.6	124.3	200.0	20.8	129.9	200.0	0.008	143.1
BTU/scf, HHV dry	......	......	1,025.3			1,025.0	......	..	976.5

about 120° F. and 895 psig.

The gas is cooled in feed residue exchanger E-101 by exchange with cold residue gas from the contactor residue gas scrubbers.

The residue gas enters the pipeline at about 90° F. The cool inlet gas next flows through feed-gas scrubber V-101 where condensed water is removed. After scrubber V-101, the gas is equally divided among four absorbers, V-102A, B, C, and D. There are two parallel rows of two absorbers each, and each row of two absorbers shares a common residue-gas scrubber, V-103A and V-103B.

Each absorber is packed with 1-in. Koch Flexirings. Cold lean Selexol enters at the top and countercurrently contacts the gas. The cold residue gas leaving each contactor contains about 3.5 mol% CO_2. The rich Selexol from each absorber, in addition to absorbed CO_2, contains both absorbed and entrained methane.

The rich Selexol flows by gravity through a common header to the sump tank, V-104. The sump tank merely provides a few minutes hold time to allow entrained methane to separate and be returned to the contactors.

A small, inline centrifugal blower, C-101, is used to return the methane vapors. The sump tank/methane blower indirectly controls the level in the four contactors. The only level control in this section of the plant is that in the sump tank.

From the sump tank, the rich Selexol flows to E-102, the solvent chiller, where it is cooled by a mechanical refrigeration unit. This is just enough to overcome the heat of pumping and heat leaks, thus providing a controlled temperature level for the lean solvent.

The chilled Selexol temperature is controlled automatically by regulating the inlet vanes of the refrigeration compressor. The temperature of the vacuum flash drum is indirectly controlled by this trim refrigeration.

After solvent chiller E-102, the rich Selexol flows through the high-pressure solvent turbine, T-101A, and then enters the high-pressure flash drum, V-105, at 252 psig. The work done in turbine T-101A is just enough to drive one 50% process pump, P-101A.

Back pressure is held on the lean Selexol downstream of pump P-101A so as to maximize the work by turbine T-101A. Pump P-101A, therefore, pumps against the minimum line pressure commensurate with getting solvent back to the contactors.

The vapor phase from the high-pressure flash drum is returned to the plant inlet via C-102, the recycle compressor. This is a centrifugal machine powered by a 2,000-hp electric motor.

The semi-rich Selexol leaves flash-drum V-105 and flows to T-101B, the low-pressure solvent turbine. The turbine power developed is not adequate to run a 50% process pump hence turbine T-101B is coupled with a full-power electric motor. The motor "coasts" at reduced output under normal conditions and the combined turbine T-101B/motor power drive pump P-101B.

Normally, pumps P-101A and P-101B

Table 3

Design material balance

Mol/hr Mitchell plant Selexol conversion				
Stream No.*	1	2	3	4
Composition	Feed gas	Residue gas	Atmospheric flash gas	Vacuum flash gas
N_2	71.1	71.0	0.1	....
CO_2	4,398.3	395.7	3,490.5	513.0
H_2S	0.2	....	0.2	....
C_1	11,110.2	10,721.4	388.1	0.7
C_2	131.1	112.3	16.5	2.3
C_3	34.9	0.5	30.7	3.7
iC_4	7.9	0.2	6.4	1.3
nC_4	12.7	0.4	10.0	2.3
iC_5	4.8	0.3	3.2	1.3
nC_5	4.8	0.4	3.1	1.3
C_6+	17.5	2.2	10.5	4.8
H_2O	37.6	0.2	3.9	1.7
	15,831.1	**11,304.6**	**3,963.2**	**532.4**
Lb/hr	383,430	195,004	164,195	23,752
Mol wt	24.22	17.25	41.43	44.53
MMscfd	144.0	102.83	36.05	4.85

*Stream numbers identify with Fig. 1.

Fig. 1

Mitchell Selexol plant process flow

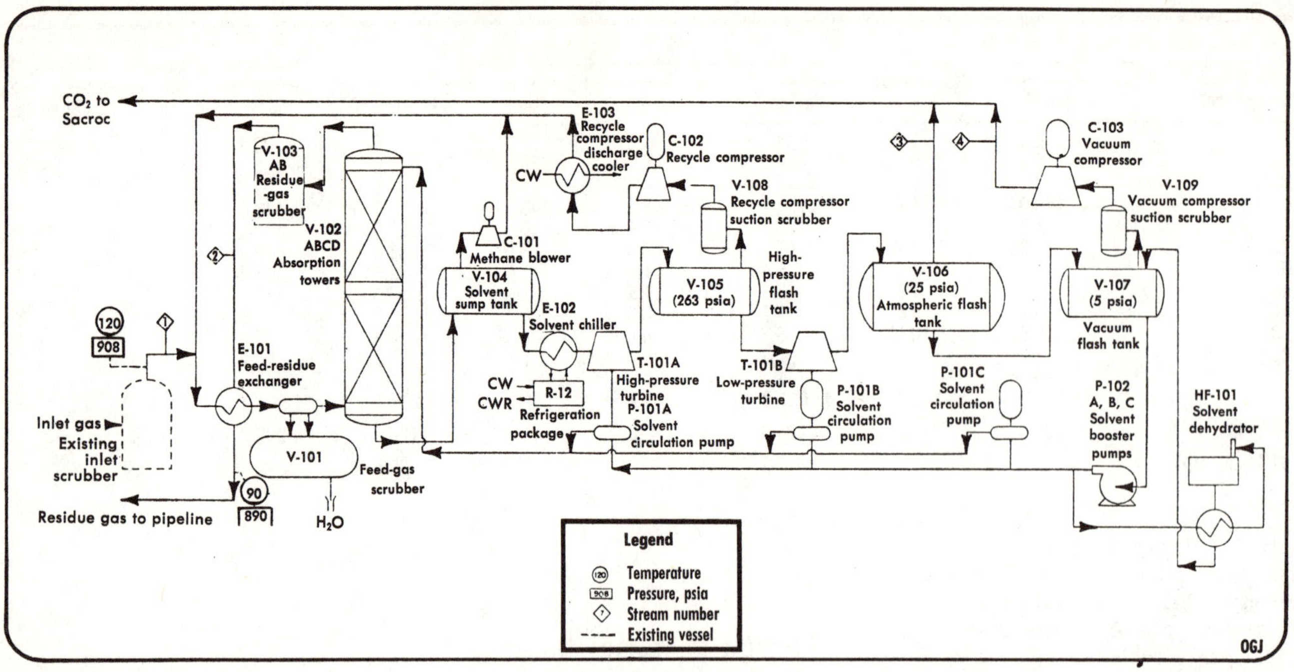

provide the total high-pressure pumping requirements. Pump P-101C (100% motor) is used only for start-up.

From turbine T-101B, the semi-rich Selexol flows to V-106, the atmospheric flash drum which operates at 25 psia. The vapor separated in drum V-106 is over 90% CO_2 and flows to the adjacent Sacroc compressor station.

The atmospheric flash-drum pressure level is adequate to move the gas to Sacroc. Drum V-106 acts as the system surge drum. There is no level control on the vessel, and the level fluctuations are observed and manually corrected.

The semi-lean Selexol next flows to V-107, the vacuum flash drum which operates at 5 psia. The vapors evolved, now over 95% CO_2, are compressed by C-103, the vacuum compressor, to 23 psia and are combined with the atmospheric flash drum vapors for delivery to Sacroc. Compressor C-103 is a centrifugal type driven by a 700-hp electric motor.

The lean Selexol, now fully regenerated, flows to solvent booster pumps, P-102A, B, and C. The vacuum flash drum is elevated to provide NPSH for these pumps.

Again, each is a 50% pump, and two are normally running to provide process capacity. Pumps P-102A and B boost the lean Selexol to about 50 psig at which pressure the solvent flows to the suction of pumps P-101A and B. From pumps P-101A and B, the Selexol enters the high-pressure solvent header and the loop is complete.

A slipstream of lean solvent is taken from the discharge of pump P-102 and charged to HP-101, the solvent dehydrator. Selexol dehydrates the gas as it removes CO_2 and thus gains water, part of which is flashed off with the CO_2 in flash tanks V-106 and V-107.

Most has to be removed by solvent dehydration. A small glycol reconcentrator was adapted to this purpose. The dehydrated slipstream of Selexol is returned to tank V-107 by pressure flow.

Process features. Numerous operating difficulties were encountered during the initial operating phase of Northern Natural's Oates plant, also near Fort Stockton, Tex. The experience gained was applied to the Mitchell plant design. Perhaps the most notable points concern process refrigeration and sulfur precipitation.

In the case of refrigeration, the Oates plant mechanical unit was not capable of removing the entire heat of pumping. The Mitchell design incorporates a 600-ton packaged refrigeration unit (R-12 refrigerant) which not only covers this but provides a small excess of mechanical refrigeration. In addition, the auto refrigeration effect is slightly greater at Mitchell because of the higher CO_2 content of the inlet gas.

Finally, two stages of hydraulic turbines are used at Mitchell, vs. a single turbine at Oates. The maximum practical use of hydraulic turbines further improves the refrigeration balance as well as providing substantial power.

The final regeneration step at the Mitchell plant is by vacuum flashing. This is a major innovation with the Selexol process. The Oates plant used air stripping in the original design, which, in the presence of sulfur, promotes sulfur precipitation. Sulfur deposits, found throughout the plant, were perhaps the most serious problem during the early phases of the Oates plant.

Various stripping alternatives were considered for Mitchell. Vacuum flashing appeared the most attractive. Design calculations indicated that an equilibrium flash at 5 psia would completely regenerate the Selexol.

To accomplish this, a centrifugal vacuum compressor was incorporated into the design. The vacuum compressor boosts the gas to 23 psia so that it can be combined with the atmospheric flash gas and delivered to the adjacent Sacroc plant.

Another major innovation is the sump-tank arrangement. In the Selexol process, a sump tank at the base of the contactor allows entrained methane to separate from the rich solvent. Normally, the methane flows back to the absorber via an equalizing line, but in the Mitchell plant, four contactors are served by one sump tank.

In addition, the sump tank is not symmetrically located with respect to the contactors. There are two parallel rows of contactors, two contactors per row. The sump tank, a horizontal vessel is located adjacent to, and parallel with, the north row of contactors. The unique layout is solely due to the lack of space.

The rich solvent from the contactors gravitates to the sump tank. Only the sump tank has a level control, and the level in this vessel thus sets the levels in the contactors. A small in-line centrifugal compressor recycles the separated vapor back to the plant inlet. This also permits a degree of level control since the sump-tank pressure can be pulled down several pounds via the compressor control.

The Selexol dehydration unit is merely a small glycol reconcentration skid that removes water (and a small amount of hydrocarbons) from a slipstream of solvent. Selexol dehydrates the process gas and thus picks up water which must be removed to maintain the system water balance.

A reboiler is ultimately required. This is the only item in the process which consumes fuel gas. It is a nominal 250,000 BTU/hr reboiler and requires practically no energy when compared to the 960,000 lb/hr stream formerly generated.

Because of the low inlet temperature, the dehydration unit reboiler has a far more efficient heat exchanger than typical glycol reconcentrators. The cold Selexol enters from the vacuum booster pump discharge and is preheated by an external, multiple hair-pin exchanger. This, again, saves energy.

A mild innovation is the use of a centrifugal compressor to recycle the high-pressure flash gas back to the plant inlet. The capacity of the machine is well under 1,000 acfm—this on the low side of capacity for centrifugal compressors. The compressor itself is dwarfed by the electrical motor.

Selexol is an expensive solvent, and precautions have been taken to prevent undue loss. First, all major process pumps have double seals with high-pressure alarms between seals to indicate leakage. Second, a large blowdown drum was installed in the relief header coming from tanks V-104, V-105, V-106 and V-107.

These vessels contain most of the plant Selexol inventory. Should a relief valve open, any solvent blown out will be collected and not lost. A sump tank collects the Selexol from the drain system. It can be pumped back to the storage tank.

The hydraulic turbines were purchased with multiple impeller/nozzle sizes. This will permit changing the internals to provide the most efficient operation as gas rates (and required solvent rates) decline.

The author . . .

Donald K. Judd

Donald K. Judd is a senior project engineer for Northern Natural Gas Co., Midland, Tex. He joined the firm in 1974 as Midland Area process specialist. Current job duties primarily involve process design, technical services, environmental problems, and economic studies. Prior affiliations include Continental Oil Co., Ponca City, Okla.; Ortloff Corp., Midland, Tex.; and Cosden Petroleum Corp., Big Spring, Tex. Judd holds a BS degree in chemical engineering from Georgia Institute of Technology (1959) and has completed 27 hours of post graduate work in chemical engineering and mathematics. He is a registered professional engineer in Texas.

All turbines and the major pumps are designed for maximum head and 50% flow rate. The major pumps have a 50% capacity common spare.

Construction. The final design was done by the Northern Natural Gas Engineering Department in Omaha. The construction was bid in three phases: civil, mechanical, and electrical. Three contractors thus worked on the plant. Dresser Construction did the civil work; Ortloff Corp. did the mechanical, which was the bulk of the construction work; and Gulf States did the electrical.

The only major equipment in the DEA plant that was reused were the four contactors and the cooling tower.

In order to build the unit around the contactors, it was necessary to fit the remaining equipment in a relatively compact area. The site was further limited by underground piping including the main line, a meter shack, and a small compressor station.

The resulting compact layout allows ready access to the equipment most likely to require maintenance. It allows reasonable access to all other equipment.

In order to facilitate the changeover, many revisions were effected during the annual shutdown in August 1977. The four contactors to be used for Selexol were permanently shut down, and the remaining contactors and accessories were modified to allow operation during final construction.

The operating plan was carefully developed so as to permit full-capacity operation of the remaining DEA plant units during construction. All operating contactors were fully loaded, and the DEA strength was increased to 32-34 mol% for the interim operation.

The CO_2 pipeline to Sacroc runs across the DEA plant to the existing CO_2 header—the distance is about 400 ft. The cooling water lines likewise run across the DEA plant for a distance of about 300 ft.

To tie these lines into the system with minimum overall downtime, it was necessary to shut down one (of six) regeneration units at the start of construction. The pipe rack associated with this unit was cleared of steam, process, and auxiliary lines servicing the regenerator.

The new CO_2 line and cooling water lines were run over this pipe rack and terminated just short of their final tie in point.

The existing inlet gas headers were modified during the August turnaround to include the inlet exchanger. This exchanger was delivered and installed before the summer turnaround. During the turnaround, the inlet piping was extended to the exchanger.

When the plant came back on after the turnaround, the inlet gas was flowing through the tubeside of the exchanger. There was no heat exchange since the residue gas connections were not made at this time and the DEA process was still in operation. However, making this connection during the turnaround saved a considerable amount of time and money.

The four contactors used for Selexol were extensively modified internally. The 20 valve trays were removed, along with the support rings. Two heavy support rings were installed in each contactor, to support the packed beds.

After the major rework had been completed, the columns were heat treated in place. The contactors were filled with 1-in. Koch Flexirings. The final step was to insulate the contactors and associated lines.

After site preparation and foundations, the large process vessels were the first major equipment items to be delivered and installed. The smaller pumps came next, followed by the refrigeration unit, large process pumps and turbines, the vacuum compressor and, finally, the recycle compressor.

The electrical work was the last scheduled phase of construction within the plant boundaries. In conjunction with the plant construction, it was necessary for Community Public Service Co. to run a power line to the area and build a substation adjacent to the Mitchell plant.

Although the transformer delivery was felt to be tight, Community Public Service completed their work on schedule in early December. The big electrical motors were first run on December 15.

A large drainage ditch at the northeast boundary of the plant site posed problems. Access to and passage through critical areas was severely restricted. A culvert was constructed across the ditch directly across from the recycle compressor site.

The recycle compressor was brought into the plant across this culvert, and the most direct access to the recycle compressor is via this culvert. Since the original plant entrance was geared to the amine-plant layout, a second culvert was built over the ditch for personnel and traffic access to the plant.

In general, the construction phase was well coordinated and proceeded smoothly. This is particularly noteworthy since a small compressor station, adjacent to the Selexol plant site, was being two-staged at the same time. There was neither conflict nor interference.

Although not related to plant construction, the Selexol charge, 95,000 gal, was manufactured and delivered in tank trucks over about a 3-month period. A large amine storage tank was converted to store the Selexol.

Start-up. Several process and instrument schools were conducted for operating personnel in the last few months prior to start-up. Two "walk-through" instruction sessions were held in December where the plant was essentially complete. Thus, the operators were able to see the equipment as it was discussed. This instrument school was invaluable since electronic instruments had not been previously used in the Midland area.

The final tie-in to the existing DEA plant water, CO_2, and electrical systems was made on December 6, 1977.

The plant was actually operated for a 5-hr period on December 21, and CO_2 was removed from the gas. The recycle compressor and vacuum compressor were not in service at this time, nor was the refrigeration unit. Hence, the operation was inefficient. The operational integrity of the system was verified and the plant actually operated as a CO_2 removal plant.

To prevent an undue outage during the transition from DEA to Selexol, arrangements were made to move the Mitchell gas up to the Gomez Complex. Excess capacity exists at that location.

During the Dec. 6 shut down, the full volume of Mitchell gas was processed at Gomez since cold weather up north required that the gas be delivered. It was also possible to manually split the flow between the old DEA section and the Selexol unit so long as the boilers and regenerators remained in commission.

Since the Selexol plant depends upon purchased electricity, a power outage will shut it down. Should a critical electric motor go down, the plant operation would be very inefficient. Hence, the capability of moving gas elsewhere for processing is decidedly advantageous.

The real start-up began on Jan. 26, 1978. At first, it was attempted to run the DEA plant while starting up the Selexol plant. Because of pressure limitations on the old DEA plant and interlocking residue gas piping, it was necessary to operate the Selexol plant at reduced pressure.

This led to very inefficient operation. It was finally decided to abandon conservation and on Feb. 11, the DEA plant was shut down. The Selexol plant alone carried the load.

There were a number of electrical and mechanical problems which plagued the start-up from this point on. There was a 15-sec power outage, throughout the entire area, that shut the plant down. A bearing failed in the recycle compressor gear box, and the plant was forced to operate without this item for a couple of weeks. The recycle gas was vented and, although some valuable methane was lost, the plant was able to pass 90 MMscfd of residue gas with less than 2.5% CO_2. END

Holmes-Stretford process offers economic H_2S removal

SRINI VASAN
Peabody Process Systems Inc.
Stamford, Conn.

ECONOMIC desulfurization of gases is becoming important in several areas. Desulfurization of sour natural gas has long been effectively achieved with amine systems followed by a Claus reactor with tail-gas cleanup systems.

Desulfurization of sour-water stripper gas and other refinery gas contaminated with hydrogen sulfide is another application.

Coal-gasification processes are currently under intense study and testing. Here also, gas streams containing H_2S must be purified to low levels of H_2S prior to using the product gas stream.

Gas-desulfurization methods. There are three alternate routes to desulfurization of gases containing H_2S:

Adapted from paper, "Economic Desulfurization of Gases with the Holmes-Stretford Process," presented at Canadian Natural Gas Processing Association meeting, Edmonton, Alta., Canada, June 9, 1977.

1. Hot-gas cleanup techniques.
2. Cold-gas cleanup methods using physical absorbents.
3. Cold-gas cleanup methods using chemical oxidants.

An advantage of the hot-gas method is that the gas is treated at the higher temperature without the need for cooling prior to desulfurization. This route is more efficient in energy consumption where the gas is directly used in gas turbines.

Four processes are currently under development: Consol Dolomite, Bureau of Mines hot iron-oxide process, Air Products' solid absorbent, and Batelle's molten salt process. None of these are commercially proven or currently available.

There are several cold-gas physical absorption processes ranging from amine-type solvents (DEA, Sulfinol, Alkazid, etc.) to modified carbonate processes Benefield, Catacarb, Vacuum Carbonate, etc.).

The physical absorbent processes have some drawbacks. They require large amounts of steam for regeneration.

Also, the regenerated gases contain H_2S along with large amounts of CO_2, hydrocarbons, etc. which have to be treated in a secondary desulfurization unit such as a Claus plant, followed by a tail-gas desulfurization unit. Thus, this approach requires three processing steps to complete removal and recovery of sulfur.

The earliest chemical oxidant developed was the iron-oxide box used by the gas industry decades ago in the United Kingdom and elsewhere. Digging out foul-smelling spent oxide from purifier boxes was the dirtiest activity in the gas industry. Too, the process is costly in capital and ground-area occupied.

The predecessors of the United Kingdom Gas Board and W. C. Holmes Ltd., Huddersfield, U.K., cooperated in developing two liquid-wash methods using chemical oxidants. The Manchester process used a suspension of hydrated ferric oxide

Fig. 1

Holmes-Stretford process

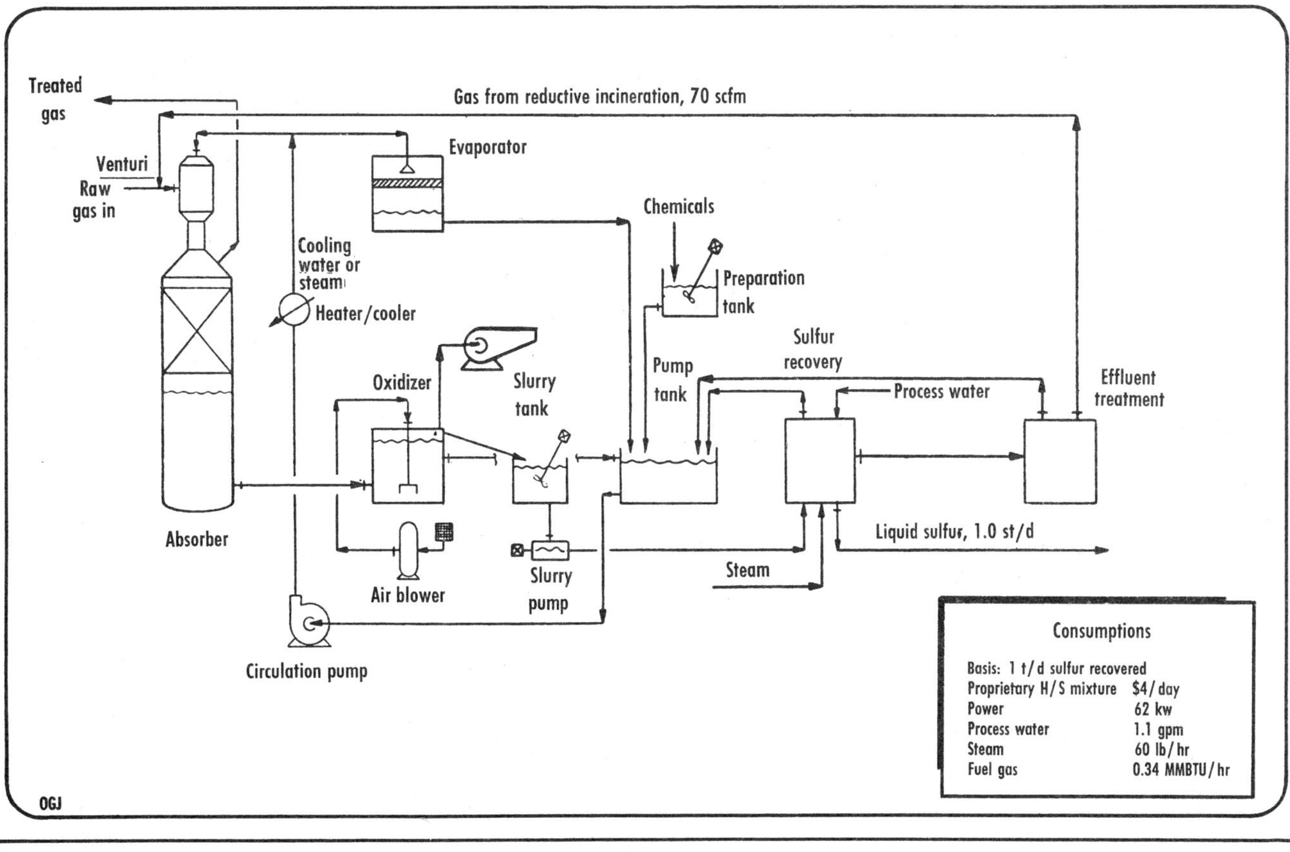

Consumptions

Basis: 1 t/d sulfur recovered	
Proprietary H/S mixture	\$4/day
Power	62 kw
Process water	1.1 gpm
Steam	60 lb/hr
Fuel gas	0.34 MMBTU/hr

and was known as the Ferrox process in the U.S.

Holmes built nine full-scale Ferrox plants from 1944 to 1951. The process suffered from slow oxidation rates and impure sulfur product.

Next came development of the Stretford process by the United Kingdom Gas Board, which has become an accepted cleanup technique for the gas industry. Holmes, appointed a licensee for the process, built many such plants which were successful in reducing the hydrogen sulfide content of the gas to less than 1 ppmv.

As new applications for the process became apparent, Holmes's experience grew. They introduced several proprietary techniques and refinements, including gas pretreatment, sulfur recovery, and waste-liquor treatment and qualify the naming of the process as the Holmes-Stretford process. Peabody Holmes has successfully commissioned 36 plants around the world by using their Stretford experience.

Another process that seems suitable for desulfurization of producer-gas is the Giammarco-Vetrocoke process. It uses an arsenate as the chemical oxidant.

However, in this process the oxidation rate is slower than that of the Stretford process, and the highly poisonous arsenate precludes passing OSHA and EPA requirements in the U.S.

Recently, several new processes based on buffered sodium sulfite are being developed. Peabody developed the Citrex process for converting H_2S into sulfur using a solution buffered with citrate. This process seems to have lower investment and operating costs than the Holmes-Stretford process. It also has no serious effluent problem, since nontoxic sodium sulfate (Glauber's Salt) is the only effluent.

Its application for desulfurization of gases from a coal-gasifier holds promise. The Citrex process, described in recent papers,[1, 2] is currently being tested in an oil refinery in the U.S. to treat sour-gas streams. The process will be offered after this demonstration unit is successfully operated for 6 months.

Holmes-Stretford process. The chemistry of the Holmes-Stretford process is quite complex. In the process, H_2S is chemically oxidized by sodium vanadate, at a pH of 9.0, into sulfur and water.

The pentavalent vanadic salt is reduced to the quadrivalent vanadous form and regenerated to its vanadic form in an oxidizer vessel by air sparging:

$$2V^{5+} + HS^{-} \rightarrow 2V^{4+} + S + H^{+}$$

$$2V^{4+} + \text{ADA (oxidized)} \rightarrow 2V^{5+} + \text{ADA (reduced)}$$

The air converts the reduced vanadium into vanadate and also acts as a flotation agent in frothing out the product sulfur. Thus, H_2S is, in effect, oxidized by air into sulfur and water with the vanadium acting as an intermediary oxidant:

$$2H_2S + O_2 \rightarrow 2S + 2H_2O$$

The use of a mechanism with ADA (anthraquinone-disulfonic acid) serves to accelerate the oxidation rate of vanadium. The Holmes-Stretford process is illustrated in Fig. 1.

Proper removal of interfering contaminants is essential for successful operation of the Stretford unit. The Holmes-Stretford process can be designed to desulfurize gases containing high concentrations of CO_2 in spite of the high pH range.

Holmes developed a proprietary venturi-absorber to handle a gas containing 84% CO_2 and 16% H_2S, where H_2S content is reduced to 4 ppm. This was proven in a plant handling about 10 tons/day of sulfur operated by a natural-gas plant of People's Gas Co. near St. Louis. In this plant, the Holmes-Stretford unit has a venturi-absorber mounted on top of the delay tank. This unit has been operating successfully for the past 5 years producing molten sulfur of high commercial quality usable in sulfuric acid plants.

Until recently, Stretford units required waste disposal of liquid effluents because of thiosulfate and thiocyanate accumulation. Also, the waste liquor contains traces of vanadium salts.

In 1973, Peabody Holmes successfully developed a new zero-discharge reductive incineration process that cracks the bleed liquor into a gas stream containing H_2S and CO_2, along with a liquid stream containing reduced vanadium salts:

$$Na_2S_2O_3 + 4H_2 \text{ (or } 2CO) \rightarrow Na_2O + 2H_2S + 2H_2O \text{ (or } 2CO_2)$$

The zero-discharge process now is employed on a large coke-oven plant, desulfurizing 16 MMcfd of coke-oven gas and producing 12 ltpd of sulfur. The problem of disposing of liquors containing vanadium has been overcome, and the zero-discharge process needs no makeup of vanadium or sodium salts.

Four additional zero-discharge units (with two in Hamilton, Ontario) have been designed for installation in the U.S., Mexico, and United Kingdom by

Table 1

Partial list of Holmes-Stretford plants

Year installed	Location	Gas flow MMscfd	Sulfur, long tons/day
Coke-oven gas units			
Typical composition: 54% H_2, 30% CH_4, 7% CO, 2.8% CO_2, 1.0% H_2S, 0.2% HCN, 5% N_2			
1965	Zeebruge, Belgium	11.0	4.0
1966	Hamilton, Canada	28.0	5.5
1967	Ebbw Vale, Wales	2.9	0.45
1968	Hartlepool, England	2.6	0.70
1969	Ravenscraig, Scotland	4.5	0.98
1970	Orgreave, England	26.0	13.0
1970	Smithywood, England	9.0	3.66
1970	Hamilton, Canada	14.0	2.63
1971	Parkgate, England	5.0	2.5
1974*	Sicartsa, Mexico	26.0	5.5
1974*	Redcar, England	43.2	18.6
Coal-gas units			
Composition: same as for coke-oven gas			
1958	Whitchurch, England	0.25	0.19
1960	Liverpool, England	6.0	3.00
1960	Bristol I, England	9.0	3.36
1961	Bristol II, England	9.0	3.36
Producer-gas unit			
Typical composition: 15% H_2, 50% N_2, 30% CO, 3.2% CO_2, 1% CH_4, 0.8% H_2S			
1965	Johannesburg, South Africa	6.0	1.78
Coal-liquefaction unit			
1975	Office of Coal Research, Cresap, W.Va.	0.40	0.67
Natural-gas sweetening (88% CO_2, 15% H_2S)			
1972	People's Gas Co., St. Elmo, Ill.	4.00	9.00
Coal-gasification			
1977*	Caterpillar Tractor Co., York, Pa.	14.00	3.5
Off-shore platform (sour gas)			
1977*	Chevron Oil Co., Santa Barbara, Calif.	25.00	3.0

*Under construction.

Fig. 2

Offshore Holmes-Stretford plant

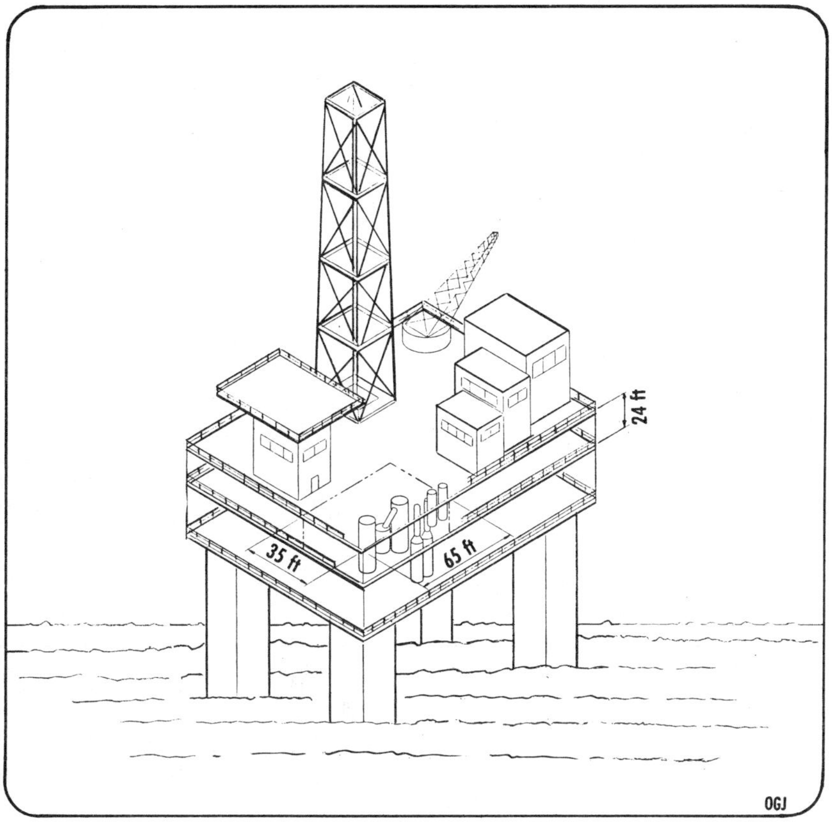

Peabody Holmes.

In 35 Holmes-Stretford plants, the sulfur product has been found to be salable and suitable for use in sulfuric acid plants. The molten sulfur produced is bright in color and free of contaminants, if proper techniques are used for feed-gas treatment prior to the Stretford unit.[3]

In some of these plants, the tonnage of sulfur recovered was low and hence disposed of as a wet-cake. A partial list of Holmes-Stretford plants built since 1956 is shown in Table 1.

The Holmes-Stretford Process removes hydrogen sulfide from gas streams at high efficiencies. Removal efficiencies of greater than 99% can be achieved along with hydrogen sulfide outlet concentrations of less than 5 ppmv.

It converts the hydrogen sulfide present in the gas into elemental sulfur in a simple sequential step at moderate temperatures. Carbon steel can be used for the most part, resulting in low capital and maintenance costs.

It has proved to be a highly efficient and specific method of reducing hydrogen sulfide to low levels even in the presence of such acid gases as carbon dioxide. An operating unit in the United States reduces hydrogen sulfide from 10% v/v to less than 15 ppmv from a gas containing 90% v/v carbon dioxide at a partial pressure of 1.1 atmospheres (16.2 psi).

The Holmes-Stretford process has no tail-gas problem such as that encountered with amine/Claus systems. Tail gases from Claus plants often contain high H_2S and sulfur dioxide concentrations and cannot be released directly to the atmosphere.

The Holmes-Stretford process can be used as part of a Claus tail-gas treatment plant, or as direct replacement of the whole amine/Claus system. The choice of plant is usually made after financial consideration.

In cases where the carbon dioxide to hydrogen sulfide ratio is quite high, the Holmes-Stretford process is the only process available capable of purifying the gas stream.

Operating costs. The chemical and utility requirements for the Holmes-Stretford unit using the zero-discharge improvement is shown in Fig. 1 on the basis of 1 ton/day of sulfur recovered. Chemical and utility requirements can be directly prorated for higher sulfur tonnages.

Operating costs of the Holmes-Stretford process are low and competitive with such processes as amine-Claus-tail gas systems. The pollution-free aspect of the process allows this type of desulfurization unit to meet any current and potential effluent standards that may be imposed.

The compact nature of a skid-mounted Holmes-Stretford unit can be visualized from a unit supplied to ERDA's gas liquefaction plant to recover about 1 ton/day of sulfur. The trouble-free operation of this process allows use of one operator per shift (except for sulfur loading).

The small venturi-absorber replacing the traditional tall Stretford towers along with Holmes's compact oxidizer and sulfur recovery systems has permitted a Holmes-Stretford unit to be specified for an offshore platform application of Chevron Oil Co.

In an offshore platform where space and labor are at a premium, a Holmes-Stretford unit has been designed to fit an area 35 ft x 65 ft with a height limit of 24 ft for recovering 3 tons/day of sulfur (as a cake) with minimum operator attention (Fig. 2).

Similarly, compact skid-mounted Holmes-Stretford units can be installed to treat sour-gas wells at remote locations with minimum operator attention and needing only power (with no need for fuel-gas or steam) and disposing of the sulfur as a cake.

Thus the economic desulfurization of sour-gas wells can be achieved with the use of low-cost skid-mounted Holmes-Stretford units requiring minimum operator attention.

References

1. Vasan, S., "The Citrex Process for SO_2 Removal," Chemical Engineering Progress, Vol. 71, No. 5, May 1975.

2. Vasan, S., "The Peabody Citrex FGD Process," paper presented at 12th Air Pollution & Industrial Hygiene Conference, Austin, Tex., Jan. 28, 1976.

3. Moyes, A. J., and Vasan, S., "Holmes-Stretford H_2S-Removal Process Proved in Use," The Oil and Gas Journal, Sept. 2, 1974, p. 56.

The author . . .

Srini Vasan

Srini Vasan is director of business development for Peabody Process Systems Inc. He is responsible for regenerable FGD systems and the Holmes-Stretford process for H_2S removal. Prior affiliations include technical director in the operation division of International Minerals & Chemicals Corp. and manager of technical and sales development for Chemical Construction Corp. Vasan also has served as a consultant to the National Science Foundation and the Environmental Protection Agency. The author of several publications and a holder of several patents, Vasan holds an MBA degree from the University of Chicago and an MS degree in chemical engineering from the Polytechnic Institute of Brooklyn, N.Y.

Holmes-Stretford H_2S-removal process proved in use

A. J. Moyes
W. C. Holmes & Co., Ltd., U.K.

S. Vasan
Peabody Engineered Systems
Stamford, Conn.

THE HOLMES-STRETFORD process has demonstrated its ability to effectively remove hydrogen sulfide from gas streams in several applications:

- In a U.S. 200-MMcfd SNG plant, the offgas containing 0.9% H_2S and 7% CO_2 is reduced to low-level ppm of hydrogen sulfide.
- When coal is gasified to produce a low-BTU gas, the process can treat the raw gas directly, eliminating an intermediate absorption step.
- By using the process, refineries can eliminate an amine absorber and a Claus unit.

How it's done. The Holmes-Stretford process uses reliable, rugged, simple technology for almost total removal of H_2S from gas streams. There are 31 successful Holmes-Stretford units operating around the world. These involve a variety of applications: towngas, coal gas, SNG, coke-oven gas, and natural-gas processing.

This proprietary process has been developed by W. C. Holmes & Co., Ltd., England, a part of Peabody Galion Corp. of New York. This process is now available for wider use in the U.S. with four installations already operating successfully in St. Louis, Honolulu, and Ontario.

The original Stretford Process was developed by the North Western Gas Board in England for use in purifying manufactured town gas. Holmes was appointed a licensee for the process and in the course of commissioning many plants around the world, Holmes introduced many proprietary techniques, including gas pretreatment, sulfur recovery, and waste-liquor treatment. These refinements qualify the naming of the process as the Holmes-Stretford Process.

The sulfur tonnage handled by individual Holmes-Stretford plants has grown over the past 14 years from 200 lb/day to 30 long tons/day in a single-stream unit now under design and construction. The volume of gas handled also has grown with recent single-train units handling 30 MMscfd.

Process features. The Holmes-Stret-

ford Process when applied to SNG, coal gasification, or natural-gas-processing plants, consists of a simple, nonplugging absorber using a nonvolatile liquor. It accomplishes near total H_2S removal down to the 4 ppm level, regardless of inlet H_2S concentration or pressure.

This process uses low-cost materials of construction and incorporates rugged, trouble-free equipment. Since most operations are handled as liquids, the process requires little operator attention. Because the Holmes-Stretford Process removes virtually all of the H_2S, a subsequent tail-gas processing step is not required (as is the case with Claus units).

HOLMES-STRETFORD process unit handles 26 MMcfd of coke-oven gas at the Orgreave works, British Steel Corp. Hydrogen sulfide concentration in effluent is very low (Fig. 2).

Many SNG and coal-gasification projects have chosen the Stretford Process because of its proven performance and trouble-free operating record. The process will meet not only current emission standards, but also any forseeable emission standards for sulfur because of its ability to remove 99.9% of H_2S from gas streams.

Process operation. The gas to be purified is counter-currently scrubbed in an absorber with an alkaline solution containing a vanadium salt along with an anthraquinone derivative.

In a single step the H_2S is converted into elemental sulfur while the vanadium salt is reduced:

$$Na_2CO_3 + H_2S \rightarrow NaHS + NaHCO_3$$

$$HS^- + 2V^{5+} \rightarrow 2V^{4+} + S + H^+$$

The reduced liquor from the absorber is fed to the oxidizers where the vanadium is restored to its oxidized form:

$$2V^{4+} + \text{ADA (oxidized)} \rightarrow 2V^{5+} + \text{ADA (reduced)}$$

Air is blown through the liquor, both

Fig. 1

Holmes-Stretford flow pattern

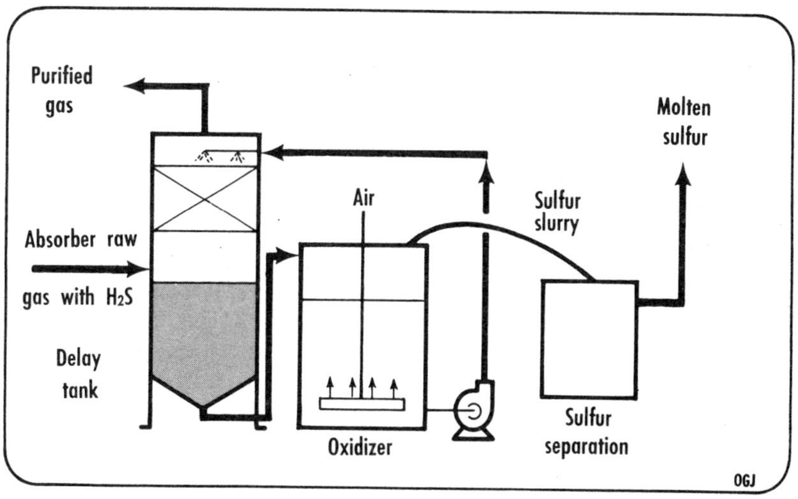

to reoxidize the solution and to separate the sulfur by flotation. The sulfur froth is either filtered or centrifuged, washed, and melted to produce high-quality sulfur.

Fig. 1 shows the process flow sheet with absorber, delay tank, oxidizer, and sulfur-melter sections.

The process produces an effluent containing thiosulfate and thiocyanate (if cyanide is present in the gas). A proprietary zero-discharge process has been recently developed by Holmes which reduces the effluent in a reductive incineration step to hydrogen sulfide and sodium carbonate. These products are recycled into the front end of the process.

Fig. 2 shows an application of the process treating 26 MMcfd of coke-oven gas, where the absorber had an allowable pressure drop of only 10 in. of water column. So there was need for tall towers with open slat packings. One of the towers here serves as a cyanide prewash tower.

Fig. 3 shows the Holmes-Stretford unit for H_2S removal from amine off gas at a natural-gas-processing plant near St. Louis. Since the gas is available at 9 psig, this gas containing 15% H_2S and 85% CO_2 is contacted in a proprietary venturi-absorber designed by Holmes.

Bottom of the venturi unit consists of the delay tank where the conversion of H_2S to sulfur is completed. The compactness of the Holmes-Stretford process could be seen from this installation.

FEED GAS for this plant is amine-offgas from a gas-processing plant near St. Louis. Feed contains 15% H_2S and 85% CO_2 (Fig. 3).

Table 1

Sulfur fixation data with HCN in the gas

	Without polysulfide wash		With Holmes polysulfide wash	
Acid gas	H_2S	HCN	H_2S	HCN
Concentration (gr/cu ft)	6.3	0.5	6.3	0.015
Assumed fixation rate (%)	1	100	1	100
Sulfur fixed (lb/day)	450	4,500	450	135
Volume of effluent (gal/day)	6,000		700	

SNG Plants. Two large SNG projects in the U.S. have adopted the Holmes-Stretford Process. One of these projects handles about 200 MMcfd of gas containing 0.9% H_2S and 7% CO_2 at 20 psig to reduce the H_2S to low ppm level in a Holmes-Stretford unit.

The process could be used directly on the raw gas without need for intermediate acid gas removal, since the process is unaffected by the pressure of the raw gas and a compact venturi-absorber could be used.

Coal gasification. In the light of the current energy situation there is a new spurt of interest in coal gasification processes. In a coal-gasification process that produces high-BTU gas, the CO_2 content of the gas must be removed by a solvent stripping step (like the Rectisol process) where the hydrogen sulfide content of the gas also is stripped along with the CO_2.

The CO_2 gas from the stripper contains up to 10% H_2S. This concentration is too dilute for handling in a Claus kiln, and the Holmes-Stretford process offers an economic and proven method that will reduce the sulfur content of the stack gas to less than 5 ppm, while recovering 99+% of the sulfur value as elemental sulfur.

In coal-gasification processes that make low-BTU gas, the Holmes-Stretford Process can treat the raw gas directly, eliminating an intermediate absorbtion step. One key problem in coal gasification processes is the presence of HCN in the raw gas. HCN will react with the Stretford liquor to form objectionable thio-cyanate along with thiosulfate which are potential pollutants.

Holmes has developed a cyanide-removal step which consists of a polysulfide wash. This step has been successfully used for over 30 years in many plants around the world. The effect of this polysulfite wash on the sulfur fixation is shown from typical plant data in Table 1.

Refinery gases. Refineries have traditionally used the Claus process to treat gas streams containing high concentrations of hydrogen sulfide. The Claus reaction is conducted as a multistage operation to get efficiencies of 94%-96% conversion.

The tail gas from the Claus kiln containing a H_2S-SO_2 mixture presents a serious pollution problem. The currently adopted method is to catalytically hydrogenate the tail gas containing sulfur dioxide into hydrogen sulfide, and the hydrogen sulfide so formed can be removed by the Holmes-Stretford process.

In some applications, the Holmes-Stretford process could handle the gases prior to the amine-absorber, eliminating the need for the amine unit as well as the Claus unit. END

Thirty-nine Catacarb-process plants are now sweetening sour natural gas

A. G. EICKMEYER
Eickmeyer & Associates
Prairie Village, Kan.

THE CATACARB process is in operation in 39 plants throughout the world, and 11 more are under construction.

The 50 plants will be capable of removing 763 MMscfd of acid gas from 4.35 thousand MMscfd of feed gas. And just announced is the 51st unit, that of Kyushu Oil's at its new Japanese hydrogen plant.

Most of the units are committed to ammonia and hydrogen-synthesis operations. Three of the plants are in 1,500-ton/day ammonia facilities, and a fourth has just been started up in a 1,700-ton/day ammonia complex.

Because of the lower heat required for regenration compared with conventional MEA scrubbing, the annual savings will be approximately 28 trillion BTU, which, taking into account boiler inefficiencies, is equivalent to saving 40 thousand MMscf of natural gas annually.

How it began. In 1963 Esso Research & Engineering Co. took a worldwide license for the Catacarb process for Standard Oil of New Jersey's affiliated companies, which now have 12 Catacarb units in operation. One of the Esso projects was the world's largest liquefied natural-gas plant, located in Libya, having a sour-gas feed which had to be treated down to low levels of residual CO_2 and H_2S.

In order to satisfy Esso that the Catacarb process was applicable to sour natural gas, and to collect some data for purpose of designing the Brega plant, a sour-gas treating plant (not Esso's) was converted to the Catacarb process, and a series of plant tests were made.

Esso furnished a crew of men and sophisticated equipment, including a CEC titrilog and a Dow precision automatic titrator for purpose of gas and solution analyses. Also, gas sample bombs were sent to Baytown for gas chromatograph and mass spectrometer analyses.

The wellhead gas, which was produced at about 3,000 psig, was let down through a choke and the temperature controlled by a heater in order to avoid hydrate formation. Hydrocarbon condensate was separated and sent to storage, and the sour gas, which was very rich in light hydrocarbons, was sent to the treating plant.

The treating plant employed a two-stage system with a direct-fired reboiler, as illustrated in the accompanying diagram.

The sour gas usually contained between 2.0 and 2.3% H_2S and 6.8-7.3% CO_2, depending upon which wells were being produced.

Absorption pressure was 950 psig and temperatures were about 210° F. in the bulk-removal section and 90° F. in the cleanup section in the upper part of the contactor. Gas leaving the contactor passed through a knockout section and a glycol dehydrator before entering the sales gas meter and line.

Rich solution was level controlled out of the contactor through a letdown valve, a flash tank, and a cartridge filter. Then it passed to the top of the stripper where the solution was regenerated in countercurrent contact with steam in two stages, yielding semilean and lean solutions. Semilean solution was withdrawn from a chimney tray and sent to a surge tank, from which it was pumped, without cooling, to an intermediate point of the absorber.

Solution overflowing the stripper chimney tray was more thoroughly stripped (because of a higher steam to solution ratio) and reboiled by direct firing before passing through an aerial cooler and goin to a surge tank. It was then pumped through a cooler to the top of the absorber. The stripper overhead acid gas-steam mixture passed through an aerial condenser, a reflux drum, and thence to a flare stack. Reflux condensate was returned to the stripper. And, since water vapor was lost with warm acid gas, the system water balance was maintained by a small evaporator generating open steam which entered the reboiler.

Treating-system performance. Purification was essential by H_2S removal at this plant. The pipeline would accept 1 grain of H_2S/100 scf, or about 16 ppm by volume, which is considered to be a liberal specification. However, before conversion to the Catacarb process, while still operating on potassium carbonate solution at full capacity, the treated gas usually had 3-6 grains of H_2S.

It was necessary to operate at less than half capacity, with the reboiler going full blast, in order to make salable 1-grain gas. This seriously lim-

Adapted and updated from paper given at Gas Conditioning Conference, Norman, Okla., 1971.

ited not only gas sales, but more important, the condensate production, since flaring the gas was not permitted. An additive for the potassium carbonate solution was tried to no avail.

After limping along for 7 months at part capacity, this plant was converted to the Catacarb process, whereupon the performance was improved. At full capacity it was possible to obtain ¼ grain residual H_2S and 0.1-0.2% CO_2 in the treated gas, while using somewhat less than the design reboiler heat input.

A detailed analysis of the test data yielded packing Kga's, substantiating data from other plants. These were then used to design the Brega plant, which had a more rigid specification for the treated gas in order to avoid CO_2 freeze-ups in the liquifaction train.

The Marsa El Brega plant was designed to treat the sour gas in two trains. Because of the political situation in Libya there was some delay in the operation of this plant under normal conditions. However, in some tests made in September 1970, on one of the Catacarb purification trains, Esso reported that the H_2S and CO_2 removal were very satisfactory.

The unit was operated at about 80% of design feed-gas rate, and the H_2S removal was from 0.79% in the feed down to 2.3 ppm (about ⅛ grain). The CO_2 removal was from 1.4% to 20 ppm by volume.

Catacarb-solution samples received from the Libyan plant were a clear yellow color with practically no suspended solids.

Catacarb flow diagram

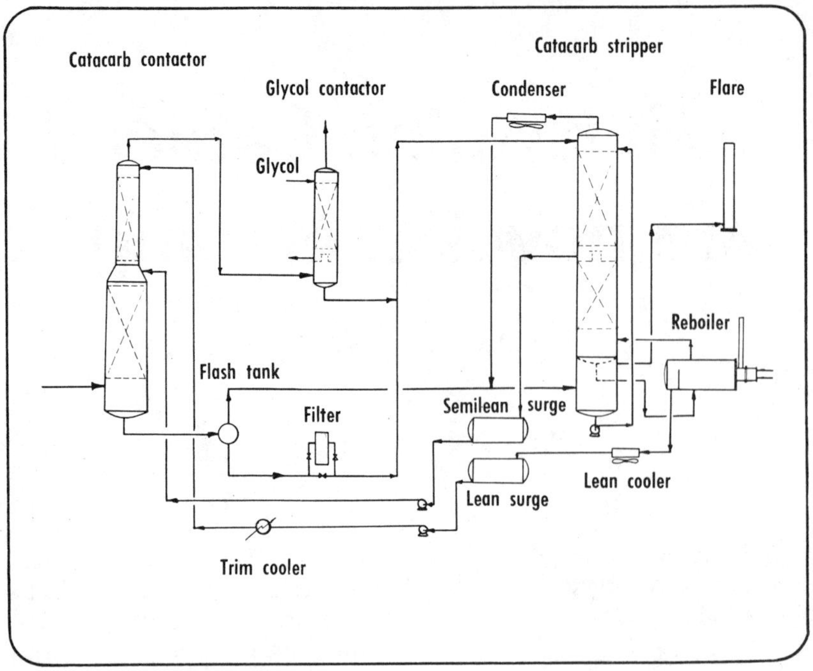

Heat requirements. The acid-gas loading of the Catacarb solution and the heat required for regeneration of the solution are both a function of the partial pressure of the acid gases in the inlet gas, the purification required, the type of tower internals selected, and the size of the towers. Obviously there are interrelated capital investment and operating cost factors which are taken into account in a design for a specific case. The higher the inlet partial pressure, the less heat required for regeneration.

Sour gas is usually treated at a pressure between 600 and 1,000 psig, and the acid gas may vary from 5 to 50%. Within this range design studies indicate that the heat requirement is ordinarily between 34,000 and 60,000 BTU/mole of acid gas, which is equivalent to 90,000 to 160,000 BTU/Mscf of acid gas removed.

The regeneration heat requirements are at a low temperature level. So it is possible to integrate this with the pumping power requirement, thus resulting in good efficiency. END

Status of SNPA-DEA

Lawrence W. Dailey
Ralph M. Parsons Co.,
Los Angeles

THE SNPA-DEA process for natural gas sweeting is receiving increased acceptance in all areas of the world where raw gas containing appreciable quantities of H_2S and CO_2 is to be sweetened at pipeline pressure. Installations are currently operating or planned in western Canada, Europe, and the Middle East.

This process deserves consideration where the total acid-gas content exceeds 10% or so, in an installation where the recovery of high-quality elemental sulfur is contemplated.

The basic flow scheme for this unit can be similar to that used for any regenerable solvent system such as a conventional DEA or an MEA unit.[1] Because of the higher solution loadings employed, in the cases where this process is applicable, the primary differences will be in equipment sizing, DEA circulation rate, and utility consumption.

Development

This process was developed by SNPA (Societe Nationale des Petroles d'Aquitaine) personnel during the late 1950's while seeking an economic sequence for processing sour gas at Lacq, in southern France. The detailed steps in this development, as well as a brief history of SNPA itself, have been the subject of past publications.[1 2 3]

A variation of the conventional DEA process was selected by SNPA after pilot, semicommercial, and commercial testing of many combinations of processes including MEA, water wash, hot carbonate, tripotassium carbonate, and solvents such as propylene carbonate and tributyl phosphate.

For the situation at Lacq, SNPA concluded that DEA, when employed at certain conditions pioneered by them, was the most attractive sweetening process—primarily on the basis of low-corrosion rates and reduced-foaming tendencies.

In addition, makeup rates of DEA were substantially below those experienced in the same units when on MEA, due largely to degradation of MEA by the COS present in the Lacq gases.

The specific conditions determined by SNPA relate to:

1. Solution circulation rate and rich solution loading.
2. Contacting temperature.
3. Solution strength, in terms of weight percent DEA.
4. Stripping steam requirements as related to gas purity.

Results of the work were to prove, on a commercial scale, that DEA solution could be employed to make pipeline specification gas having an H_2S content of ¼ grain/100 scf or less.

Commercial applications

Existing units, plus those currently in an engineering or construction phase, are capable of processing 2,580 MMscfd of sour raw gas. The geographic distribution of this capacity is given in Table 1.

Table 1

SNPA-DEA sour-gas-treating capacity

	Sour gas capacity, MMscfd Installed	Building	Total
Canada	840	492	1,332
Europe	1,080	—	1,080
Middle East	—	168	168
	1,920	660	2,580

In addition, SNPA has recently been awarded a contract for a plant with a capacity of 1,620 MMscfd which will involve export of know-how to the U.S.S.R. When completed, the total worldwide capacity will be 4.2 billion scfd of sour raw gas.

Raw-gas compositions. The analyses of sour-gas streams entering the contactor of several SNPA-DEA units are given in Table 2. Total acid-gas content ranges from a low of about 10 to a current high of 35 vol %. This latter figure represents the highest commercial application to date, but not an upper limit for the capability of the process.

In these commercial plants the ratio of H_2S to CO_2 varies from a high of 34 to a low of about 0.65. Again, these are indicative of installation to date, and not necessarily any limitation on the process.

Table 2 also shows the trace contaminant levels of COS, CS_2, and mercaptan sulfur contained in the raw-gas stream entering the contactor, and their hydrocarbon distribution.

Treated gas purity. Two basic specifications for the sulfur content of residue gas stream from a plant are normally encountered in North America. These are:

1. 0.25 grain H_2S/100 scf, 0.2 grain mercaptan sulfur/100 scf, and a total sulfur content of 1.0 grain/100 scf.
2. 1.0 grain H_2S/100 scf and a total sulfur content of 20 grains/100 scf.

Table 2

Raw-gas compositions

Component Mole %	Middle East		France	
H_2S	34.0	25.0	15.5	6.4
CO_2	1.0	11.0	10.0	9.9
(Total acid gases)	(35.0)	(36.0)	(25.5)	(16.3)
N_2	—	—	Trace	0.3
C_1	63.7	62.8	69.4	77.4
C_2	0.7	0.7	2.9	3.2
C_3	0.3	0.2	0.9	1.0
C_4	0.1	0.1	0.6	0.8
C_5	0.1	0.1	0.3	0.4
C_6+	0.1	0.1	0.4	0.6
ppmv				
COS	1,200	1,200	200	60
CS_2	Nil	Nil	50	Nil
RSH	300	300	750	Nil

Component Mole %	Alberta, Canada plants					
H_2S	26.3	9.8	17.9	13.3	15.7	26.6
CO_2	4.8	3.8	3.3	2.5	3.5	4.1
(Total acid gases)	(31.2)	(13.6)	(21.2)	(15.8)	(19.2)	(30.7)
N_2	0.7	4.9	0.8	9.3	1.4	0.7
C_1	64.4	75.0	73.8	66.2	67.8	51.7
C_2	1.2	3.1	1.4	3.6	7.3	10.2
C_3	0.7	1.5	0.8	2.2	2.4	3.9
C_4	0.8	1.4	0.9	2.1	1.2	2.0
C_5	0.3	0.3	0.3	0.5	0.4	0.6
C_6+	0.7	0.2	0.8	0.3	0.3	0.2
ppmv						
COS	70	—	—	NR	—	—
CS_2	Nil	—	—	400	—	—
RSH	30	—	—	NR	—	—

A variation of (1) permits a total sulfur content of 5.0 grains/100 scf, with the H_2S and mercaptan limits remaining as given.

While these specifications also may serve as a basis for sulfur purity in other areas of the world, more severe requirements may be encountered.

For example, in France the H_2S content is limited to 0.15 grain/100 scf scf with a total sulfur requirement of 1.3 grains/100 scf maximum.

Nearly all recent plants have been designed with the capability of meeting the pipeline specification of 0.25

Presented at NGPA annual convention, Denver, 1970

grain H_2S/100 scf, although many still deliver to systems permitting 1 grain of H_2S/100 scf. In practice, values of 0.05 to 0.15 grains are attained on a day-in, day-out basis.

The CO_2 content of the treated gas from the contactors operating at 850 to 1,000 psig has been reported from 100 to 1,000 ppmv; this is more than adequate for the normal requirements of either 1 or 2 vol % maximum.

Removal of mercaptans by the SNPA-DEA process can be predicted once their distribution and boiling point are known. This removal ranges from 10 to 60% on the basis of both solubility and chemical reaction.

As a result, the pipeline specification permitting 0.2 grain of mercaptan sulfur per 100 scf may not always be met directly off the DEA contactor. In these cases, even moderate downstream chilling for propane recovery (or refrigerated oil absorption) will produce a residue gas meeting this specification with raw-gas mercaptan contents of up to several hundred grains per 100 scf.

Since some degree of propane recovery is normally required to comply with the dew-point specification, mercaptan level will also be met.

The effect of COS and CS_2 in the raw-gas stream can be considered together, since the manner of their removal by DEA is believed to be similar. Contrasted with the irreversible chemical reactions between MEA and these components, which lead to consumption of the solvent through the formation of degradation products, DEA, probably because of its lower alkalinity, is relatively inert.

The removal of COS and CS_2 in SNPA-DEA units is believed to be largely through hydrolysis and physical solubility. The latter is borne out by the apparent occurrence of these components in the overhead stream from the regenerator. We say apparent, because of the obvious difficulties in analyzing for trace COS or other sulfur-containing compounds in the presence of large quantities of H_2S and CO_2.

Hydrolysis is believed to be the mechanism by which the major removal of COS and CS_2 occurs. Higher operating temperatures in the contactor are thus conducive to more complete removal.

These higher operating temperatures are normally associated with an SNPA-type unit, because of the lower circulation rates employed. Thus, high removal of COS and CS_2 can be attained with DEA without high solution losses through chemical degradation.

Acid-gas quality. As the percent H_2S in the raw gas increases, the importance of the return from the elemental sulfur produced takes on greater economic significance. In any plant it is mandatory to consistently produce salable sulfur, concurrent with long runs on the sulfur plant catalyst.

In this respect the acid-gas-removal facilities must be regarded as the sulfur plant feed-gas-conditioning unit, in addition to its prime requirement for natural-gas sweetening.

Since the SNPA-DEA process employs only an aqueous solution of DEA, the question of hydrocarbon absorption in the contactor, with subsequent release in the regenerator to the sulfur plant is minimized. Further, the absorption is not a function, to any degree, of the hydrocarbon type present in the raw gas.

The lower-solution circulation rates employed with this process also tend toward a reduced takedown of hydrocarbons, and their subsequent release in the flash tank.

The raw-gas compositions listed in Table 2 are all producing an acid-gas stream normally containing less than ½% hydrocarbons, and in many cases substantially so.

For this reason, the operator can be assured of bright yellow sulfur production, without the need for a separate acid-gas conditioning step, and with a minimum of sulfur plant operational adjustments.

Makeup requirements. The solution makeup requirements represent a total of chemical-reaction losses, entrainment, and mechanical losses and, thus, include a measure of plant housekeeping. Using the conventional terms of reporting—pounds pure DEA per MMSCF of sour raw gas treated—values from 2 to 9 have been reported on SNPA-DEA units to represent annual average rates.

One 300-MMscfd plant recently converted to SNPA-DEA from another process sequence reported annual losses at 4 lb/MMscf.

These makeup figures are from plants treating raw gas containing from 13 to 31% acid gas, and with appreciable quantities of COS, CS_2, and mercaptans present. For these high acid-gas applications, we have been better able to correlate makeup requirements on the basis of pounds pure DEA per MMscf of sour raw gas treated, for percent acid gas in the sour raw gas.

On this basis, long-term operating values of 0.15 to 0.20 lb/MMscf/% acid gas are being achieved in several units. During the first few months of operation following start-up of a new or converted plant, makeup may be twice these values or more, but can be reduced by adequate attention to housekeeping.

The results of a survey of solution makeup requirements on still another basis were recently reported.[7] In this comparison, the solvent losses from three currently popular treating systems were compared on the basis of solvent cost per MMscf of acid gas removed. The values are shown in Table 3.

Table 3

Solvent Cost

Solvent	Cost per MMscf of acid gas removed, $
DEA	3 to 8
Sulfinol	8 to 20
MEA	7 to 27

With the ever-increasing size of individual plants, one or more of the split-flow variations can often be economically incorporated in the design of an SNPA-DEA unit. Split flow of lean solution to an absorber operating at pipeline pressure is probably the most easily justified since a mimimum of additional piping and instrumentation is required. A somewhat greater throughput will be needed to support the increased complexity of a doubly split-flow design.

In such a design, a partially regenerated or semilean solution is produced from an intermediate point of the regenerator and circulated through the lower section of the contactor where bulk removal of acid gases is occurring. Stripping of this semilean solution is carried out at reduced stripping steam rates while some fraction of the total circulation is severely stripped to produce an extremely lean solution for use in the top section of the absorber.

Another sequence that has been successfully employed—particularly when converting an existing unit from some other treating solution to SNPA-DEA—at the same time that an overall plant capacity increase is being made, involves the use of separate bulk and trim acid-gas-removal sections.

For revamping high-acid-gas-content plants where this is economic, the bulk tank unit serves to reduce the acid-gas content from a starting value of 25 to 35% down to an intermediate level, following which specification gas is produced in the trim unit.

In these designs the lean solution circulated in the bulk unit is again only partially regenerated and can be considered analogous to the semilean solution in a doubly split-flow unit. The lean solution circulated in the trim unit is regenerated in a separate facility to the degree necessary for the treated gas-purity requirement.

Solution conditioning. The most effective step in maintaining a clean treating solution of any variety is to minimize the amount of contaminants entering with the raw gas. Once liquid hydrocarbons, sand, pipe scale, glycol, or other wellhead additives, and corrosion products from upstream units enter the treating unit, their removal is an order of magnitude more difficult.

A reduction of 1 ppm by weight of material continuously entering the contactor of a typical 100 MMscfd plant means that 1 ton/year of contaminant does not have to be removed from the solvent by other means.

For this reason an adequately sized knockout drum or filter-separator on the raw-gas stream is a prerequisite for minimizing foaming, corrosion, solution degradation, and losses in the downstream treating unit. An SNPA-DEA unit is no different from other regenerable solvent processes in this regard.

Since 100% removal of potential contaminants is not economically possible, slipstream filtration of the DEA solution is employed to keep the solution free of dissolved and solid impurities. In an SNPA-DEA unit, this filtration is normally carried out in a pressure leaf-type filter using conventional filter aid plus pulverized activated carbon as the precoat material.

This filtration step removes solids that may be products of corrosion, or that may have come in with the raw gas. Addition of the activated carbon serves to absorb trace liquid hydrocarbon compounds and amine degradation products, to reduce foaming and corrosion tendencies.

Elimination of contact between DEA solution and oxygen is necessary to avoid gradual generation of degradation products that would otherwise require removal. Gas blanketing of the running and pure solution storage tanks has proven effective in this regard, plus gas purging of the solution sump.

The use of steam condensate for solution makeup is also recommended for two reasons. First, this gain eliminates a possible source of oxygen contact; and second, the dissolved solids that may be present in the raw or treated water can be a source of decomposition products, or release the protective iron-sulfide film, requiring its removal via the filter and possibly resulting in increased corrosion.

Even with these precautions, solution foaming with attendant loss of production can be encountered during upsets. Thus, all SNPA-DEA units are built with antifoam-injection facilities. Use of either the heavy alcohol-type or silicone-base antifoam have been successfully used in most plants at different times.

The two basic types of prime offenders with regard to foaming are hydrocarbons (either entering with the raw gas, or condensed within the contactor) and glycols or additives used to flush the well flow lines and casing.

Condensation within the contactor can be avoided by maintaining a proper differential between DEA temperature and the gas dew point, while any truly commercial process must be capable of coping with the necessary well-treatment steps.

In the case of units being converted from another treating solution to an SNPA-DEA operation, it is mandatory that the unit be thoroughly flushed to remove all of the prior treating solution. For example, any residual MEA remaining following a conversion will still be capable of reacting with COS and CO_2 to form permanent degradation compounds.

Since the SNPA-DEA process does not require reclaiming, there is no effective way, other than bleeding solution, to remove these materials that may prove conducive to corrosion.

Corrosion experience. The corrosion experienced to date has been extremely low on new plants specifically designed for the SNPA-DEA process, as well as on a majority of the converted units. However, it is necessary to separate the corrosion experience on completely new units from that on converted units, for sometimes we find that a certain section, or sections, of existing units may not have been originally designed with the parameters that are now used on a new unit.

If the conditions of severity of an existing unit are close to, but somewhat in excess of, those that would be selected for a new plant, it oftentimes appears more economic to leave the existing equipment in place and see how it performs, rather than arbitrarily call for replacement at the time of conversion.

All areas of any type of acid-gas removal units are normally subject to varying degrees of corrosion. These include the rich-solution piping, rich-solution letdown valve, rich-lean solution exchanger, regenerator vessel and internals, regenerator reboiler, lean-solution piping, lean-solution pump, lean-solution cooler, regenerator overhead piping, regenerator overhead condenser, and accumulator.

Each of these areas will be covered in the discussion below insofar as SNPA-DEA units are concerned, since each is a potential source of corrosion for these units as well. We specify stress relieving for all equipment and piping in contact with lean DEA solution operating at temperatures of 180° F. and above, and all rich-solution piping.

These precautions, plus code requirements for heavy-walled vessels, mean that nearly the entire unit is stress relieved. Where this procedure may not have been followed during the initial construction of a converted unit, premature failure can be expected.

The contactor and regenerator of a new plant, each containing valve-type trays, were reported to show no corrosion or tray defects after about 3 years of operation. The interior of the columns on another unit (which had previously been in MEA service for about 5 years) was described as dirty from iron sulfide after 3 years on an SNPA-DEA type operation.

Also, the bottom 10 trays in the regenerator required replacement. Although scaled, the vessel walls were not severely corroded from a metal-loss standpoint and are still in operation today (4 years after above report).

The operators of still another converted unit reported that the contactor, regenerator, and trays (sieve type) were in good condition 3 years after conversion, and that there was no indication of fouling in these columns. These columns had also been in MEA service before conversion.

Progressive pitting of a bypass line around a letdown turbine in the rich-solution piping, immediately downstream of a large pressure reduction, has been experienced in a new unit after 3 years of operation. This leads us to expect a life of 3 to 4 years for this section of piping in those plants having such a line.

If a hydraulic turbine is not a part of the plant, then the letdown valve can be located directly on the flash tank. Where rich-solution piping is under pressure, only nominal corrosion has been experienced.

One new unit has reported pitting of the high-differential letdown valve in SNPA-DEA service after 3 years of service, due to the flashing occurring. Similar valves in other units have operated for up to 3 years with no reported excessive corrosion or erosion. The preferred metallurgy of this angle-type valve is a carbon-steel body with stellited plug and seat.

Our experience regarding corrosion of rich-lean solution exchangers also falls into categories; namely, those with carbon-steel tubes and those with type 304 SS tubes. In all of these units the lean solution is on the shell side and rich solution on the tube side. It is desirable to locate the control valve, holding liquid level in the flash drum, on the outlet line from the rich-lean solution exchanger to minimize vaporization and, hence, velocity in the heat exchanger.

The velocity of the mixed-phase rich solutions, as calculated at the outlet operating conditions of some of the SNPA units at Lacq, is in excess of 5 fps following debottlenecking to increase capacity.

The tubes in these units have been changed to type 304 SS. All of our new units employ carbon-steel tubes since the operating velocities do not approach this value. A life of over 4 years has been attained on a new unit in Alberta with a computed tube velocity of 3.4 fps at operating conditions.

In other units undergoing conversion to SNPA-DEA, we have calculated that operating velocities in excess of this would result, but have been unable to justify discarding an existing carbon-steel bundle, for replacement with 304 SS, at the outset of a conversion. Rather, it has proved reasonable to leave the carbon steel bundle in place and gain maximum life from it.

As points of reference, a stress-relieved carbon-steel bundle operating at an outlet velocity of 5.25 fps failed after approximately 5 months in a converted unit, while a nonstress-relieved replacement lasted only about 11 weeks. Use of type 304 SS tubes corrected this problem area in the converted unit.

Hydrogen blistering was also reported in certain areas of the channel of the unit operating at 5.25 fps. Weld-overlay material was applied to the areas affected as a precaution against premature failure.

On the basis that an aqueous phase, hydrogen sulfide, and hydrogen must be present for hydrogen blistering to occur, it was reasoned that this step, plus the use of a type 304 SS bundle, would mitigate the hydrogen attack, since hydrogen (apparently initially present as a by-product of tube corrosion) would no longer be present.

Corrosion of the shell or on the shell side of the tubes has not been of any concern on units built to date, although fouling is encountered in some plants to the extent of limiting heat transfer after a year of onstream time.

All of SNPA's reboiler bundles at Lacq are stress-relieved carbon steel U-tube units in kettle-type reboilers. The reboiler steam is desuperheated and the reboilers are instrumented so that condensate cannot flood the tubes. One unit in Alberta experienced severe reboiler corrosion following conversion from MEA to SNPA-DEA. Contributory causes for this situation included:

1. Use of superheated steam during some periods of the year.
2. Use of raw water rather than condensate in preparation of filter precoat.
3. Condensate flooding of the tubes.
4. Insufficient tube submergence (weir height) in the kettle as originally installed.
5. Peculiarity of the reboiler design whereby reboiler feed was directed to a splash plate above the tube bundle and overflowed to the bundle compartment, rather than entering at the bottom of the reboiler.
6. Tube spacing insufficient to permit passage of the vapors generated in the reboiler.

Changes were made in the unit to correct the above problems, and caustic was added to the DEA solution to maintain heat-stable salts at a value of 0.2 wt % or less.

Two other converted units in Alberta, plus one new unit, (the latter representing about 4 years of operation) have experienced no appreciable reboiler corrosion, pitting, or baffle cutting. By this it is meant that no bundle replacement or tube plugging has been required for this period of time.

Our conclusion is that stress-relieved carbon steel is the proper choice of material for a reboiler that is designed properly from a hydraulic and mechanical standpoint, provided the solution filter is operating on schedule.

In the rare case that some impurity in the raw gas causes the accumulation of a corrosive degradation product, the addition of soda ash appears justified. The quantity of soda ash involved has affected neither treated gas purity nor the fouling tendencies of the solution.

The overhead condensing systems of solution regenerators for all plants employ air coolers with carbon-steel tubes, followed in some cases by trim water coolers, while those at Lacq mainly have water coolers for the entire service. Connections are normally provided for the periodic injection of amine or other corrosion inhibitors into this area.

Of the SNPA-DEA type plants now operating in Alberta, most have indicated only nominal corrosion in the regenerator overhead piping, condenser, and accumulator, while one has detected some loss of metal in the regenerator outlet vapor line only, but the rate of loss does not indicate failure in the near future.

This represents a service life of 3-10 years on the units showing no corrosion, while the one with detectable metal loss has been in DEA service for about 3 years following about 3 years in MEA service.

References and bibliography

1. Wendt, C. J., Jr., and Dailey, L. W., "Gas treating: the SNPA process," Hydrocarbon Processing, Vol. 46, No. 10, October 1967, 155. (Paper presented at the June 1967 meeting of the Canadian Natural Gas Processing Association, Edmonton, Alberta).
2. Bailleul, Michel, "Development of SNPA-DEA process," Canadian Gas Journal, Gas Processing/Canada, Winter 1969. (Paper presented at the September 1969 meeting of the Canadian Natural Gas Processing Association, Calgary, Alberta).
3. Chute, A. E., "Lacq gas processing and sulfur removal plant," Proceedings of the 34th annual fall meeting, California Natural Gasoline Association, Oct. 8-9, 1959.
4. Younger, A. H., "Fundamental and applied work needed," Canadian Gas Journal, Gas Processing/Canada, Spring, 1969. (Paper presented at the October 1968 6th annual conference of Canadian Research Management Association).
5. Operators List 7, "Natural gas processing plants in Canada, January 1969," Mineral Resources Division, Department of Energy, Mines, and Resources, Ottawa.
6. Berlie, E. M., "The Canadian gas processing industry." (Paper presented at the 49th annual conference of the Chemical Institute of Canada at Saskatoon, Saskatchewan, June 6, 1966).
7. Beddome, J. M., "Current natural gas sweetening practice." (Paper presented at the 19th Canadian Chemical Engineering Conference, Edmonton, Alberta, Oct. 19-22, 1969).

Glycol choice merits close study for gas dehydration

SILVANO GROSSO
Dow Chemical U.S.A.
Freeport, Tex.

DIETHYLENE (DEG), triethylene (TEG), tetraethylene glycol (T_4EG), and blends of these and other glycols have all found application in dehydration of natural gas. As a class, glycols have high hydroscopicity, good thermal stability, and low vapor pressure.

Individual physical properties, however, vary considerably (Table 1). As a consequence, there is a great variation in the performance of these fluids for natural-gas dehydration.

Choice of a glycol for either new or existing equipment must be based on close scrutiny of the initial cost of the fluid, and plant operating conditions and gas specifications. By considering these factors, the overall performance, operating costs, and reliability of the dehydration equipment can be optimized.

In most cases, TEG has an overall performance advantage over other fluids. When a DEG blend can dry gas without sacrifice in dew point, it is advisable to consider use of pure DEG since it may be both less expensive and of more uniform quality. Tetraethylene glycol blends, provided they are of adequate quality, can provide performance comparable to TEG.

Dehydration principles. Dehydration is accomplished by countercurrent contact of wet gas with a relatively pure or "lean" glycol solution, i.e., a solution that contains as little water as possible. The contactor is either a packed or trayed tower.

Water is absorbed from the gas phase into the glycol so dry gas exits at the top and glycol rich in water at the bottom. The rich glycol must then be regenerated to its original conditions to repeat the cycle in a continuous process.

Most pipeline-gas specifications require a water content not to exceed 7.0 lb H_2O/MMscf. The gas dryness can also be expressed as the water dew point, which is that temperature to which the gas must be cooled before condensation will start to occur.

The dew-point depression will then be the difference between the temperature of the saturated gas and the dew point after dehydration.

Dew point depends principally on the partial pressure of water over the glycol in contact with the gas at the top of the tower, which in turn depends on the glycol concentration.

The regeneration step determines the ultimate concentration of lean glycol attainable, hence the dew point of the dry gas.

Dehydration capacity. Fig. 1 shows the relationship between the dew point that can be obtained at various contact temperatures with aqueous solutions of DEG and TEG. These curves represent composite data from references 1 and 2 for DEG, and references 1, 2, and 3 for TEG.

Note that the higher concentration, the lower the resulting dew point. For concentrations greater than 98%, TEG gives lower dew points than DEG. For example, at 100° F., a 99% wt DEG solution gives a dew point of 14° F., but a 99% wt TEG gives a dew point of 8° F., or an advantage of 6° in dew-point depression.

Fig. 1 shows the dew point that can be obtained under equilibrium conditions. In practice, limits on equipment design (number of contact stages, circulation rates, etc.) permit only an approach to equilibrium ranging from 5° to 15° from the dew points shown.

Similar relationships as those shown in Fig. 1 have been developed for ethylene glycol and tetraethylene glycol.[1,2] Over the same concentration and temperature ranges, EG gives slightly lower dew points than DEG and tetraethylene glycol slightly higher dew points than TEG.

If there were no limits to the concentration of lean glycol that could be obtained, one could achieve the desired degree of dehydration with any of the glycols or any blend of them. A choice of fluid could then be based exclusively on initial cost.

Unfortunately, there are practical limits to the purity obtainable in the regeneration of the various glycols. Thus the cheapest fluid is not necessarily the one that will give the lowest dehydration cost.

Glycol regeneration. In the commercial production of glycols, a mixture of aqueous glycols is obtained at one point in the process. Purification of the single components, EG through T_4EG, cannot be accomplished by atmospheric distillation since the glycols decompose as the temperature approaches their boiling points.

Successive vacuum distillations are necessary to get pure EG, DEG, TEG, and T_4EG. This explains why cost is higher for the higher glycols and why unpurified mixtures are cheaper.

Similarly, in the reboiler of a glycol dehydrator, the glycol cannot be heated to its boiling point to remove all the water. Otherwise, decomposition will occur. The products of decomposition, which are acidic, would result in a quite corrosive solution.

Both laboratory data and field experience have helped establish certain temperature limits which, in conjunction with other precautions, permit safe thermal regeneration of the various glycols. Table 2 shows the maximum temperatures to which the glycols can be safely heated, together with the corresponding concentrations and the approximate dew point depression obtainable.

The boiling-point curves of the various glycols (Figs. 4-6) can be used to determine the glycol concentration corresponding to any given temperature. These temperature limits are easily exceeded at the glycol/fire-box interface.

High metal skin temperatures can rapidly decompose the glycol even though the bulk temperature is below the decomposition limit. For this reason, it is important to minimize skin temperature by proper fire-box design and by providing sufficient surface area so the average heat flux does not exceed a maximum value of 7,000 BTU/hr/ft².

When high dew-point depressions are required, glycol solutions can be regenerated to quite high concentrations by specialized techniques such

Fig. 1

Equilibrium water dew points

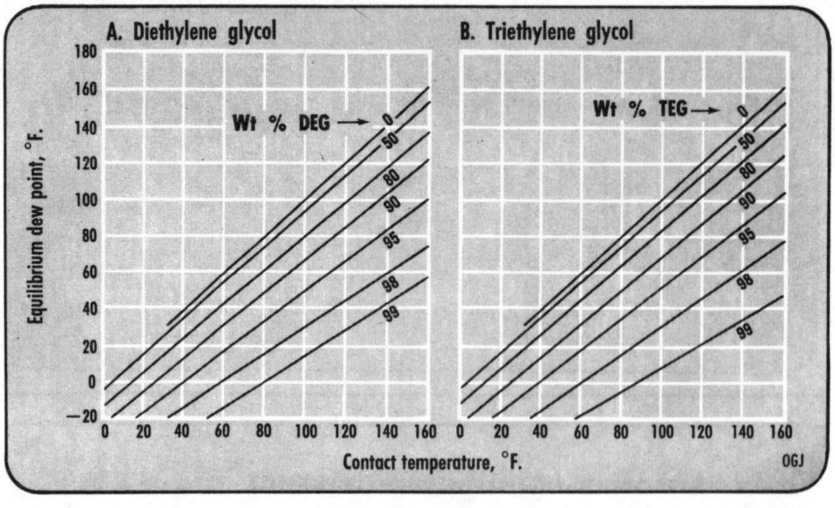

Table 1

Properties of commercial glycols[1]

Property	Ethylene glycol	Diethylene glycol	Triethylene glycol	Tetraethylene glycol
Formula	HOC_2H_4OH	$HOC_2H_4OC_2-H_4OH$	$HO(C_2H_4O)_2-C_2H_4OH$	$HO(C_2H_4O)_3-C_2H_4OH$
Molecular weight	62.1	106.1	150.2	194.2
Boiling point of pure compound at 760 mm Hg, °F.	387.1	472.6	545.9	597.2
Vapor pressure* at 77° F. (25° C.), mm Hg	0.007	0.002	0.0004	0.00005
Density, g/ml				
At 77° F. (25° C.)	1.110	1.113	1.119	1.120
At 140° F. (60° C.)	1.085	1.088	1.092	1.092
Pounds/gal @ 77° F. (25° C.)	9.26	9.29	9.34	9.34
Freezing point, °F.	8	17	19	22
Pour point, °F.	....	−65	−73	−42
Absolute viscosity, cp				
At 77° F. (25° C.)	16.5	28.2	37.3	39.9
At 140° F. (60° C.)	5.08	7.6	9.6	10.2
Surface tension at 77° F. (25° C.), dyne/cm	47	44	45	45
Refractive index at 77° F. (25° C.)	1.430	1.446	1.454	1.457
Specific heat at 77° F. (25° C.), BTU/lb/°F.	0.58	0.55	0.53	0.52
Flash point, °F.†	240	280	320	365
Fire point, °F.†	245	290	330	375

*Extrapolated values. †Cleveland open cup.

as hot gas stripping, vacuum regeneration, or the Drizo process, a patented process licensed by Dow Chemical Co. Limitations of temperature and heat flux, however, must be observed.

TEG vs. other glycols. If dew-point depression were taken as the only measure of performance of a glycol fluid, it is apparent from Table 2 that T_4EG would be the best choice. T_4EG can be thermally regenerated without decomposing to higher concentrations than the lower glycols and thus give higher dew-point depression.

However, when all costs are taken into account, it can be shown that TEG is the most cost effective of the pure glycols. For this reason, most modern equipment for glycol dehydration is designed specifically for TEG. Reliability is also greatest for TEG. These three factors together—dew point depression, operating cost, and reliability of operation—can be used to define performance of a glycol fluid.

In general, both initial cost and performance are lower for DEG than TEG. Tetraethylene glycol finds some application in cases where the dew-point requirements fall between conventional TEG units and units using TEG regenerated to high concentrations by ways mentioned previously.

Blends of various glycols usually, but not always, cost less than the pure glycols by themselves, since they may consist simply of distillation "bottoms" with little or no further puri-

Fig. 2

Dew-point comparison of TEG and DEG/TEG blend

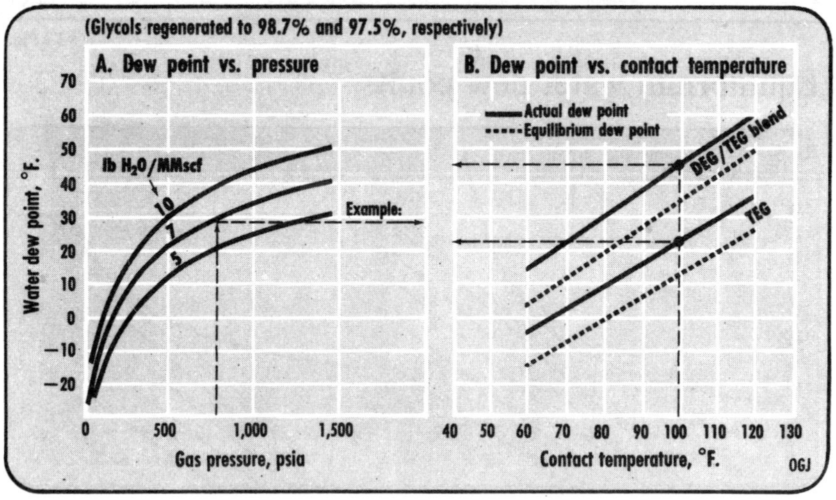

Fig. 3

Glycol absorber equilibrium diagram

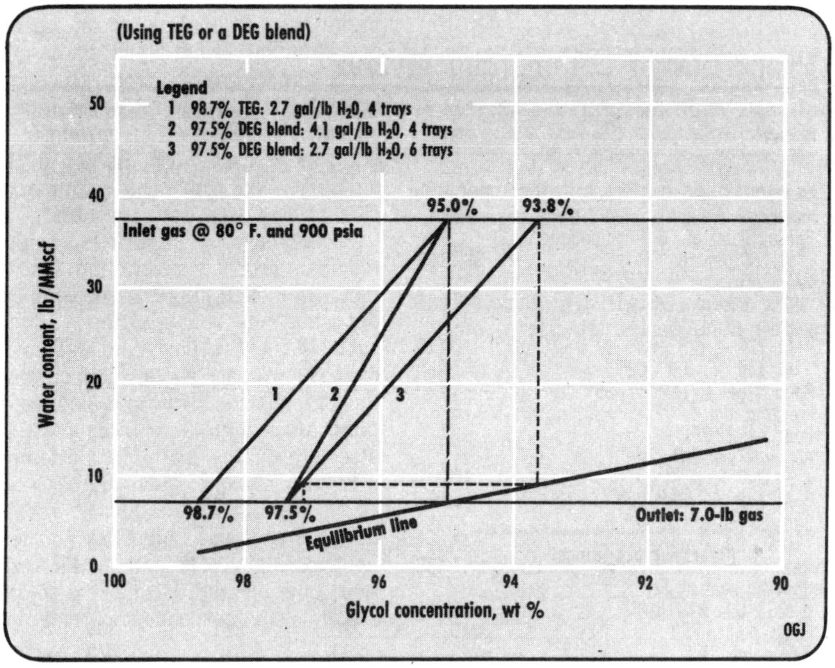

fication. Their performance, therefore, is variable and highly dependent on the type and quality of the blend.

The type of blend depends on whether it consists of TEG and lower glycols or TEG and higher glycols. Each blend type may also be susceptible to considerable variation in quality, depending on its source. Quality of the pure glycols (and some blends) is uniform since it is regulated by production and sales.

DEG blends. In this type of blend, there may be 15-30% TEG, the rest being principally DEG. Smaller amounts of EG and other glycols may be present. Also, water may have been added to lower the freezing point in storage. Some blends of this type may be formulated into a complete product including antifoam, inhibitor, and dye.

Since this type of fluid contains mainly DEG, the maximum operating temperature is limited by the decomposition of DEG, 350° F. EG, however, which can be present at levels as high as 2-3%, will decompose and volatilize even at these conditions.

It is a serious mistake to assume that a DEG blend can replace TEG directly. Unless the equipment design, the capabilities of the fluid, and the dehydration requirements are closely matched, straight substitution of TEG with a DEG blend can be expected to result in any or all of the following: (a) pH and corrosion problems, (b) excessive glycol losses, and (c) insufficient dehydration.

Fig. 2 shows the dew points that can be obtained with this type of fluid over a range of contact temperatures. In some instances, the blend cannot give a 7.0-lb gas even if circulation rate and contact stages are increased until an approach to equilibrium closer than 10° F. is obtained.

The example (Fig. 2A) shows that, for a gas at 800 psia and 100° F., a 7.0-lb gas must have a dew point of 28° F. In standard equipment, the DEG/TEG blends can only give a dew point of 45° F. while the equilibrium dew point, that is the lowest possible, is 35° F.

In this case, the blend cannot give 7.0-lb gas unless the glycol is regenerated to higher concentrations by either gas stripping, the Drizo process, or vacuum regeneration.

In contrast, TEG regenerated at 400° F. will give a dew point as low at 22° F. in standard equipment.

Table 3 shows the operating conditions of a DEG blend vs. pure TEG for dehydrating gas at a temperature and pressure that will permit the attainment of 7.0-lb gas.

As shown in Fig. 3, the required dew point can be achieved with either a circulation rate of 2.8 gpm of TEG (2.7 gal/lb H_2O absorbed) or with 4.3 gpm of the DEG blend (4.1 gal/lb H_2O). If the circulation rate is set at 2.8 gpm in both cases then the use of the blend would require two additional absorber trays.

Table 3 also shows that both the reboiler duty and vaporization losses are higher for the DEG blend than for pure TEG. The calculations assume that the reboiler feed temperature is equal in both cases.

If the equipment was designed for TEG, it is likely that the reboiler feed is at a lower temperature when a DEG blend is used, in which case yet more fuel would be necessary to keep the reboiler at 350° F. It does not follow, that lowering the reboiler temperature will automatically decrease the operating costs.

There may be cases when this type of glycol blends can give the desired dew point without increasing the circulation rate. However, these blends are not as versatile as TEG.

Consider that equipment modifications may be required if the dehydrator was designed for TEG. There may be a need, for example, of additional heat-exchanger capacity, more con-

Figs. 4-6

Boiling points of aqueous glycol solutions

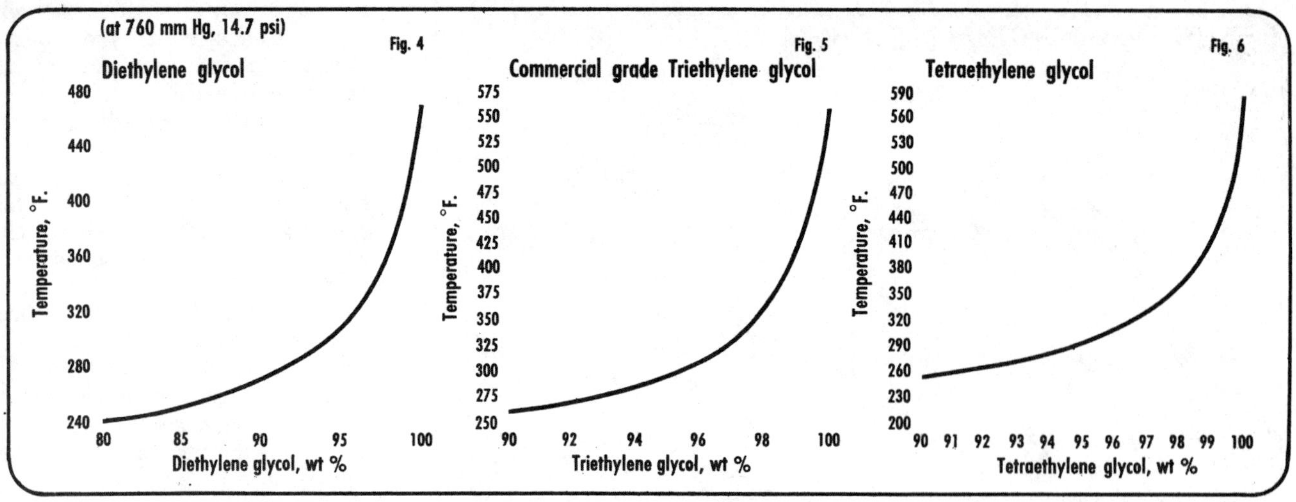

tact trays in the absorber, and increased reflux capacity in the reboiler still to minimize loss of the more volatile components.

In some cases, an external overhead condenser for DEG is necessary. As the well pressure decreases, the dew point requirements for a 7.0-lb gas may increase beyond the capability of the blend so that a complete change of glycol inventory (back to TEG) may be necessary.

Tetraethylene glycol blends. This type of blend consists mainly of T_4EG with 20-30% TEG. Reboiler temperature must be no higher than 400° F. Otherwise, TEG will decompose. At these conditions, the total glycol concentration will be about 98.7% so that dew-point depressions comparable to pure TEG can be obtained.

Table 2

Dew point depressions

Glycol	Max. temp.,* °F.	Glycol purity,† % wt.	Max. dew point depression,‡ °F.
EG	325	95.8	53
DEG	350	97.0	60
TEG	400	98.7	85
T_4EG	435	99.3	95
DEG blend	350	97.5	60
T_4EG blend	400	98.9	90

*Temperature above which decomposition increases rapidly. Assume heat flux ≦ 7,000 BTU/ft²/hr and oxygen excluded. †Maximum concentration obtained in atmospheric reboiler. ‡A slight increase in dew-point depression can be achieved by increasing glycol circulation rate.

Tetra blends suffer from some of the limitations of DEG blends. But the fuel requirements for regeneration are no greater than for pure TEG.

With no advantage in dew point depression or cost of regeneration, one must weigh the initial savings in purchase price against any operational problems and replacement costs.

As mentioned earlier, product specifications for the glycol blends are not as stringent as for the pure glycols. Therefore, such impurities as salt, lower glycols (EG, DEG), glycols of the propylene series, chlorides, etc., can be present in sufficient quantities to cause continuing problems.

Not all blends contain all of those impurities. A lot depends on whether one is buying crude glycol "bottoms", a reclaimed product from glycol still bottoms or from some other type of operation such as an aromatic extraction process. Also, as the degree of purity of the blend approaches that of TEG, so does the cost.

Problems with impurities. Impurities can produce several adverse effects. For example, salt contaminants deposited on heat-transfer surfaces, such as the fire box, can lead to the formation of hot spots, tube burn-out, and expensive down-time and repair bills. Along with other impurities, salt also aggravates foaming problems.

Lower glycols, such as EG and DEG, decompose readily at reboiler temperatures used for TEG and tetra contributing to coke and sludge formation. Low pH and a corrosive solution result with consequent metal loss,

Table 3

DEG blend vs. TEG

(Operating conditions)	DEG blend	TEG
Gas flow rate, MMscfd	50	50
Gas temp., °F.	80	80
Pressure, psia	900	900
Water in, lb/MMscf	37.0	37.0
Outlet water spec., lb/MMscf	7.0	7.0
Dew point, °F.	31.0	31.0
Absorber trays (bubble caps)	4	4
Glycol units, gal/lb H_2O absorbed	4.1	2.7
Glycol circ. rate, gpm	4.3	2.8
Lean glycol conc., % wt	97.5	98.7
Reboiler temp., °F.	350	400
Hot rich glycol temp., °F.	250	250
Reboiler duty, BTU/hr	232,000	222,000
Glycol vaporization loss, lb/day	2.0	0.4

The author . . .

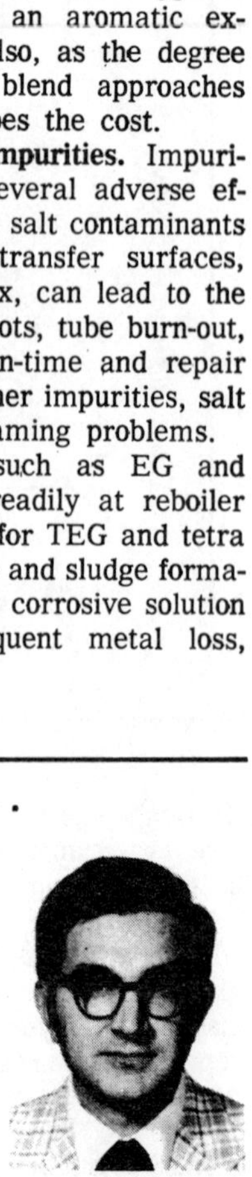
Silvano Grosso

Silvano Grosso is a gas-conditioning specialist with responsibilities for product and process development in gas dehydration and treating for Dow Chemical U.S.A., Freeport, Tex. He joined Dow's Michigan Division in 1972. Grosso transferred to the Texas Division in 1975. He graduated from Bob Jones University in 1969 and from Northern Illinois University in 1974 with an MS degree in inorganic chemistry.

pitting (especially if chlorides are present), and fouling. The higher vapor pressure of the lower glycols translates into losses by vaporization that are higher than normal.

The presence of propylene glycol (PG) and/or dipropylene glycol (DPG) can cause severe foaming. This effect can be readily lab demonstrated.

In the standard foaming test,[1] as little as 0.5% DPG will cause a fresh sample of triethylene glycol, by itself completely nonfoaming, to be whipped into a stable foam in just a few minutes. The problem is even more severe with tetra blends, since they have a higher viscosity than TEG. Replacement of glycol lost by foaming can become expensive very quickly!

Another disadvantage of the blends is that one is never assured of a continued supply of a good-quality product. The availability of the blends depends on the market demand for the pure glycols both in gas conditioning and in other applications.

References

1. Gas Conditioning Fact Book, Dow Chemical Co., Midland, MI, 1962.
2. Polderman, L. D., Proceedings of 1957 Gas Conditioning Conference, University of Oklahoma, Norman.
3. Worley, M. S., Proceedings of 1967 Gas Conditioning Conference, University of Oklahoma, Norman.
4. Pearce, R. L., Proceedings of 1973 Gas Conditioning Conference, University of Oklahoma, Norman.
5. Ballard, D., Hydrocarbon Processing, April 1977.
6. Campbell, J. D., Gas Conditioning and Processing, Norman, Ok, 1976.

Molecular sieves can sweeten fuel gas

LORENZ V. KUNKEL
JAMES W. CHOBOTUK
Amoco Production Co.
Tulsa

WITH PROPER DESIGN, an economical and reliable molecular-sieve unit can be used to sweeten natural gas with low H_2S content. However, actual project costs vary with each unit and depend on amount of H_2S and CO_2 in the feed gas.

In 1970-71, Amoco Production Co. developed two large reserves of natural gas between British Columbia, Yukon, and Northwest Territories. Economics favored drying this gas in two separate plants, one at Beaver River and the other at Pointed Mountain.

Gas was pipelined to Fort Nelson, B.C., to be sweetened. Each plant required some 885 Mscfd of fuel gas. Since no sweet gas was available at either location, fuel-gas sweetening units were required.

Molecular sieves were selected because of potential advantages in overall cost and reliability. They also can remove H_2S and require little maintenance and operating attention.

However, a molecular sieve to further reduce COS conversion with high H_2S gases must be developed if low-sulfur gas is required. Presently, further experience and molecular-sieve development is expected to better define safe limits of molecular-sieve use.

This article abstracted from a paper presented by the authors at NGPA's annual convention, Mar. 26-28, Dallas.

Dehydration plant. Feed gas to the treater is taken from the residue scrubber downstream of the triethylene glycol dehydrator, Fig. 1.

The dehydrator lowers water in the feed to the molecular-sieve treater to less than 1 lb/MMscf. Entrainment of triethylene glycol into the system is controlled by mist extractors located in both the absorber and the residue-gas scrubber.

A treated gas containing a maximum of 3 grains of H_2S/scf was specified, Table 1. The nature of molecular-sieve treaters is such that the 3-grain limit would be reached only at the end of the cycle.

Essentially complete removal of H_2S could be expected so that the fuel gas would average less than ¼ grain H_2S/100 scf. The molecular sieves would also remove essentially all mercaptans present in the feed gas.

Design. The molecular-sieve fuel-gas treaters were each designed to sweeten 1,770 Mscfd of dehydrated natural gas at 1,200 psia and 112° F. Fuel-gas requirement at each plant was 885 Mscfd with the remaining half of the treated gas used for bed cooling and regeneration.

The design provides for downflow for treating and cooling and upflow for regeneration, Fig. 2. The 885 Mscfd of treated gas used for regeneration is first used to cool the second bed which had been regenerated on the previous cycle.

Gas from the cooling bed is then heated to about 600° F. in a salt-bath

Table 1

Inlet gas*

Component	Beaver River, mol %	Pointed Mountain, mol %
He	....	0.05
N_2	3.57	1.30
H_2S	0.57	0.47
CO_2	6.93	10.77
C_1	88.62	87.35
C_2	0.31	0.06
C_3+	Trace	Trace
	100.00	100.00

Treated fuel-gas specification

H_2S content — Maximum of 3 grains/100 scf
water vapor content — Maximum of 3 lb/MMscf

*Beaver River and Pointed Mountain plants

Fig. 1

Beaver River, Pointed Mountain plants

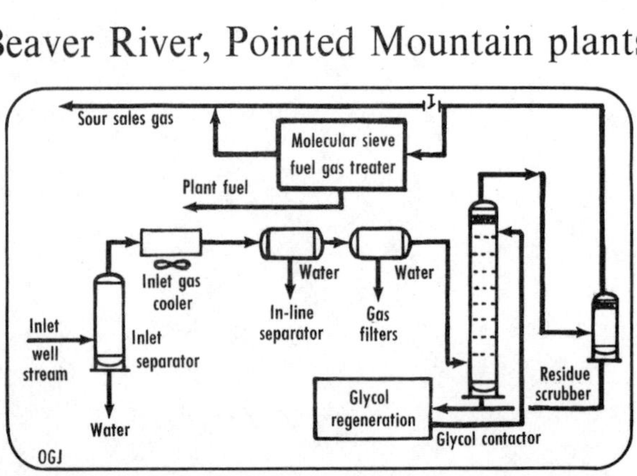

Fig. 2

Molecular-sieve fuel-gas treater

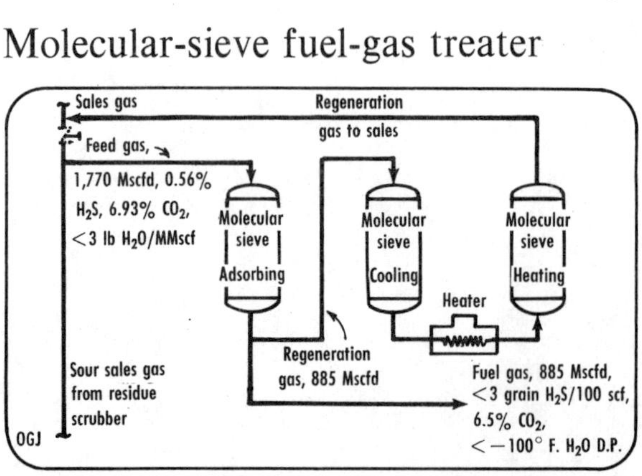

Fig. 3

Isotherm

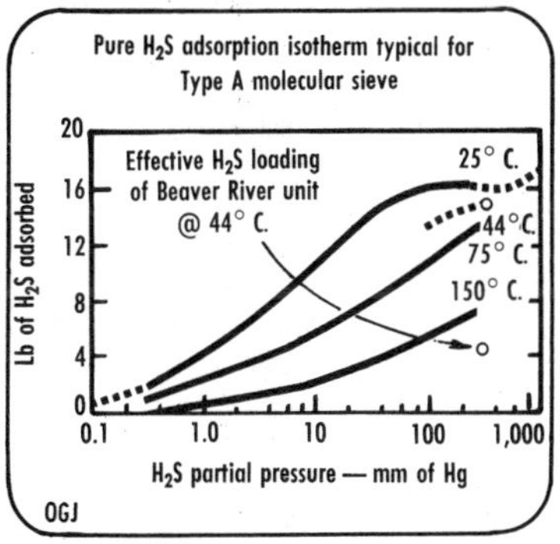

heater and used to regenerate the third bed that was previously on adsorption. By introducing a 12-psi pressure drop in the sales-gas line, regeneration gas at 1,188 psia can reenter the sales-gas line downstream of the back-pressure-control valve.

A three-bed system such as this saves fuel as the sensible heat from the hot regenerated bed is recovered by the cooling.

The 42-in. OD vessels are lined with 3-in. refractory insulation. They were originally charged with 10-ft beds of 1/16-in. diameter, pelletized Type A molecular sieve weighing 3,300 lb/bed. The unit was designed to operate on a 4-hr cycle.

Beaver River plant, designed to deliver 240 MMscfd, went on stream in September 1971. Pointed Mountain plant, designed for 180 MMscfd, went on stream 1 year later. Design operating conditions for the installation were about the same as the actual operating conditions, Table 2.

Sieve traits. Union Carbide selected a specific composition of its Type A molecular for Beaver River. Selection was based on capacity and mass-transfer traits combined with particle strength and crystal stability in acid-gas-sweetening service.

Later, another Type A sieve was installed which had similar characteristics to the first. But it had smaller granulated form and increased bulk density.

Both types have an internal surface area between 500 and 700 sq m/g with adsorption of nitrogen at minus 196° F. from 0.1 to 233 mm mercury pressure. The nominal pore diameter of 4A sieve is 4 A. and 5A sieve is 5 A. Either type can be used to selectively adsorb H_2S or CO_2.

Average loading of the sieve was 4.48% by weight of H_2S at an H_2S partial pressure of 348 mm of mercury and a temperature of 112° F. This loading was set well below the equilibrium loading isotherm for pure H_2S shown in Fig. 3. This allows for interference by coadsorption, for partial effectiveness in the transition zone, and for loss of sieve activity with time.

Problems and solutions. The Beaver River fuel-gas treater failed to produce specification sweet gas on start-up in September 1971. Treated gas contained approximately 100 to 300 ppm H_2S at the beginning of the treating cycle. But it increased to about 2,000 ppm H_2S at the end of the 4-hr cycle.

Early H_2S breakthrough showed mechanical problems. And the high H_2S concentration at the end of the cycle showed a possible sieve-capacity problem.

Valve leakage, a frequent problem with this type of equipment, is normally one of the first items to be checked. Effect of leakage can be to preload the cooling bed with H_2S and to reduce the regeneration efficiency of the bed being heated.

Where necessary, switching valves were removed and adjusted until they were found bubble tight by bench testing. Where possible, valves were reinstalled in the direction such that the pressure differential during operation helped seat the closed valves.

Larger pneumatic actuators were installed to provide higher pressure for better valve seating. Tests showed no significant valve leakage, but performance of the unit did not improve.

Water contamination. Because of continued poor performance, the Beaver River molecular-sieve unit was shut down in November 1971. Analyses showed that the sieve contained 1 wt % excess water and, after drying, retained about 92% of the new sieve H_2S capacity.

Each bed was regenerated for 36 hr at 65 psia with an inlet gas temperature of 600° F. to remove any water present. Data taken after the regeneration showed no great improvement.

Low-pressure operation. Performance improved when operating at about 465 psia for both adsorption and regeneration instead of 1,200-psia design pressure.

H_2S content of the treated gas at the start of the treating cycle was 200 ppm or about the same as at design pressure operation. But at the end of the 4-hr cycle, the H_2S content increased to only about 400 ppm compared with about 2,000 ppm when operating at design pressure. Temporary permits were obtained to allow flaring of regeneration gas during low-pressure operation.

Refractory lining leakage. After switching to the adsorption cycle, the treated gas contained about 200 ppm H_2S. This indicated that some feed gas was bypassing part of the sieve bed.

Tower linings were inspected in January 1972 and found to be porous in certain areas with numerous long hairline cracks. The original two layered, trowel-applied lining was removed, and all of the towers were relined with a single gunited layer.

After the tower was relined, the two beds with new sieve produced specification fuel gas throughout the 4-hr cycle at 465 psia. This showed that the lining leakage problem had been solved.

There was also no H_2S breakthrough with the old sieve until after 3½ hr. For the first time since start-up, specification fuel gas could be produced from the unit on a routine basis at a lower temperature.

Table 2

Design vs. actual*

	Design	Normal operating range
Adsorption cycle		
Pressure, psia	1,203.7	1,104 - 1,288
Inlet temperature, °F.	112	83 - 112
Flow rate, Mscfd	1,770	1,497 - 1,820
Fuel-gas rate, Mscfd	885	682 - 900
Regeneration cycle		
Pressure, psia	1,203.5	1,104 - 1,288
Inlet temperature	550	605 - 550
Outlet temperature	500	520 - 450
Flow rate, Mscfd	885	815 - 920

*Beaver River molecular-sieve-treater operating conditions.

Fig. 4

Inlet distributor designs

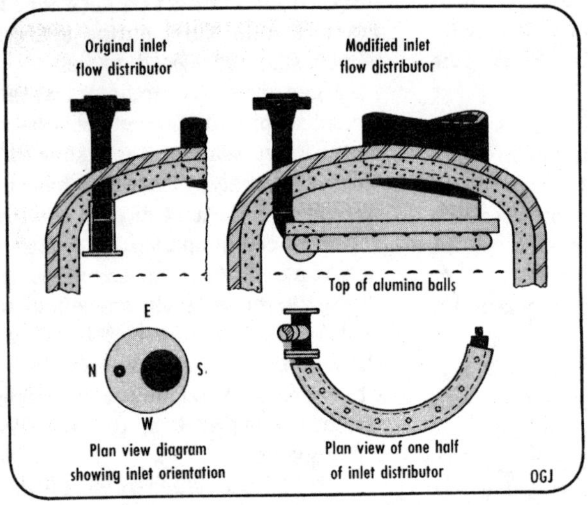

Fig. 5

Sieve-bed support

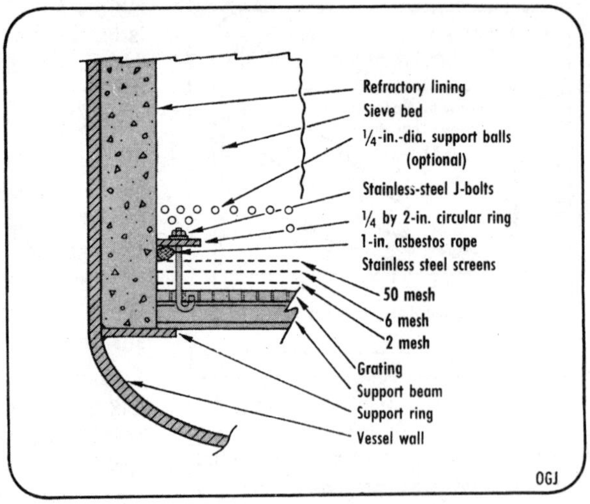

Table 3

H_2S concentration*

Hours into treating cycle	H_2S content of treated gas in ppm — Date: 2-22-72	2-23-72	4-1-72
1.0	0	0	0
2.0	0	0	0
3.0	50	0	5
4.0	2,000	150	...
Cycle time, hr	4	4	3
Pressure, psia	1,155	465	465

*After tower relining.
Beaver River plant Tower 3 containing original sieve
Inlet temperature: 105° F.
Treating rate: 1,600 Mscfd

Table 4

Design and low-pressure operation*

Hours into treating cycle	H_2S content of treated gas in ppm — Outlet	Center	Probes — North	East	West	South
Tower 1 at 1,195 psia						
0.5	0	0	50	0	0	0
1.0	20	0	70	0	0	0
1.5	160	0	200	0	0	0
2.0	250	0	2,000	0	0	0
2.5	450	Trace	2,000+	Trace	0	0
3.0	800	250	2,000+	20	0	0
3.5	2,000	2,000	2,000+	750	Trace	50
4.0	2,000+	2,000+	2,000+	2,000	Trace	350
Tower 1 at 465 psia						
1.0	0	0	0	0	0	0
2.0	0	0	0	0	0	0
3.0	0	0	0	0	0	0
4.0	0	0	0	0	0	0
5.0	0	0	0	0	0	0
5.5	0	0	0	0	0	0
6.0	200	Trace	200	0	0	Trace
6.5	....	55	1,400	0	0	350
7.0		1,700	2,000+	200	200	2,000

Beaver River plant data (upper, 2-23-73; lower, 3-12-72)
Inlet ltemperature: 105° F. @ 1,195 psia; 85° F. @ 465 psia
Treating rate: 1,700 Mscfd

*H_2S concentration

At design pressure of 1,200 psia, there was also no breakthrough of H_2S at the start of the cycle. However, the treated gas built up to 2,000 ppm H_2S or higher after 4 hr, Table 3.

H_2S capacity of sieve beds. During relining of the towers, five probes were installed in the bottom 5 in. of the sieve bed in tower 1. The probes were arranged to sample each quadrant near the edge of the wall and center of the bed. Sieve beds were tested for H_2S capacity at the design pressure of 1,200 psia and a lower pressure of 465 psia, Table 4.

At design pressure, H_2S concentration does not meet specification after only 1½ hr into the adsorption cycle. There is also general breakthrough of H_2S in four of the five probes after 3½ hr. This indicates a deficiency in H_2S capacity.

At 465 psia, an improvement in performance is observed. There was no breakthrough of H_2S in the treated gas until 5½ hr of adsorption at design rates. This corresponds to a 138% increase in capacity above that required for a 4-hr cycle.

Flow distribution. Probe data in Table 4 show that the H_2S capacity problem at 1,200 psia may be due to flow maldistribution. A rapid buildup of H_2S in certain areas of the bed was seen long before general H_2S breakthrough.

Nature of the H_2S breakthrough was not a distinct, massive front with the H_2S concentration rapidly building up to the feed concentration as would be expected. Flow maldistribution was more pronounced at 1,200 psia than at 465 psia.

The molecular sieve treater was modified in July 1972 to determine if the problem was mechanical in nature or could be attributed to the sieve. The original and modified inlet distributor designs are shown in Fig. 4.

Table 5

H_2S concentration after modifications

Hours into treating cycle	H_2S content of treated gas in ppm Bed 1	Bed 2	Bed 3
1.0	0	0	0
2.0	300	40	trace
3.0	650	600	5
3.8	950	....	
4.0[2]	1,350	2,000+	8
Type of sieve	Original pelletized Type A	Original pelletized Type A	Granulated Type A
Modified inlet flow distributor	Yes	No	Yes

(1) Treater will produce specification fuel gas at 465 psia operation with 4-hr cycle and identical conditions.
(2) Difference between H_2S values of Bed 1 and 2 is within the normal reproducibility of the data and is not significant.
Beaver River plant data, date 9-12-72
Inlet temperature 110° F.
Treating pressure 1,270 psia[1], 4-hr cycle
Rates: treating 1,458 Mscfd, regen 748 Mscfd

Results at 1,200 psia indicate that the new inlet distributors did not improve bed performance, Table 5. The new sieve, however, did improve performance as would be expected because a 20% greater mass of molecular sieve was available.

At the Pointed Mountain plant, after 3 months of operation, specification fuel gas could be produced with a 3-hr cycle at 1,200 psia. But it could not be produced on the design 4-hr cycle.

Average H_2S concentration of treated gas at Pointed Mountain at the end of a 4-hr cycle was about 140 ppm. The unit was operated satisfactorily for several months on a 3-hr cycle at design pressure.

New compressors. After the high sieve H_2S capacity at low-pressure operation was established, two actions were taken. Efforts to correct the original capacity limitation at design pressure were continued and a regeneration gas cooler and a recompressor were designed.

These facilities were installed at a cost of $75,000 to each plant to provide greater flexibility. This was done to ensure sufficient H_2S removal capacity in the existing sieve units. However, recompression would not necessarily have been the most economic alternative for a new design.

The increased sieve capacity at low pressure permits a longer cycle time. And sieve life is lengthened with longer cycles. So operating costs of the recompression facilities appear to be offset by reduced sieve replacement.

Recommendations

Sieve capacity. Information on H_2S capacity and coadsorption is largely proprietary. So, the operator must rely on the sieve supplier for an adequate design. Data should be based on actual plant experience or on laboratory work at similar H_2S, CO_2 and hydrocarbon partial pressures.

In setting treated-gas specifications, total sulfur content as well as maximum H_2S content should be limited to acceptable values. The allowable conversion of H_2S to COS and to other sulfur compounds should also be specified and guaranteed by the sieve supplier.

Process and mechanical design. Process and mechanical design details critical to the performance of the sieve units should be reviewed and approved by the sieve vendor. These should include tower dimensions, internal lining specifications, bed supports, and inlet distributor design.

Velocities, temperatures, and pressure drops in treating, cooling, and regeneration should also be approved by the sieve vendor. The maximum permissible upflow velocities should be specified to avoid lifting the bed.

Pressure drops. Proper gas-flow distribution is necessary for successful operation of molecular-sieve units. Tower dimensions and size of sieve will determine pressure drop across the sieve bed for given flow-rate conditions.

Position of the inlet nozzle, plenum space in the top of the tower, and design of a distributor ring or baffle are also important. A pressure drop to distribute the flow evenly is desirable. Too high a pressure drop, however, can lead to sieve attrition.

Contamination. During purging and pressuring, care should be taken to avoid water or methanol contamination of the sieve. High water content is difficult to remove and delays successful operation of the unit. Testing and start-up with dry gas is recommended if it can be made available.

Normal design sieve life is about 3 years. Longer cycles enhance sieve life because loss of activity is related to the number of cycles. Contaminants in the inlet gas can substantially shorten sieve life.

Clean inlet gas should be taken downstream of an adequate gas scrubber. A special feed-gas filter or separator should be installed if there is a danger of contamination due to upsets or entrainment.

Valves. Switching-valve leakage can be troublesome. Some valve manufacturers supply special valves for such high-temperature, fast-cycle service. Type 416 stainless-steel seats were used in the high-temperature valves. Soft seats are recommended for the low-temperature valves.

Valves should be bench tested for positive shutoff prior to installation. Valves should be installed so that the pressure drop during operation helps seat the closed valve, where possible.

Refractory. Internal refractory insulation can reduce both regeneration-gas-heater size and fuel-gas requirements. But it can also be a source of feed-gas leakage if improperly applied. A gun-applied refractory was found to be more impervious than a trowelled application.

A gunned refractory mix similar to Haydite-Lumnite mixture was used on the Beaver River-Pointed Mountain molecular-sieve units. Refractory anchors were installed on 9-in. maximum centers to prevent the lining from separating from the vessels.

Bed supports. A bed support must be designed to prevent loss of sieve and permit proper flow distribution. It must also have the strength to support the bed.

Bed supports can contract as much as ¼-in. in diameter on cooling. Fig. 5 shows the design used at Beaver River and Pointed Mountain. It uses a compressed asbestos rope to prevent sieve leakage.

Test probes. Test probes extending into each quadrant and the center of the sieve bed about 5 in. from the bottom are recommended. They enable testing for flow distribution and checking for leakage through the lining. The probes can be installed much more easily before startup. END

Quick design charts for diethanolamine plants

DON LOVE
Williams Bros. Process Services Inc.
Tulsa

DIETHANOLAMINE (DEA) is being used more and more in the field and in gas-processing plants to remove acid gases from natural-gas streams.

Reason is that DEA is comparatively unreactive with carbonyl sulfide and carbon disulfide, impurities which appear in greater quantities as more and more sour gas is produced in the field.

Historically, monoethanolamine (MEA) has been used for removal of the common acid gases (hydrogen sulfide and carbon dioxide). MEA removes these compounds effectively, and less circulation is required to maintain an equivalent molal solution.

But the problem of dealing with carbonyl sulfide grows more acute every year, and the need for DEA systems increases.[1]

The two DEA design charts, Figs. 2 and 3, may be used for quick and accurate estimation of DEA systems, though more detailed calculations may be needed for final design.

Following is a discussion of the conditions on which Figs. 2 and 3 are based and example of use.

Solution strength and loading. Although various amine-solution strengths have been and are being used in treating plants, it has been found that a 20 wt % solution is the recommended design basis for new plants. Solutions stronger than 22% increase corrosion rates.

The maximum solution loading is based on 0.333 moles of acid gas/mole of DEA. It has been common in DEA plants not to exceed this loading due to corrosion problems, although some plants operating in western Canada use 0.667 moles of acid gas/mole of DEA. The designer may use any loading he desires by ratioing the circulation rate:

$$\text{New circulation rate} = \frac{(\text{Rate from charts})\ (0.333)}{(\text{Desired loading})}$$

The new circulating rate may then be used by starting at the respective value and proceeding accordingly. Fig. 1 shows an example of how this is done. The same procedure may apply for a different solution strength.

Example of solution-loading adjustment

Fig. 1

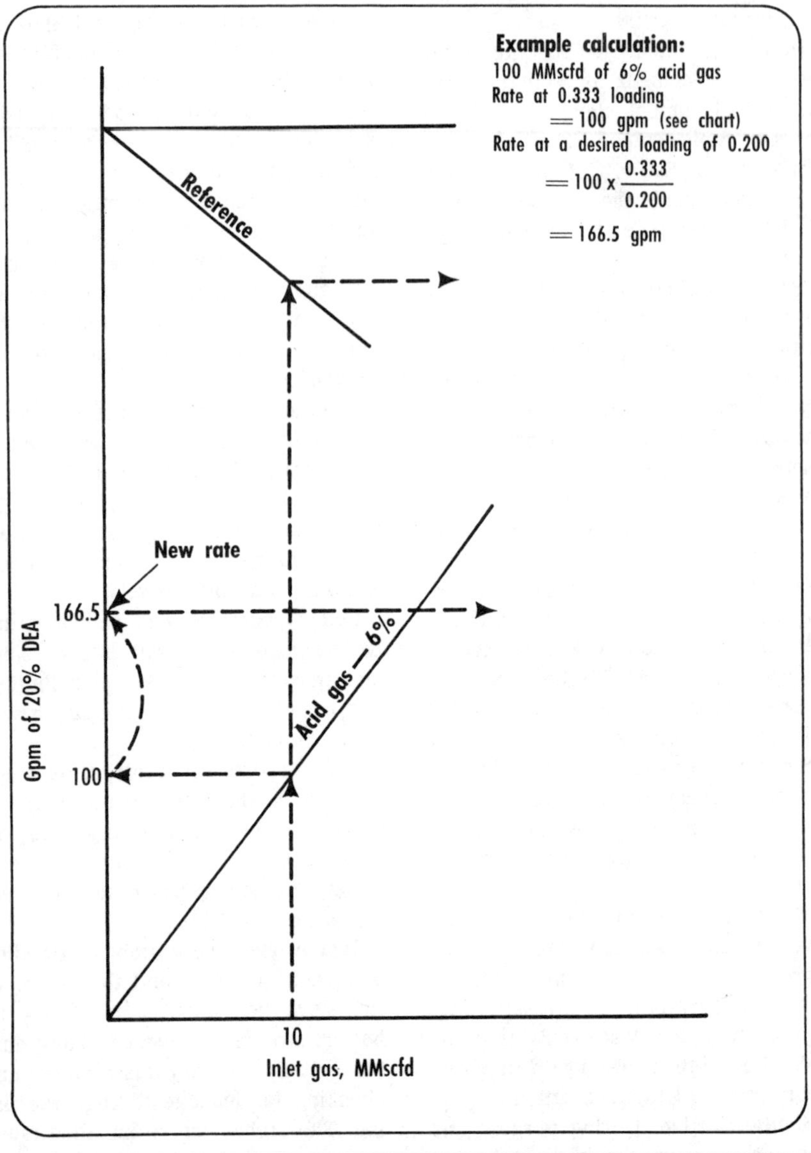

Tower diameter. Study of these charts shows that absorber diameter is based on the vapor loading, and the regenerator on liquid loading. In almost every case this is true. But the designer should verify this before making a final design.

Regenerator reflux ratio. The quantity of water pumped back to the regenerator as reflux is a measure of the quantity of stripping steam coming up the column per quantity of acid gases, based on moles. The higher the reflux ratio, the less likely the corrosion on the trays and in the reboiler.[2]

This ratio varies from 1.0 to 3.0, depending on the designer's preference. Generally, a reflux ratio less than 2.0 is very lightly loaded and subject to more corrosion.

Tower trays or packing height. For design estimates, experience is perhaps the best answer in evaluating the number of trays or height of packing required. Calculation of the number of theoretical stages of contact required is felt by some authors to be meaningless.[3] Most installations contain 20 trays for the absorber and 22-25 trays for the regenerator, or an equivalent amount of packing.

References

1. Pearce, R. L. Arnold, J. L., and Hall, C. K., "Studies Show Carbonyl Sulfide Problem," Hydrocarbon Processing, Vol. 40, No. 8, 1961.
2. Beck, J. E., "Basic Acid Gas Removal Plant Design and Operation," Gas Training Conference, 1966.
3. Connors, J. S., Perkins, C. L., and Vandeveer, F. F., "Removal of Sulfur Compounds and Carbon Dioxide from Some Natural Gas," Gas Engineers Handbook, Chap. 9, The Industrial Press, New York, 1965.

Absorption system designed for 20% DEA

Fig. 2

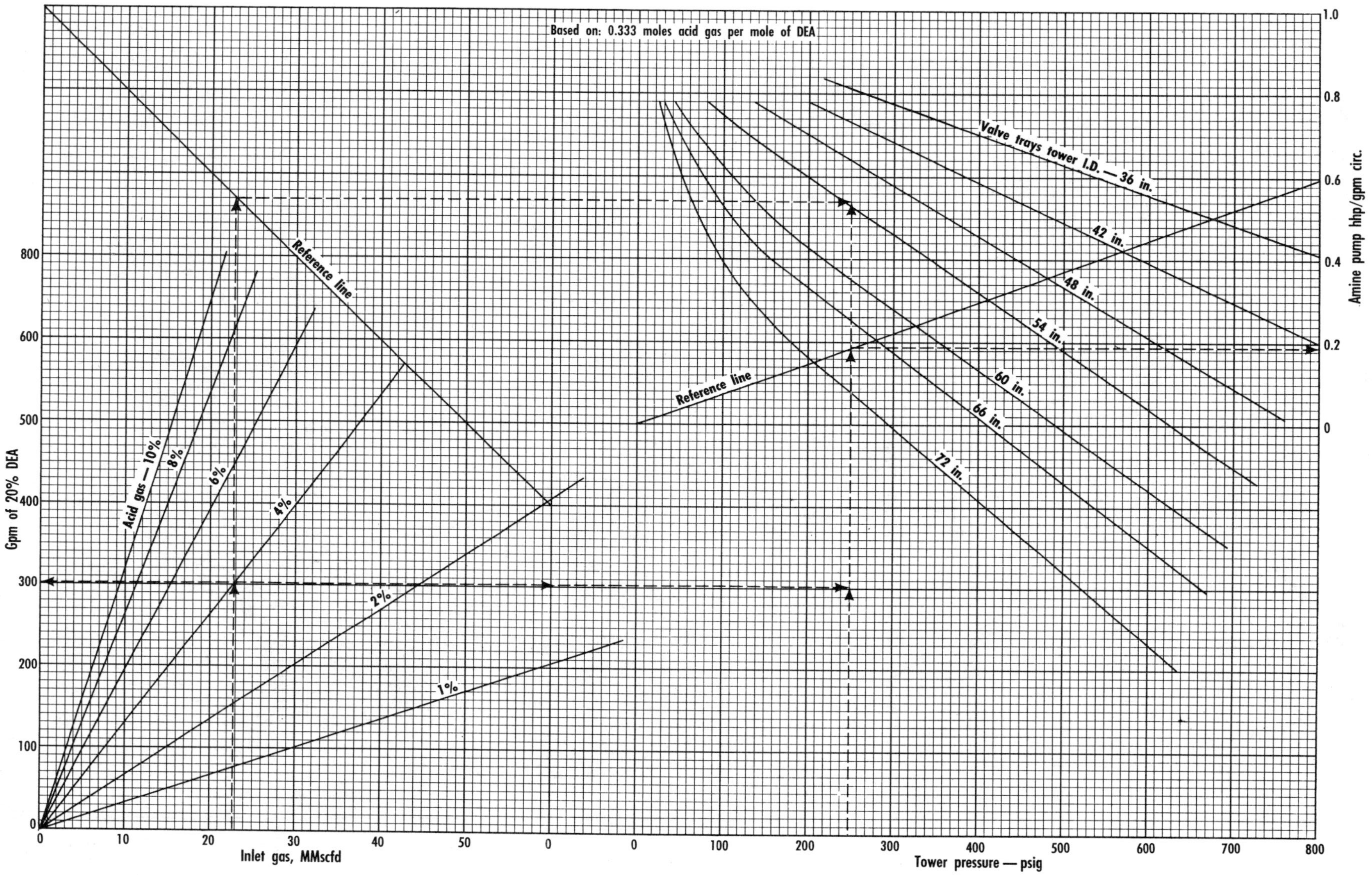

Fig. 3

Regeneration-system design for 20% DEA

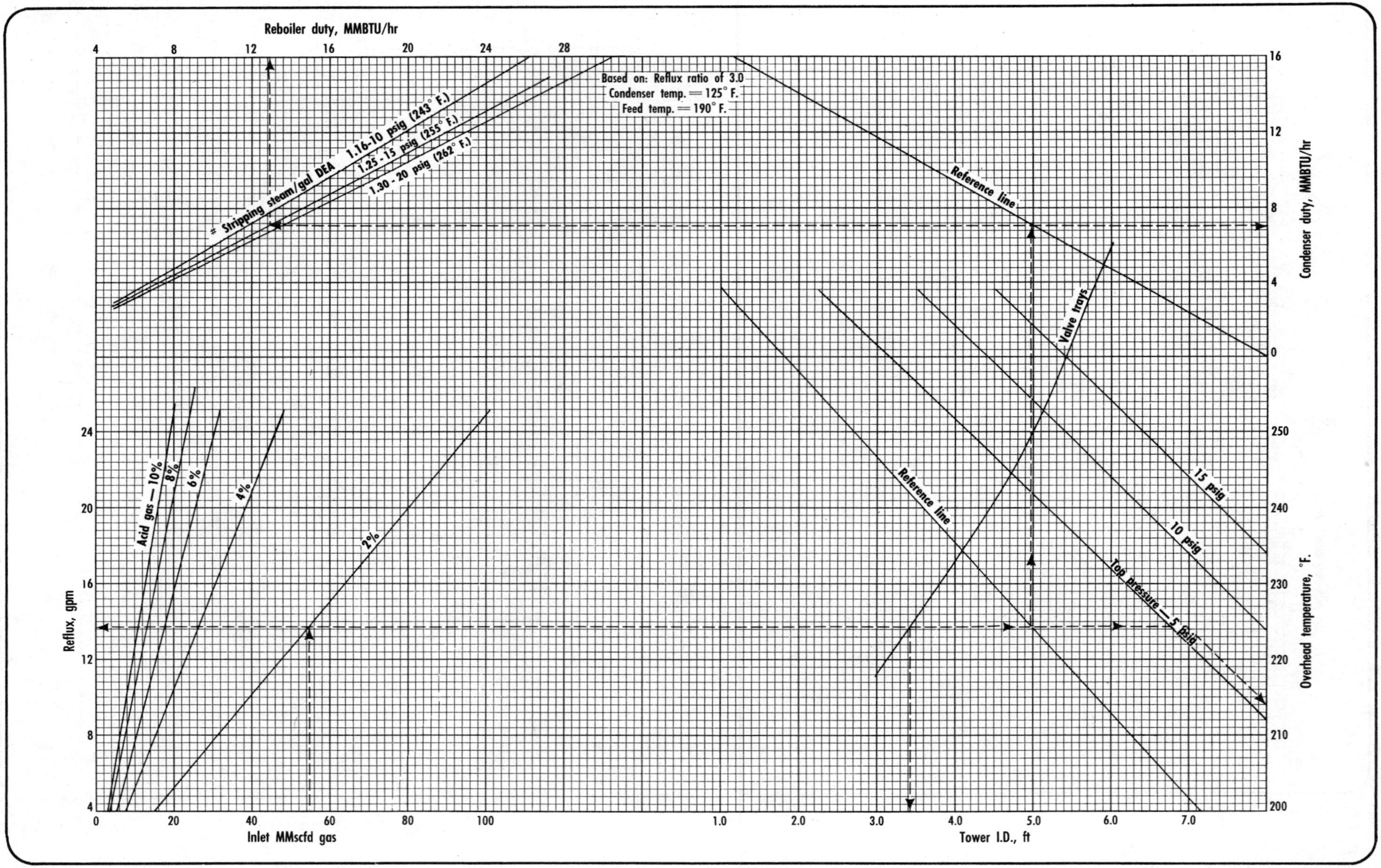

Studies of the performance of dehydration systems pay off

ARTHUR J. SCHROEDER, JR.
Amoco Production Co.
New Orleans

WITH the value of natural gas steadily rising, Amoco Production Co. is continuing to take a close look at reducing internal energy consumption and operating costs at company-operated gas-processing plants.

One area of investigation within the processing plants recently examined was gas dehydration. Detailed evaluations or case studies were performed on the dehydration equipment at four refrigerated lean or light-oil absorption plants located in the Gulf Coast region.

In one instance, the system examined was determined to be operating at its economic optimum. But modifications and recommendations for improvement at the other three plants should provide gross after-payout savings of over $130,000/year.

By evaluating the dehydration systems' economic performance in relation to overall plant operation, several worthy projects have been instigated. As fuel, glycol, and NGL-product prices continue to increase, reevaluation will be necessary to balance these increased costs with application of existing and new product technology.

Study objectives. Each dehydration system was first examined to determine its ability to mesh into the total plant operating scheme and then as a separate economic unit.[1 5] The primary focus of the investigation was to: (1) determine if the system was regenerating glycol sufficiently to meet process specifications and (2) what effect the quality of regeneration had on the total process.

In all instances, the quality of regeneration was not up to the manufacturer performance specifications. But, in most cases, this was not necessary to meet current process-operating conditions.

Where dehydration equipment did prove to be a factor in plant performance, an economic analysis was performed to determine the attractiveness of modifying the system.

The second portion of each system review focused on reducing total operating expenses and minimizing energy consumption. This involved balancing glycol, fuel, and operating expenses against the costs of equipment additions and/or modifications.

Low gas prices were by far the single largest deterrent in economically minimizing energy consumption and upgrading system efficiency. The prevailing low gas prices coupled with soaring equipment costs still tended to place the more exotic methods of regeneration, such as vacuum distillation,[7] azeotropic distillation methods,[3 6] and the Coldfinger method,[8] out of economic reach.

As the case studies attest, a complete dehydration-system evaluation can prove beneficial in many cases. From the four plant studies presented, the modifications and recommendations made should have a gross after-payout savings of over $130,000/year.

Plant 1. The first plant examined originally used a diethylene glycol injection system which was converted in the mid-1960's to a TEG contactor system. The 100+ MMcfd unit was designed to dehydrate a saturated 90° F. inlet stream at 1,050 psig to provide an outlet dew-point gas of −20° F.

The unit featured a 12-tray contactor and a standard regeneration section equipped with facilities to inject stripping gas into the reboiler.

The process at this plant doesn't require a dew point as low as −20° F. for which the glycol unit was designed. Thus the dew-point depression the dehydration system was capable of delivering was never actually needed. Also, the gas prices at this particular location have been, and still are, rather low.

These conditions, combined with normal operating costs for the unit and just slightly above normal glycol losses, indicated very little potential for improvement. However, as part of an overall plant-efficiency study, it was determined that one facet of the dehydration operation could economically be improved.

Unlike most skid-mounted regeneration units built today, this particular unit was not equipped with a final lean glycol cooler (Fig. 1). Regenerated glycol at 400° F. flows over a weir out of the fired reboiler and into an accumulator directly underneath.

The accumulator serves to provide a suction head for the glycol pump, and also as a preheat for the entering water-rich glycol. However, the temperature of the lean glycol leaving the accumulator and being pumped to the contactor was found to be 230° F. This high temperature was also found to be heating up the plant inlet-gas stream by about 10° F.

As a result of the abnormally high lean-glycol temperature, the heat duty on the refrigeration system was increased and the processing plant's recovery factors lowered.

The plant operating conditions were as follows:

Category	With cooler	Without cooler
Date	11/10/77	11/9/77
Lean glycol after pump, °F.	234	230
Lean glycol after cooler, °F.	102	230
Inlet gas to absorber, °F.	75	77
Inlet gas after absorber, °F.	76	85
Gas out of chiller, °F.	−9	−6
Reduced heat load on refrigeration system	424 MBTU/hr	
Increased NGL recovery	300 gal/day	

Paper titled "Case Study Reviews of Optimizing the Economic Performance of Gas Dehydration Units," presented at 28th Annual Gas Conditioning Conference, University of Oklahoma, Norman, Mar. 6-8, 1978.

Figs 1 - 5

Plant revamp schemes

Plant 1 — glycol cooler

Fig. 1

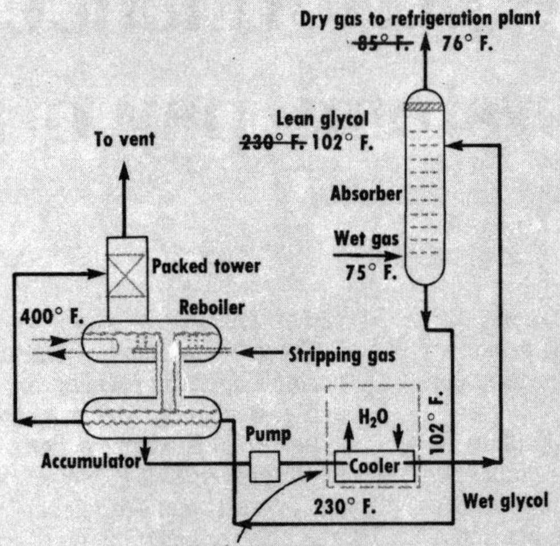

Lean glycol cooler addition:
Investment: $6,500
Savings: $15,000/year

Plant 2 — azeotropic distillation

Fig. 2

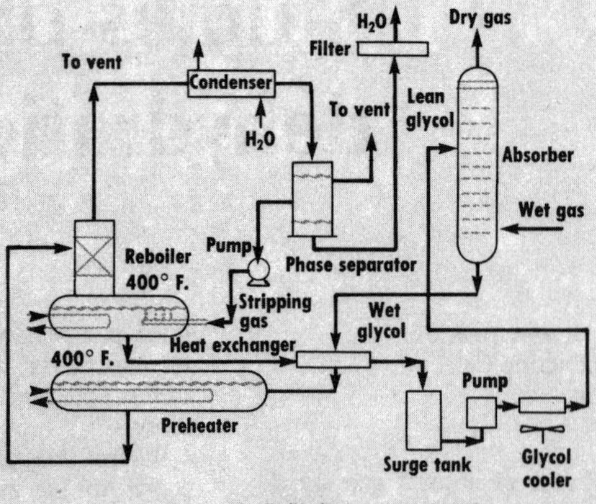

Azeotropic distillation system rejected:
Investment: $21,000
Savings: $6,000/year

Plant 3 — glycol recycling

Fig. 3

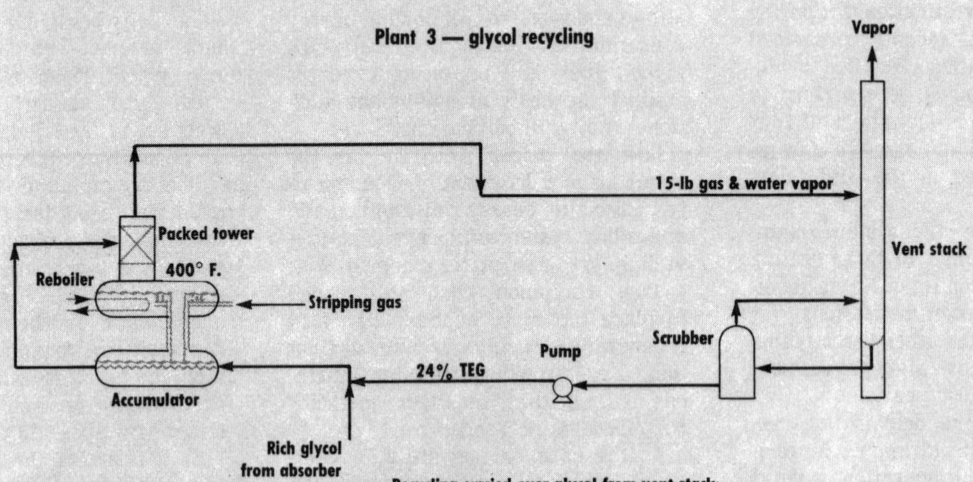

Recycling carried-over glycol from vent stack:
Investment: $1,500
Savings: $30,000/year

Plant 4 — azeotropic distillation

Fig. 4

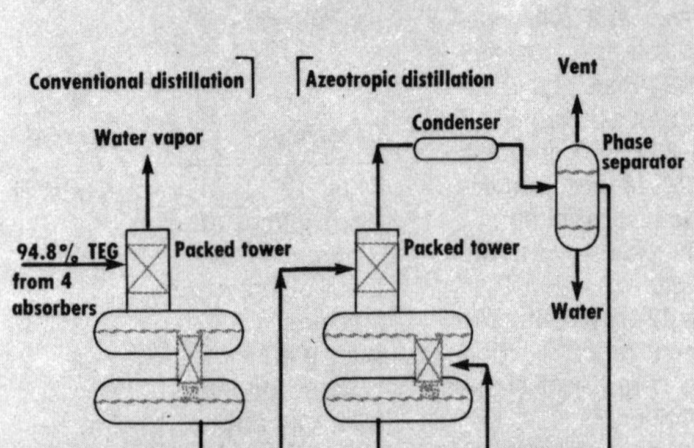

Proposed configuration using azeotropic distillation:
Investment: $33,000
Savings: $86,000/year

Plant 4 — vapor recovery

Fig. 5

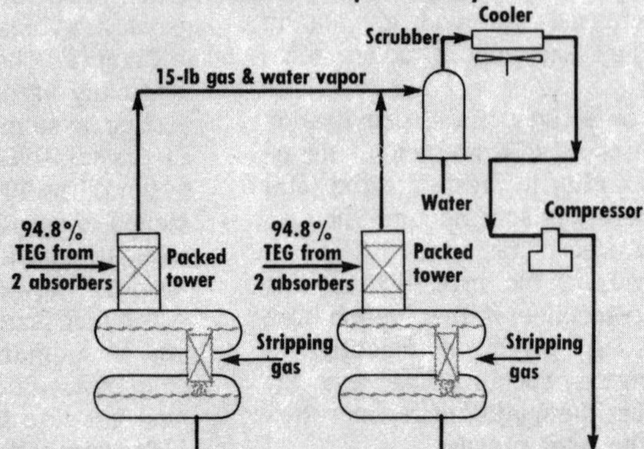

Vapor-recovery system rejected:
Investment: $31,000
Savings: $60,000/year

By installing a lean glycol cooler on this hot glycol stream, vaporization losses over the contactor were minimized, and the plant's recovery rate was increased.

Calculations showed, and plant tests confirmed, that this $6,500 project paid out in less than a year and should provide over $15,000/year in additional revenue.

By examining the entire dehydration system in relationship to the plant's total economic picture, an original design deficiency was profitably corrected.

Plant 2. The second plant examined had a design capacity of over 100 MMcfd, but was only processing about 50% of its designed throughput. The TEG dehydration system (Fig. 2) was designed to provide a −15° F. dewpoint outlet with a saturated 90° F., 1,050-psig inlet.

Unit design varies slightly from other company-operated units in that it is equipped with both a fired reboiler and a fired preheater. This arrangement permits the small reboiler on top to operate at close to isothermal conditions, and thus the higher efficiency of countercurrent gas stripping is obtained.

Because of the slightly unusual physical configuration of the system, adaptation of an azeotropic distillation system was examined. Using design data furnished by a licensor of the process,[4] equipment and material costs for conversion were estimated at $15,000. This included a water-cooled condenser; phase separator; two pumps (one spare); charcoal filter for water drawoff; and all piping, valves, and fittings for a complete installation.

Labor charges for on-site modifications to the present unit and installation were estimated at $7,000. To determine the potential benefits of the proposed azeotropic system, operational data were collected and heat-duty calculations performed to determine the present energy consumption.[2,3]

Plant operating conditions were as follows:

Date	8/16/1977
Inlet gas rate, Mcfd	68,620
Gas inlet temperature, °F.	80
Gas inlet pressure, psig	875
H_2O content of dried gas, lb/MMcf	1.2
Glycol circulation rate, gpm	7
Stripping gas rate, cfm	4.7
TEG concentration (rich), %	97.0
TEG concentration (lean), %	99.6

The author . . .

A. J. Schroeder, Jr.

Arthur J. Schroeder, Jr., is currently an operations engineer for the onshore producing department of Amoco Production Co., New Orleans (La.) Division. Until recently, he was a member of the plant and mechanical department where one of his responsibilities was optimizing the dehydration systems at the company-operated NGL plants. Before joining Amoco, Schroeder was employed by the research department of the Coca Cola Co. while in school. A member of the SPE and AIChE, he holds a BS degree and an MS degree (obtained while working under a grant from the EPA) in chemical engineering, both from the Georgia Institute of Technology.

At these conditions, total energy requirement for the system was calculated at 870 MBTU/hr or the equivalent of 21 Mcfd. At the time the data were taken, process conditions within the plant required only a −10° F. dew point which was easily obtained.

Because of variable process conditions, the dehydration system is occasionally required to provide a −20° F. dew point to prevent hydrate formation within the plant. The existing dehydration equipment is capable and does provide this lower dew point when the glycol circulation rate is increased to 10 gpm and the stripping-gas rate to about 20 cfm. Under these more-stringent conditions, total energy requirement increases to approximately 1,800 MBTU/hr or an equivalent of 43 Mcfd.

Using test and design data furnished by the licensor, energy requirements were calculated for providing the −20° F. dew point. By employing an azeotropic distillation process, an estimated total energy savings of 1,100 MBTU/hr or about 26 Mcfd could have been realized.

Even though the process offered a potential energy savings of 40%, it was not economically attractive. With relatively low gas prices, on an after federal-income-tax basis, the payout on the system was over 5 years after royalty charges to the licensor were included.

At the higher dew point (−10° F.), which is the minimum required under most conditions, the potential savings dropped even further.

In summary, using the relatively efficient design concepts built into the existing system and considering the modest dew-point requirements, the dehydration system at this plant is considered to be at its economic optimum.

Plant 3. The third case study reviewed a dehydration system designed for a saturated 60 MMcfd, 90° F. inlet stream at 1,050 psig. The skid-mounted package can consistently deliver dehydrated gas with 0.8 to 1.0 lb H_2O/MMcf water content with fairly low stripping-gas rates.

Even though the unit was designed to dehydrate gas to 0.5 lb H_2O/MMcf, present performance is considered satisfactory for the plant processing conditions. A major problem which has plagued the unit since installation though has been high glycol losses.

Shortly after installation, speculation on the reasons for the excessive glycol losses centered around suspected carry-over problems. In an attempt to remedy the problem, since tower inspection indicated no tray damage, a heat exchanger was added to cool the lean glycol going to the contact tower.

A larger still column was also installed to reduce the velocity of the vapors going overhead and to provide for more packing.

To recover some of the inlet gas glycol entrainment losses, a cyclotube separator was installed downstream of the contactor.

With these modifications and close and persistent operator attention, losses have occasionally dropped down to 0.6 or 0.7 gal of TEG/MMcf, but the long-term average has remained above 1.0. However, recent modifications have significantly reduced these losses.

The overhead vapors from the packed section of the regenerator are routed through a 3-in. pipe, 50 ft away where they are vented to the atmosphere.

At the base of the vent stack, a small scrubber knocks out any entrained or condensed liquids.

An analysis of this liquid, which is normally pumped to the plant sump, showed 24% glycol by volume. To immediately recover this glycol-rich solution, a small pump was installed to return the liquid back to the rich-glycol stream coming from the contactor (Fig. 3).

This modification resulted in a slight fuel increase but an immediate and drastic reduction in glycol losses. In an attempt to correct the problem at the source, an alteration was made in the still-feed location which resulted in an overhead temperature drop from 250° F. to 190° F. This action subsequently reduced the percentage of glycol in the scrubber liquid to about 7% by volume.

To further reduce this carry-over, recommendations have been made to install a mist extractor in the top of the still section, repipe the distributor on the stripping gas line for better distribution, and use an emulsion-breaker solution to facilitate better separation in the three-phase separator.

At this particular location where gas prices are relatively low, a small fuel savings has been temporarily sacrificed to gain a 30-gal/day reduction in glycol losses. The recycling of the scrubber water and glycol was initiated quickly and cheaply to immediately reduce large glycol losses while a more permanent solution to eliminate the source of the problem was being instigated.

Plant 4. In the final plant examined, the dehydration facilities offered a couple of possibilities for increased efficiency and decreased operating costs. The plant design throughput is 300 MMcfd, and process temperatures require a gas dew point of −40° F.

The present dehydration facilities include four contact towers, each containing 12 trays, and two complete regeneration units, one 1.5-MMBTU/hr unit and one 1.25-MMBTU/hr unit. The minimum dehydration requirements are as follows:

Inlet gas rate, MMcfd	300
Inlet gas temperature, °F.	80
Inlet gas pressure, psig	900
Inlet gas H_2O content, lb/MMcf	36
Required dew point, °F.	−40

To achieve the required dew point, a lean TEG concentration of 99.95% is required. This is presently accomplished by countercurrently injecting stripping gas after the reboiler.

To recover or eliminate waste of this gas (about 120 Mcfd), two projects were evaluated: use of an azeotropic distillation process (Fig. 4) and installation of a vapor-recovery unit (Fig. 5). As a result of this evaluation, a recommendation has been made and plans are being finalized to use azeotropic distillation to achieve the required process condition and to reduce energy requirements by about 70%.

The estimated cost for an installed stripping-gas recovery system was $31,000. The equipment list included a multivane rotary compressor, a 25-hp electric motor, a suction scrubber, and a heat exchanger as well as all of the required piping, valves, and fittings.

Installation was considered to be relatively simple with only a maximum of 8 hr downtime expected for tie-ins and start-up. The project was calculated to have an after-tax payout of less than a year with gross savings of about $60,000/year.

As an alternative to the vapor-recovery unit, an evaluation, as previously mentioned, was conducted to determine the economic attractiveness of using an azeotropic distillation system.

To minimize equipment costs and to obtain additional fuel savings, the two regeneration units would be set up in a series configuration, as indicated in Fig. 4. All modifications to the regenerators are so planned that an operator could quickly and easily revert the system back to stripping gas.

With this additional capability, should service be required on one unit, the other could operate independently and still provide sales-specification gas.

Using design and test data provided by the licensor and considering the present fuel-gas consumption rates, a total energy savings of 7 MMBTU/hr was calculated.[4] A significant portion of the savings was due to the elimination of stripping gas which is now being vented to the atmosphere.

Total project economics for the azeotropic distillation process indicated a gross saving of $86,000/year with an after-tax payout of less than a year.

Even though the incremental net savings associated with the azeotropic process over the vapor-recovery system were relatively small, as gas prices start to more accurately reflect the true value of this energy source, the economic advantages of azeotropic distillation should rapidly increase over the vapor-recovery system. This installation will also provide Amoco Production with field-test support for consideration in evaluating future applications.

References

1. Ballard, D., "How to Operate a Glycol Plant," Reprint from Hydrocarbon Processing, 1966.
2. Campbell, J. M., Gas Conditioning and Processing, Campbell Petroleum Series, 1976.
3. Fowler, A., "Super-Drizo, The Dow Dehydration Process," University of Oklahoma Gas Conditioning Conference, 1975.
4. Grosso, Silvano, Private communication, Dow Chemical USA, 1977.
5. Pearce, R. L., "Fundamentals of Gas Dehydration with Glycol Solutions," University of Oklahoma Gas Conditioning Conference, 1973.
6. Pearce, R. L., Protz, J. E., and Lyon, G. W., "Drizo-Improved Regeneration of Glycol Solutions," University of Oklahoma Gas Conditioning Conference, 1972.
7. Polderman, L. D., "The Performance of Vacuum-regenerated Glycol-type Natural Gas Dehydration Plants," University of Oklahoma Gas Conditioning Conference, 1957.
8. Reid, L. S., "Coldfinger, An Exhauster for Removing Trace Quantities of Water from Glycol Solutions Used for Gas Dehydration," University of Oklahoma Gas Conditioning Conference, 1975.

Regeneration gas heaters require better controls

W. A. Lewis
G. M. Franklin
K. R. Sheetz
Amoco Production Co.
Chicago

Fig. 1

Fired heater controls

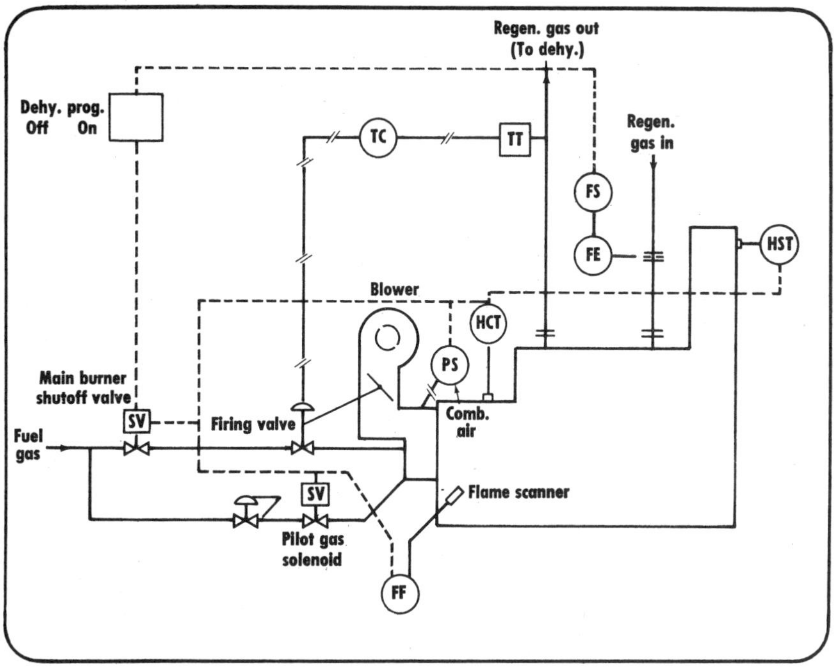

THE WIDESPREAD adoption of molecular-sieve units for superdehydration in cryogenic plants and the pressing need for higher thermal efficiencies have led to increased use of mechanical-draft, convection type of regeneration-gas heaters. Many of these heaters are subject to cyclic operations imposed by the heating and cooling periods.

Cyclic operation of these heaters can subject the tubes to high operating temperatures and stresses under certain conditions. Amoco Production has learned that careful attention is required to proper application of controls and safeguards to avoid tube ruptures caused by creep at elevated temperatures.

Design considerations. The gas-processing industry has used conventionally designed fired heaters for a long time. But the energy crunch, with its pressing need for improved thermal efficiencies and the advent of cryogenic gas processing, have brought about the need to reconsider some of the past design and operating practices for fired heaters.

Within the past 4 years, Amoco Production Co. has acquired experience with a number of forced-draft and natural-draft direct-fired heaters which are used to heat regeneration gas in cyclic service. Amoco Production has had tube failures because it did not anticipate the need for certain controls which it now requires.

Some of those failures were due to operator errors. Others were due to inoperative safety instruments, and some were due to maladjusted firing valves.

In at least two cases, the failures were caused by defective tube material.

In a well-designed heater, the thermal and mechanical designs proceed hand-in-hand because of their interdependence. The allowable operating stress in a heater tube is a function of the tube-metal temperature, and, at the same time, the tube-metal temperature depends upon the ability of the process fluid to transfer heat from the tube to the fluid.

The heat-transfer mechanism is a function of flowing velocity and heat-transfer properties of the process fluid. In addition to these heat-transfer criteria, the heater manufacturer must provide enough space between the burner and the tube bundle to preclude flame impingement

Paper titled "Pertinent Considerations in the Design and Operation of Regeneration Gas Heaters in Cyclic Service," presented at the Annual Gas Processors Association Convention, New Orleans, Mar. 20-22, 1978.

Figs. 2-5

Heater control schemes

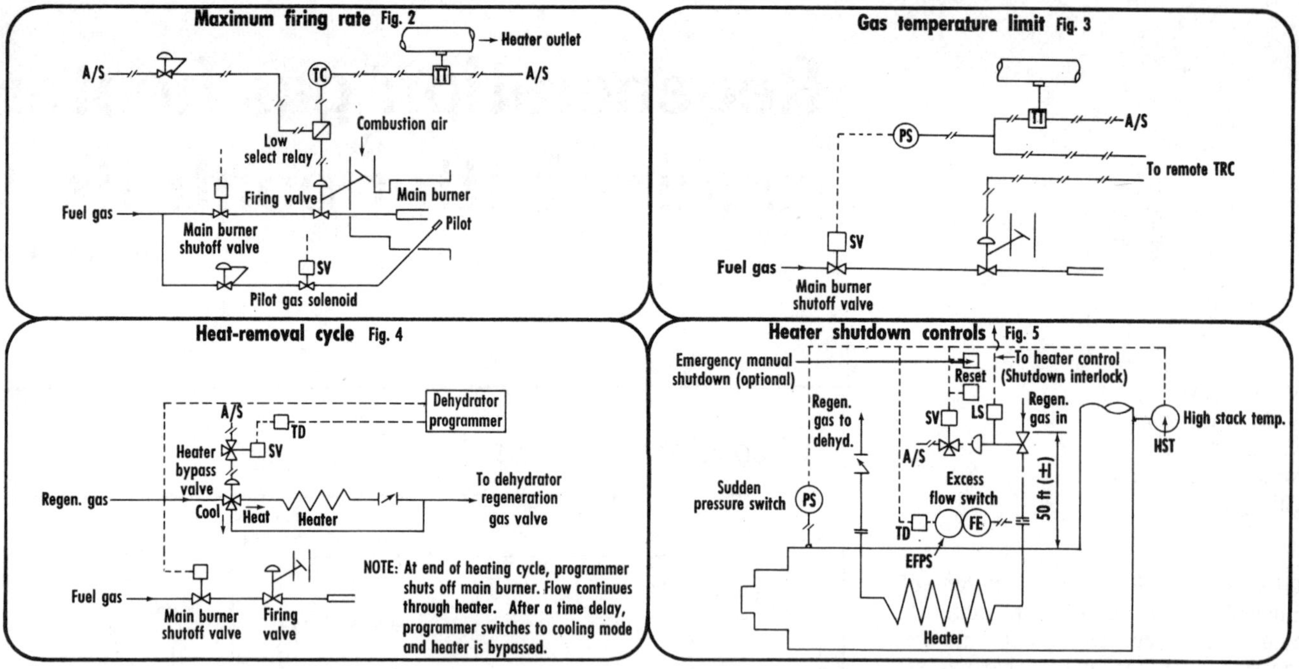

when the unit is firing at full capacity.

In some cases, the process design makes it necessary to operate the heater in a cyclic manner to provide the necessary heating and cooling periods for the molecular-sieve units. Mechanical-draft, convection-type heaters readily lend themselves to such operations while providing favorable thermal efficiencies.

These are fine attributes for a heater, but the cyclic service conditions can result in tube failures if improper applications are made. Amoco Production has found that refinements and additions to the heater controls are necessary to minimize the effects of cyclic service on heater tubes.

Heater controls, safeguards

Providing mechanical-draft heaters with adequate controls and safeguards for cyclic operation requires that several special features be included in their design and installation. The automatic starting sequence of events must include sufficient purge before lighting the pilot, followed by automatic ignition of the pilot, and proving the pilot light-off by a flame scanner.

These heaters are normally operated with a continuous pilot, and the main burner is arranged to light-off when the mol-sieve unit programmer switches to the heating mode in the regeneration cycle.

Regeneration-gas flow must also be established before the main burner will light-off. If flow drops below a minimum safe value or stops completely, the main burner shuts off. Fig. 1 shows a simplified diagram for typical fired-heater controls.

Maximum firing limit. When the mol-sieve programmer initiates a heating cycle, the heater outlet-gas temperature controller will be at full firing output, and the burner will go to maximum firing. The addition of controls to limit the maximum firing rate has proved beneficial.

One method of limiting the maximum firing rate is to connect the firing-valve control-signal line to the output of a low-select relay (Fig. 2). One input to the relay is from the temperature controller and the other input is from a regulated preset source.

The preset signal is set slightly higher than normal level required to provide adequate heat for regeneration of the mol-sieve. On initial start, the relay will select the lower preset input.

As gas temperature increases, the controller output will drop. When it becomes lower than the preset signal, the relay will select that signal and the valve is then on normal temperature control.

Low regeneration-gas flow. On any heater, the main burner should be shut off if flow of the material being heated stops for any reason. A switch sensing flow to the heater can be used for this purpose.

In regeneration-gas service, the switch should normally be set to shut off the main burner if flow drops to about 40-50% of normal flow required for regeneration.

Maximum gas-temperature limit. If the regeneration-gas temperature controller is equipped for "manual-auto" operation (such as a remote board mounted TRC), it is desirable to prevent overfiring if the controller is inadvertently left in the "manual" position.

The desired protection can be accomplished by a pressure switch on the temperature transmitter output to the controller (Fig. 3). The switch should be set to actuate at an output corresponding to a temperature about 50° F. higher than the desired temperature and would shut off the main burner.

When the temperature dropped, of course, the burner would again light-off, and this procedure would be repeated until the controller was switched back to the "auto" position.

Heat-removal cycle. On completion of a heating cycle, the mol-sieve programmer shuts off the main burner. If the regeneration-gas circuit bypasses the heater in the cooling cycle and flow through the heater is shut off at that point, the tubes will be subjected to the residual heat in the cabin with only the combustion-air blowers to remove the heat from the cabin.

The addition of a heat-removal

cycle (Fig. 4) to the programmer control which allows the gas to continue to flow for a period of time after the main burner shuts off, before switching to the cooling cycle, will remove the residual heat and the tubes will not be subjected to excessive temperatures.

Experience has shown that 7 to 10 min is adequate to reduce cabin temperature to a satisfactory level before flow through the tubes is stopped in heaters having low thermal inertia.

Emergency shutdown devices

The purpose of all these measures is to prevent damage to the heater in normal service. In the event of a tube failure, however, it is imperative that the heater be shut down and regeneration gas be shut off for the protection of personnel and to minimize damage to the unit.

A fast-closing shutoff valve should be installed in the regeneration-gas line to the heater and a check valve installed downstream of the heater to prevent backflow through the rupture from process pressures.

The shutoff valve should be located a safe distance—say 50 ft—from the heater, and the control should be arranged such that after the trouble is cleared it will have to be manually reset to open. The valve would be closed by any or all of the actions described here and a limit switch on the valve should be provided as a safety interlock to prevent heater start-up unless the valve is open.

Fig. 5 illustrates the application of several other shutdown controls which are as follows:

High stack temperature. This shutoff feature is workable provided the amount of gas released in the cabin will cause the stack temperature to rise to a point high enough to actuate the sensing device.

It has been our experience that, on a tube rupture, the volume of gas and the cabin pressure become so great that gas is forced up the stack and burns several feet above the stack.

Under such conditions, the thermocouple sensor does not see a temperature high enough to actuate the shutoff valve.

Excess flow. If a tube ruptures, the flow of high-pressure gas in the line will be considerably greater than normal. A flow switch in the line to the heater, set to actuate at a flow somewhat higher than normal, can be used to actuate the shutoff valve. But this arrangement has some shortcomings.

When gas flow is established at the start of a heating cycle, the flow may exceed flow setting due to line packing so it is necessary to time delay the shutoff until a normal flow is established.

If the switch is set higher to overcome initial flow conditions without a time delay then it may well be set too high to sense an actual excess flow due to a tube rupture.

While this scheme will function if properly designed and adjusted, it does not provide a fast enough shutoff due to the time delay introduced and can be the cause of nuisance shutdowns.

Sudden pressure shutoff. A ruptured tube will generally cause an excessive pressure buildup in the heater cabin as evidenced by the blow-torch action. A more-reasonable approach to fast shutoff of regeneration gas and heater shutdown is to use a "sudden pressure" switch on the heater cabin.

The pressure switch senses cabin pressure in the same manner as the combustion-air pressure switches and should be set at a value higher than the normal maximum cabin pressure experienced with the combustion-air blowers running when the main burner lights-off.

Natural-draft heaters

Forced convection heaters with low thermal inertia and high thermal efficiencies are well suited for cyclic operation and are readily adapted to automatic ignition and programmed firing. However, cabin-type heaters are also being built for cyclic gas-heating services where high thermal efficiency and low thermal inertia are essential.

These cabin heaters present a somewhat different problem with regard to tube safety and protection of operating personnel.

Automatic ignition of a natural-draft heater requires special considerations since purging depends on natural draft. Amoco Production uses remotely operated spark-ignition, flame monitoring to prove the pilot and enable the main burner valves to open.

Many of the safety features used on forced convection units such as low fire, high stack temperature, flame failure, etc., are also applicable to natural-draft heaters.

There are other safety hazards with natural-draft heaters which must be considered. Ignition of this type unit has, in times past, been by torch. Amoco Production recently adopted electric-spark ignition as its standard, arranging the system so ignition of all pilots can be accomplished from a panel at a safe location.

In both natural-draft and forced-draft fired heaters in regeneration-gas service Amoco Production uses thicker tube walls than theoretically required just to provide an additional margin of safety. When lead time permits, the company also ultrasonically inspects tubes before fabrication to assure minimum wall thickness and to minimize defects.

Acknowledgment

We thank Amoco Production Co. for permission to present this paper and to acknowledge with appreciation the assistance from several of our associates.

The authors . . .

Willard A. Lewis

George M. Franklin

Kenneth R. Sheetz

Willard A. Lewis is plant engineering supervisor of the mechanical engineering group for Amoco Production Co. in the firm's general office in Chicago. He joined Stanolind Oil & Gas Co. (now Amoco Production Co.) in 1946 and has held several positions in the plant engineering group. Before joining Amoco, Lewis worked for a major contractor in the field and in their New York office. He holds a BS degree in mechanical engineering from Texas Technological College, is a licensed professional engineer in Texas, and a member of ASME.

George M. Franklin joined Stanolind Oil & Gas Co. (now Amoco Production Co.) in 1948 and became an engineering specialist in Amoco's Plant Engineering Group, Mechanical Section. He was Amoco's heat exchanger, fired heater, and pressure-vessel specialist and a staff engineer SG at his retirement in 1977. Franklin has coauthored several papers, contributed to industry developments in heat exchangers and sulfur plants, and was a representative for the Standard Oil of Indiana affiliate to the Heat Transfer Research Institute. He attended Texas A&M University and Southern Methodist University.

Kenneth R. Sheetz joined Standard Oil of Indiana in 1951 in Whiting, Ind., and transferred to Amoco Production Co. in 1965. He became an engineering specialist in Amoco's Plant Engineering Group, Mechanical Section and was Amoco's electrical engineering specialist and a senior staff engineer at his retirement on Jan. 1, 1978. Sheetz was active in IEEE, is a registered engineer in Indiana, and a graduate of Rose Polytechnic Institute with a BS degree in electrical engineering.

Processors counterattack amine-unit corrosion

Gary E. Congram
Processing Editor

TO RELIEVE amine-unit corrosion, gas processors are trying these major innovations:

- Reducing the acid-gas loading ratio.
- Using tougher alloys.
- Stress relieving critical components.
- Watching regeneration temperatures of the amine solution.
- Keeping the reboiler-steam temperature down to prevent degradation of the amine solution.

Other processors are showing success with corrosion inhibitors, solids-removal side streams, filtration, and pH monitoring of the amine solution.

The causes. The CO_2 and H_2S in sour gas combine with water and form acids. These in turn cause corrosion, degrade the amine solution, and retard the efficiency of the units. Corrosion usually occurs in high-temperature areas and at areas of extremely high-flow turbulence and stress.

Gas-plant operators have found most corrosion occurs in the amine solution heat exchanger, the stripper, and the reboiler (Fig. 1). In addition to corrosion, operators often find that the impurities from corrosion cause foaming which in turn cause poor contact between the gas and amine solution. In severe cases, foaming results in entrainment of the amine solution and carry-over from the absorber with substantial pressure drops across the absorber.

Corrosion reduction. Gas processors are pinpointing their causes of corrosion and keeping temperature under control. Most operators are holding regeneration temperatures under 250° F. except where CO_2 content is very high.

In this case, a temperature not exceeding 255° F. is sufficient for effective CO_2 removal.

Acid-gas loadings. Severe corrosion occurs when acid-gas loadings reach 0.40 mols of gas/mol of amine solution and gas treaters are keeping these loadings lower. Most are holding loadings of 0.34 mols/mol of amine or less to ensure minimal chances of attack by corrosion.

Alloys. Stress-corrosion cracking in bubble-cap yokes in the regeneration section is being eliminated with type 304 stainless steel. Processors are also using 304 stainless in pressure let down valves and in the main impellers of the amine-circulation pumps.

Stress relieving. Stress cracking has been halted by the processor by stress relieving many critically exposed areas of high turbulence in piping and process units. Failures from stress cracking have been noted in the hot, reactivated amine lines between the regenerator and the lean-rich amine exchanger train.

Steam temperature. Operators are also watching reboiler-steam tempera-

Corrosion, foam, and amine-loss sites

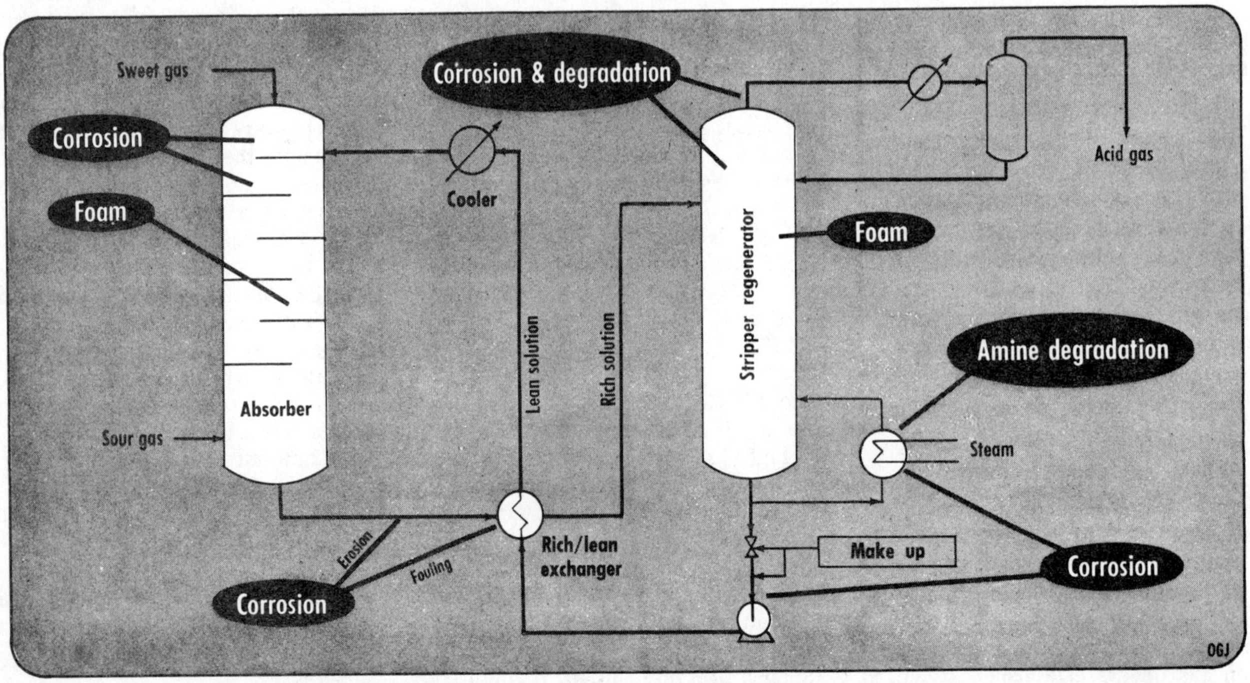

tures to prevent degradation of the amine solution. Most are trying to hold the temperature at less than 300° F. Although not a common problem, high temperature can break down the amine and cause corrosive compounds to be formed.

Other means. Corrosion inhibitors have been used successfully in many sour-gas-treating units. Primary requirement is that the inhibitor be water soluble rather than the conventional hydrocarbon-soluble inhibitors. Several patented processes using corrosion inhibitors are currently available. Most carry royalty fees and secrecy agreements but have proven successful in bringing corrosion rates down. An example of such a process is Union Carbide's "Ucar" process.

Since foaming can be a major problem, processors are using filtration and solids-removal side streams. In this way, potential foam-causing contaminants are removed and the amine solution is cleaned. Reclaimer filters can remove contaminants as small as 5 μ.

New foam-control methods also include the use of water-soluble silicone antifoams. These compounds help reduce foaming by lowering the surface tension of the amine solution. The antifoam compounds are usually added to the circulating amine solution and to any make up amine. Continuous addition is not necessary; batch treatment is usually sufficient. Usage rates for antifoams are as low as 10 ppm based on total circulation capacity and have proved effective in preventing foam in bubble cap or packed columns.

Other gas processors monitor pH of the amine solution to detect any changes in alkalinity. They find that if the solution becomes overloaded with acid gas, the amine will begin to neutralize and then swing the pH towards the corrosive acid side.

Some plant operators take grab samples of the amine solution for visual inspection. Many claim they can diagnose corrosion occurrence by the color of the amine solution. For example, the amine solution may be an off-white color normally, but if corrosion occurs, it may change to a dirty gray or black because of iron oxide or iron sulfide contamination. Others have seen the characteristic brownish red solution indicating a higher oxidized iron oxide.

Some gas processors are inert-gas blanketing amine-storage tanks to prevent oxygen from entering the system and promoting amine degradation or corrosion. Others are taking special care to keep outside air leaks in equipment from occurring.

Sour-gas processing has been plagued with many problems in the past, but today's processors are finding that by careful attention to operating variables, many difficulties can be overcome. END

Part 2
Sulfur
Recovery

Tighter control of Claus plants needed by TGCU system

B. Gene Goar
Goar, Arrington & Associates Inc.
Tyler, Tex.

THE CLAUS process has been extensively used to recover elemental sulfur from H_2S-bearing acid-gas streams. Until recently, most Claus plants were considered of significant importance in reducing SO_2 emissions to the atmosphere.

Now, most Claus plants are considered potential major offenders of air-pollution regulations.

The Claus process is recognized as a very reliable and very "forgiving" process. Industry has been able to operate Claus plants with a relatively small amount of attention and a large amount of fluctuations in tail-gas composition. The incinerator would handle large swings in tail-gas flows and composition.

Tail-gas cleanup (TGCU) processes have been developed to clean up the tail gas from Claus plants. With the advent of TGCU systems, much better process reliability and closer process control will be required in the future. The TGCU systems are not quite so "forgiving" as the parent Claus plants.

When a TGCU system is planned, it may be advisable to install a Claus-plant tail-gas analyzer (possibly with feedback control) to monitor the H_2S/SO_2 ratio and permit adjustment of the combustion air flow to achieve the optimum ratio of 2/1 at all times.

Before improved design and operations of Claus plants can be achieved, an understanding of the basic components of a Claus plant and the chemical reactions involved are essential.

Basic Claus process. Fig. 1 shows a simplified flow diagram of a typical Claus-type sulfur recovery unit. An H_2S-bearing acid-gas feed stream passes through an inlet separator and flows into the burner of the reaction furnace.

In the furnace, enough air is added to burn one-third of the H_2S in the acid gas to SO_2 and all hydrocarbons present.

Adapted from paper titled, "Fundamentals of Sulfur Recovery by the Claus Process," presented at University of Oklahoma Gas Conditioning Conference, Norman, Mar. 7-9, 1977.

Fig. 1

Typical Claus plant

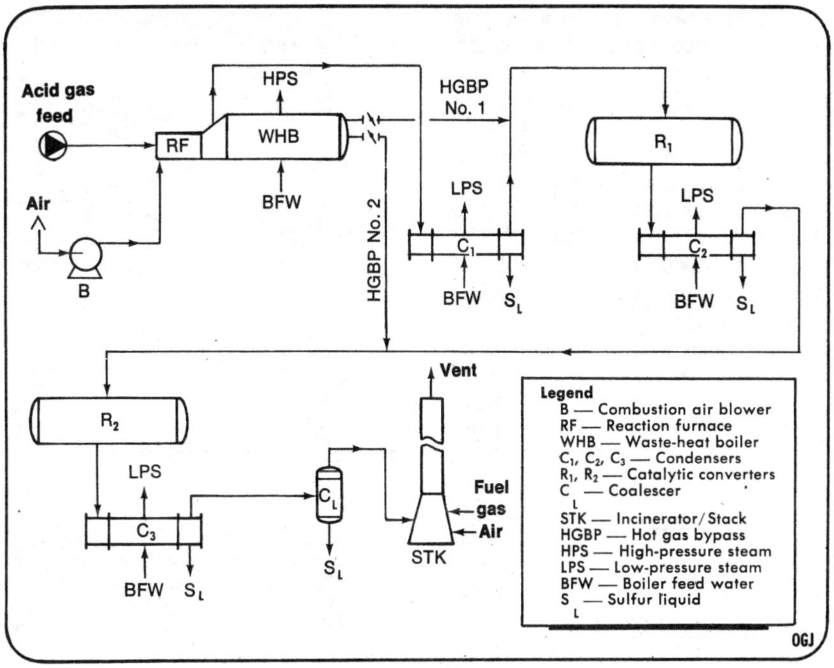

Fig. 2

Reaction furnace, waste-heat boiler

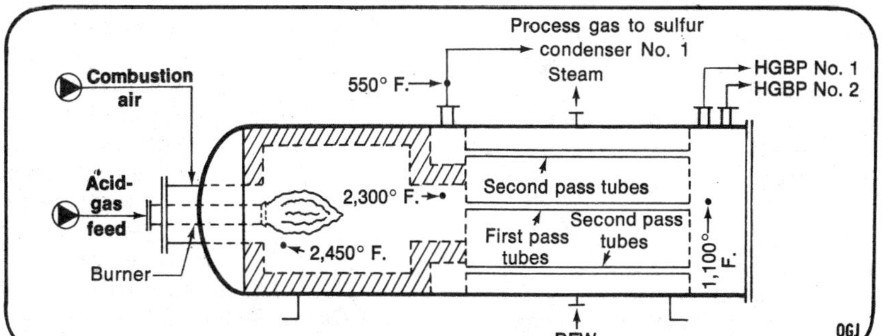

The flow diagram (Fig. 1) shows a once-through flow scheme. It uses two catalytic converters, two hot-gas bypass (HGBP) reheats, and three sulfur condensers. Inside the reaction furnace, unburned H_2S in the acid gas reacts with the SO_2 formed to yield elemental sulfur vapor. A conversion reaction occurs between the H_2S and SO_2 in the reaction furnace of the waste heat boiler (high-temperature, thermal zone), and the sulfur converters (lower-temperature, catalytic zone).

A sulfur condenser is provided after each conversion step to allow sulfur vapor to be condensed and the liquid sulfur formed to be separated from the main gas stream. The gas streams from sulfur condensers C_1 and C_2 must be reheated before going to a sulfur converter. This is necessary to prevent sulfur condensation and subsequent catalyst fouling on the converter catalyst beds.

Both reheating steps are accom-

Fig. 3

Theoretical conversion of H_2S to elemental sulfur

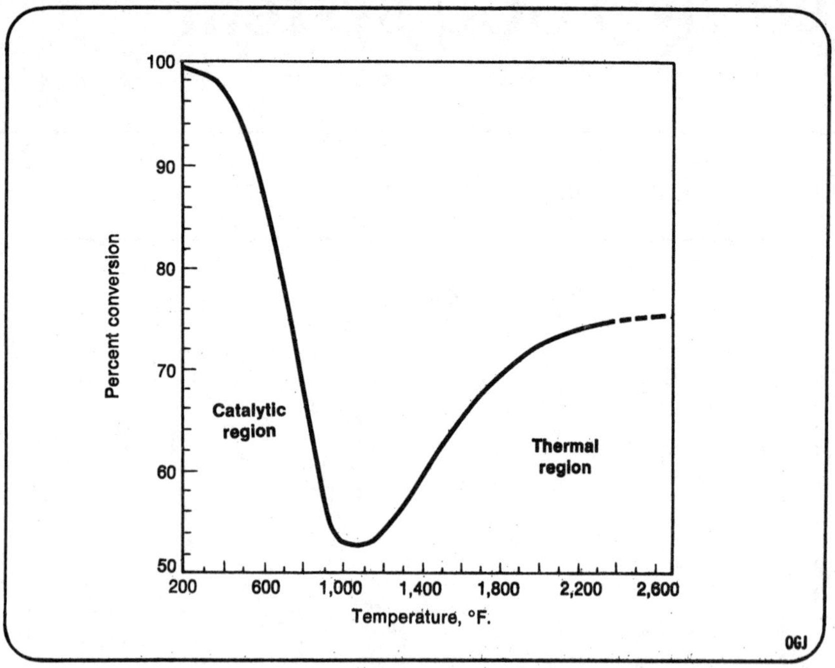

Fig. 4

Equilibrium

(between major molecular sulfur species in vapor phase)

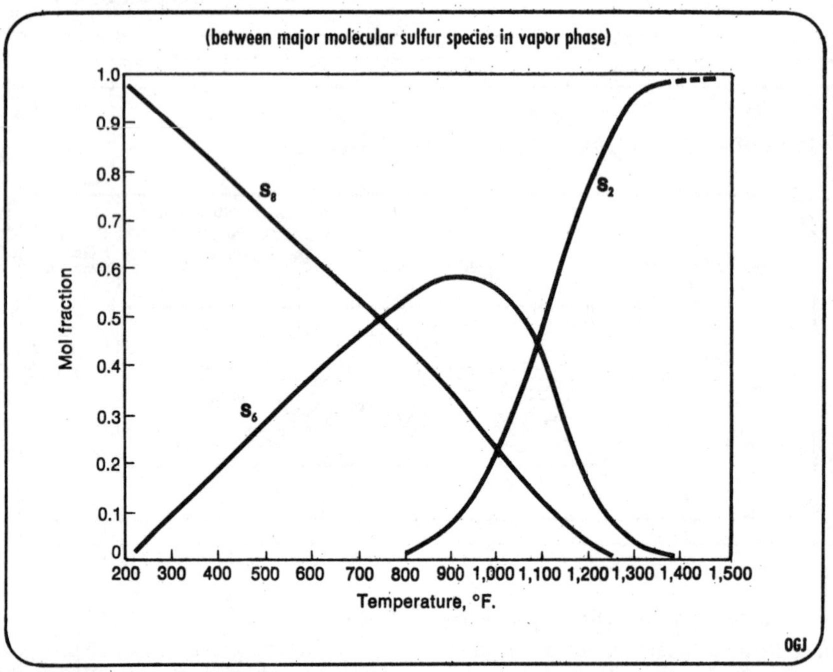

Fig. 5

Claus process

Possible side reactions

$CO_2 + H_2S \rightarrow COS + H_2O$

$CO + 1/2 S_2 \rightarrow COS$

$CH_4 + SO_2 \rightarrow COS + H_2O + H_2$

$CH_4 + 2\ S_2 \rightarrow CS_2 + 2\ H_2S$

$2\ CO + S_2 \rightarrow CO_2 + CS_2$

$2\ COS \rightarrow CO_2 + CS_2$

$COS + H_2S \rightarrow CS_2 + H_2O$

$C + S_2 \rightarrow CS_2$

$2\ COS + 3\ O_2 \rightarrow 2\ SO_2 + 2\ CO_2$

$2\ COS + O_2 \rightarrow 2\ CO_2 + S_2$

$2\ COS + SO_2 \rightarrow 3/2\ S_2 + 2\ CO_2$

$CS_2 + SO_2 \rightarrow 3/2\ S_2 + CO_2$

$H_2S \rightarrow H_2 + 1/2\ S_2$

$SO_2 + 1/2 O_2 \rightarrow SO_3$ (with excess air)

$1/2 S_2 + 3/2 O_2 \rightarrow SO_3$ (with excess air)

$1/2 S_2 + O_2 \rightarrow SO_2$

$C_n H_{(2n+2)} + \left(\frac{3n+1}{2}\right) O_2 \rightarrow n\ CO_2 + (n+1)\ H_2O$

$C_n H_{(2n+2)} + \left(\frac{2n+1}{2}\right) O_2 \rightarrow n\ CO + (n+1)\ H_2O$

OGJ

plished by HGBP streams, which allow the proper amount of hot gas (at, say, 900°-1,200° F.) to mix with the condenser outlet gas to give the desired gas temperature to the converter (say, 400°-450° F.).

All sulfur rundown lines pass to seal pots and on to the sulfur storage pit. The final tail gas from the last condenser (C_3) passes through a tail-gas coalescer (to de-entrain liquid sulfur droplets), and on to a TGCU unit or the incinerator stack.

After the initial burning step, the Claus process consists of a repetition of three basic steps: (1) reaction, (2) cooling and condensing, and (3) reheating.

This sequence may be repeated as many times as desired. However, one thermal conversion step and two to three catalytic conversion steps are normally considered about optimum.

If a TGCU process is to follow the Claus plant, probably only two catalytic converters would be used. If no TGCU unit is planned, then three or four catalytic converters might be used. This depends upon the size of the plant and the overall sulfur recovery that must be achieved.

Reaction furnace, waste-heat boiler. It is extremely important to understand what happens inside the reaction furnace and the waste-heat (WH) boiler. Fig. 2 shows a typical arrangement. Sufficient air is mixed with the acid-gas feed to burn one-third of the H_2S to SO_2 and all hydrocarbons present.

If the acid gas contains 80-90 mol % H_2S, the resulting flame temperature might be 2,450° F. Then the unburned H_2S reacts with the SO_2 formed in accordance with the Claus reaction, in an endothermic manner to form sulfur vapor (S_2 specie). A decrease in temperature to about 2,300° F. would be realized by the time the hot gases pass to the rear of the reaction furnace.

The hot gases then enter the first-pass tubes of the WH boiler where they are typically cooled from 2,300° F. to 1,100° F. The rear chamber of the WH boiler serves as the source of hot gas for both HGBP lines.

The remaining hot combustion gases

Figs. 6-8

Claus plant once-through processes

A. Using hot-gas bypass, acid-gas inline burner

Fig. 6

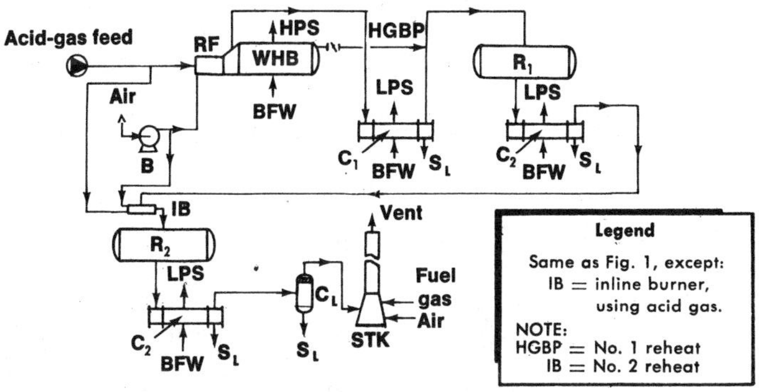

B. Using hot-gas bypass and gas/gas exchanger

Fig. 7

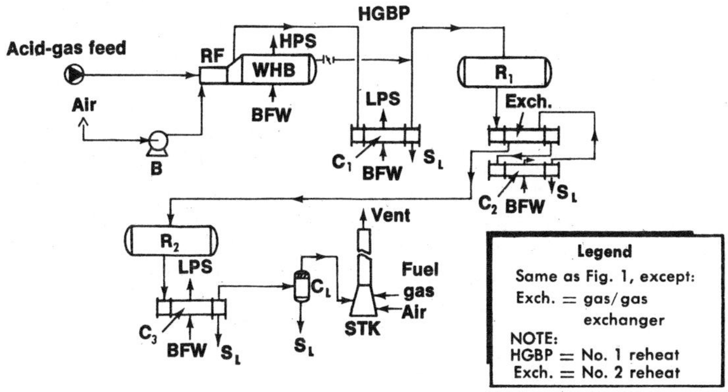

C. Using two steam-heated indirect heaters

Fig. 8

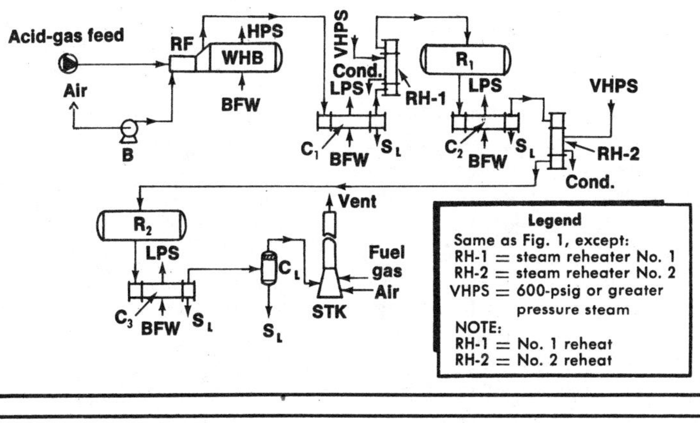

OGJ

enter the second pass of the WH boiler, where they are typically cooled from 1,100° F. to 550° F. Heat is removed from both passes in the WH boiler by generation of steam on the shell side of the boiler.

Reactions occurring in the reaction furnace and WH boiler may be represented as follows:

1. In the burner, let X = mols/hr of H_2S in acid-gas feed. One-third of X undergoes a "burn" as,

$$H_2S + 3/2\ O_2 \rightarrow SO_2 + H_2O \quad \text{(exothermic)}$$

2. In the reaction furnace, the remaining two-thirds of X undergoes the Claus reaction as,

$$2H_2S + SO_2 \rightarrow 3/2\ S_2 + 2H_2O \quad \text{(endothermic)}$$

The SO_2 generated in the "burn" step is the same SO_2 that enters into the "Claus reaction" step. At the high temperature encountered here, only S_2 sulfur vapor specie exists.[1 2]

Fig. 3 shows the theoretical conversion of H_2S to elemental sulfur by the Claus reaction.[1] Notice that in the "thermal region," percent conversion is enhanced by higher temperatures. In the "catalytic region," percent conversion improves with a reduction in temperature. At a pressure of 1 atm, the overall reaction is:

$$H_2S + \tfrac{1}{2}\ O_2 \rightarrow S + H_2O$$

3. In cool-down through the WH boiler, the S_2 sulfur-vapor specie formed in the reaction furnace shifts to other species as the temperature is reduced as,

$$S_2 \rightarrow 1/3\ S_6 \text{ (exothermic)}$$
$$S_2 \rightarrow 1/4\ S_8 \text{ (exothermic)}$$

The cooler the temperature, the less S_2 specie and the more S_6 and S_8 species are present, down to about 800°-900° F. Then below 800° F., all S_2 specie is gone, S_6 specie begins to decrease, and S_8 specie begins to increase.

Other species are known to be present in minor amounts, but the major species are S_2, S_6, and S_8. Fig. 4 shows the approximate equilibrium between the major molecular species of sulfur in the vapor phase at various temperatures.[1]

Other side reactions involving CO_2, hydrocarbons, H_2S, SO_2, COS, and CS_2 may occur in the reaction furnace. Typical side reactions are shown in Fig. 5. If the acid-gas feed contains 80-90% H_2S, few of these side reactions go to any significant degree of completion.[3]

Catalytic converters, sulfur condensers. The Claus reaction proceeds to a further degree of completion in the presence of Claus catalyst in the converters. The reactions involved are,

$$2H_2S + SO_2 \rightarrow 3/6\ S_6 + 2H_2O \quad \text{(exothermic)}$$
$$2H_2S + SO_2 \rightarrow 3/8\ S_8 + 2H_2O \quad \text{(exothermic)}$$

After the initial "burn" step in the reaction furnace, the optimum ratio of H_2S/SO_2 of 2/1 is established and is present for reactions in the rest of the plant. Of course, each conversion step can only make some reasonable approach to theoretical equilibrium and ultimate total reaction. Therefore, there is always some unreacted H_2S and SO_2 (in a ratio of H_2S/SO_2 = 2/1) in the effluent gas

The author . . .

B. Gene Goar is a consulting chemical engineer with Goar, Arrington & Associates Inc., Tyler, Tex. His specialties are gas processing, gas treating, sulfur recovery, and stack-gas cleanup technology. Goar is the author of several articles and contributor to books on those subjects. He was previously associated with Howe-Baker Engineers Inc., Stearns-Roger Corp., Pan American Petroleum Corp., and Struthers Scientific & International Corp. Goar is a graduate of the University of Texas in chemical engineering.

B. Gene Goar

from each conversion step.

Usually 60-70% of actual conversion of H_2S to elemental sulfur (S_2 vapor specie) is achieved in the "thermal conversion" step.[1,3] Then the overall conversion is improved each time a "catalytic conversion" step is performed.

With a rich acid-gas feed, 94-96% overall conversion can be realistically achieved with one thermal and two catalytic conversion steps. For the same feed gas, 95-97.5% overall conversion might be achieved with one thermal and three catalytic conversion steps.

In the sulfur condensers, the sulfur species are shifted and sulfur vapor is condensed to liquid as follows:[1,3]

1. Sulfur vapor specie shifts:

$S_2 \rightarrow 1/3\ S_6$ (exothermic)
$S_2 \rightarrow 1/4\ S_8$ (exothermic)
$1/3\ S_6 \rightarrow 1/4\ S_8$ (exothermic)

2. Sulfur condensation:

$S_6 \rightarrow 6S_1$ (exothermic)
$S_8 \rightarrow 8S_1$ (exothermic)

The outlet gas stream from a sulfur condenser normally operates at 320°-340° F., with the final condenser operating at as low as 265° F., if it is being used as a boiler-feedwater preheater. The sulfur condensers typically produce 45-70 psig steam on the shell side.

Methods of reheat. The various methods of reheating gas to a catalytic converter are:[2,3,4]

- Hot-gas bypass.
- Inline burners.
- Gas-gas exchangers.
- Indirect heaters, with fuel-gas firing or steam heating.

The above reheat methods are shown in general order of increasing cost and in general order of increasing overall sulfur conversion efficiency. Various methods of reheats are shown in Figs. 6, 7, and 8.

Fig. 9

Claus plant split-stream process

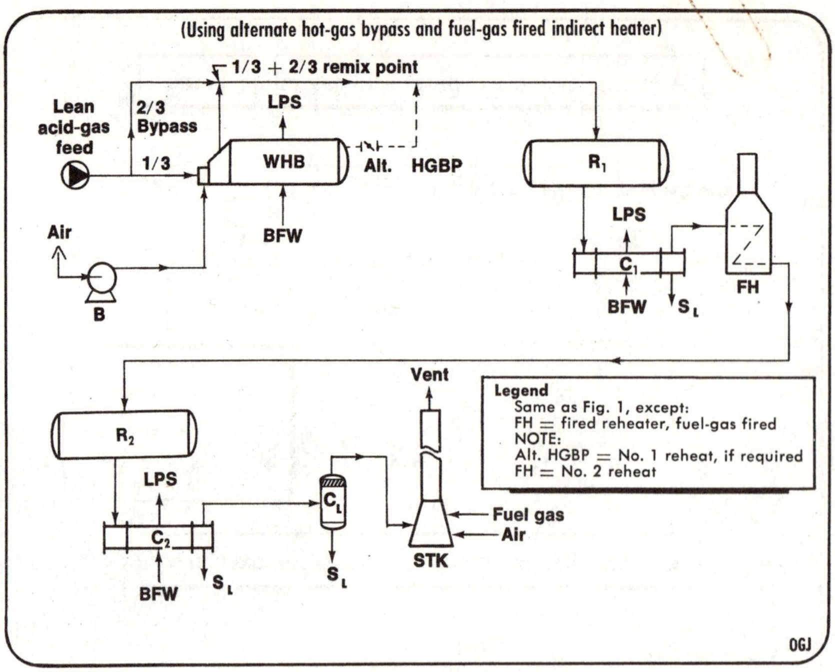

Lean H_2S acid-gas streams. Acid-gas feed streams containing less than 25-35 mol % H_2S are considered "lean H_2S" streams and usually require the use of a split-stream flow scheme (Fig. 9). The more CO_2 and hydrocarbons present in the stream, the greater the amount of COS and CS_2 formation is realized.[6]

Also, more of the side reactions shown in Fig. 5 go to a higher degree of completion. Excessive quantities of feed gas impurities can cause severe problems in a Claus plant.[6]

Major design considerations. Many good design considerations for a Claus plant are discussed in the literature.[1,3,4,7,8] The major variables and criteria to be considered are:

1. Composition of acid-gas feed.
2. Combustion of acid gas.
3. For a once-through scheme, retention time of combustion gases at elevated temperatures.
4. Catalytic-converter feed-gas temperature.
5. Optimum reheat scheme(s).
6. Space velocity in converters.
7. Sulfur condensing temperatures.
8. Coalescing and separation of entrained sulfur from condenser effluent gases.
9. Turndown of plant throughput.

Major operating control variables. The major operating control variables for a Claus plant are:[2,5,7]

1. Control of the H_2S/SO_2 ratio at the optimum value of 2/1.
2. Catalytic converter feed-gas temperatures. The inlet temperature must be kept high enough to prevent sulfur condensation in the catalyst bed.
3. Catalyst maintenance.[9]
4. Sulfur condenser outlet temperatures.
5. Elimination of hydrocarbons and other impurities in the feed gas.[6]
6. Proper operation of instrumentation at severe turn-down conditions.

References

1. Gamson, B. W., and Elkins, R. H., "Sulfur From Hydrogen Sulfide," Chemical Engineering Progress, Vol. 49, No. 4, April 1953, pp. 203-215.
2. Goar, B. G., "Today's Sulfur Recovery Processes," Hydrocarbon Processing, September 1968, pp. 248-252.
3. Opekar, P. C., and Goar, B. G., "This Computer Program Optimizes Sulfur Plant Design and Operation," Hydrocarbon Processing, June 1966, pp. 181-185.
4. Valdes, A. R., "New Look at Sulfur Plants, Part 1: Design," Hydrocarbon Processing, March 1964, pp. 104-108.
5. Valdes, A. R., "New Look at Sulfur Plants, Part 2: Operations," Hydrocarbon Processing, April 1964, pp. 122-124.
6. Goar, B. G., "Impure Feeds Cause Claus Plant Problems," Hydrocarbon Processing, July 1974, pp. 129-132.
7. Goar, B. G., "Sulfur Recovery From Natural Gas Involves Big Investment," The Oil and Gas Journal, July 14, 1975, pp. 78-85.
8. Palm, J. W., "Watch These Trends in Sulfur Plant Design," Hydrocarbon Processing, March 1972, pp. 105-108.
9. Norman, W. S., "There Are Ways to Smoother Operation of Sulfur Plants," The Oil and Gas Journal, Nov. 15, 1976, pp. 55-60.

GPA panelist outlines Claus process improvements in sulfur recovery

THE Claus sulfur-recovery process is widely used today for converting hydrogen sulfide to elemental sulfur. This process has now advanced to the point that recoveries have increased from the 90-92% levels first visualized in 1948-50 to 1978 recoveries above 98% or 99% of inlet sulfur.[1 2]

These comments and improvements in the Claus sulfur recovery process and what these mean to gas processors handling sour natural gas were discussed by L. V. Kunkel, L. V. Kunkel Engineering, Naperville, Ill., during a panel session at the recent Gas Processors Association Convention in New Orleans. A group of noted authorities comprised the panel, which went into H_2S removal methods. Parts 1 and 2 of this series explored gas-sweetening processes.

Claus process improvements

Key elements of Kunkel's formal presentation on Claus process improvements follow:

At the same time that the capabilities of the process have improved so dramatically, innovations and process optimization have reduced investments and operating costs to substantially offset the effect of inflation on plant construction and operation.

As the need for reduction of SO_2 emissions receives greater emphasis in the future, Claus process-reaction technology will be improved to obtain improved recovery which will be adequate to meet reasonable emission limits and which will be more economical than other tail-gas-cleanup processes.

The process was originally conceived in 1883 by a British scientist named Claus, but major improvements were not made in the process itself or in the techniques of its application until a process of this type was required in the 1950's in the U.S.A. The process was then improved to meet new needs.

Many variations have been developed over the years in response to the need for an improved process when it became necessary to process natural gas containing substantial volumes of hydrogen sulfide in both the United States and Canada.

Improved technology to meet the new needs was enhanced by several developments. These include publication of the paper, "Sulfur From Hydrogen Sulfide" by Gamson and Elkins in April 1953[3], and development of computer programs which used Gamson and Elkins and other data for preparation of heat and material balances and equipment designs and design optimization.

Other contributions include entry of operating companies which devote substantial effort to sulfur-recovery research and development into plant design and licensing activities, state environmental and federal EPA regulations which require lower sulfur-dioxide emission levels and consequent higher sulfur recoveries, and competition between designers in extending application of the Claus technology toward higher sulfur recovery at lower cost.

Table 10

Process variations

Main process variations
1. Straight-through process
   a. All air and acid gas through combustion zone.
2. Split-flow process
   a. All air and 40% or more of acid gas through combustion zone; 60% or less routed to first catalytic converter.

Variations in stages
1. Flame-conversion stage
2. Muffle furnace or fire-tunnel stage
3. Catalytic converters—one to four converters
4. Tail-gas-cleanup stage.

Variations in heating methods
1. Fired preheaters or reheaters—direct injection
2. Exchange preheaters or reheaters
   a. Internal fired exchange
   b. External fired exchange
3. Bypass hot gas injection from upstream in process.

Variations in cooling methods
By steam generation
   a. High pressure—power generation and heat utilization
   b. Medium pressure—heat utilization only
   c. Low pressure—no heat utilization

Variations in separation techniques
   a. Separators—separate or combined with condensers
   b. Packed bed
   c. Mist pad separators

The principal variations of the Claus process used in the world today are listed in Table 10.

Other variations of the Claus process have been required to solve special problems such as the presence of hydrocarbons or the presence of ammonia in the sulfur-plant feed.

Hydrocarbon effects. The concentration of hydrocarbons in acid gas is usually reduced by reducing amine circulation, or by installing an amine flash tank.

A 5% increase in methane-content acid gas containing 50% hydrogen sulfide in a straight-through sulfur plant can increase air requirement by as much as 40% and can increase process gas quantity as much as 25%. If the hydrocarbon is heavier, the effect on air and process-gas volume will be greater.

For example, an increase of only 2% pentane in similar acid gas containing 50% hydrogen sulfide requires the air to be increased by 64% and the process-gas quantity to increase as much as 42%. Side-reaction losses will also increase, and catalyst life will be decreased when heavy hydrocarbons must be handled.

Ammonia is usually handled by modifying the sulfur-recovery process equipment to obtain adequate combustion of the ammonia, or by selecting operating conditions which prevent interference of ammonia with operations.

Process capabilities. The superiority of the straight-through process can

Table 11

Process comparison

Investment	Straight through	Split flow
Treating plant	Lower	Higher
Sulfur plant	Lower	Higher
Tail-gas cleanup	Lower	Higher
Operating Cost		
Fuel, maint., etc.	Lower	Higher
Recovery level	Higher	Lower
Downtime	Lower	Higher
Control of process composition	Easier	More difficult

Figs. 13 and 14

% H_2S + SO_2 vs. recovery

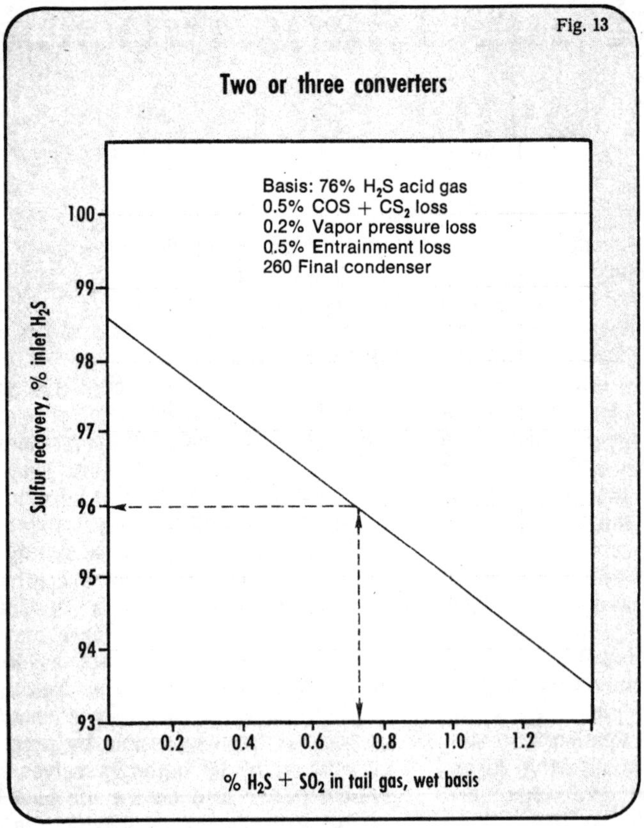

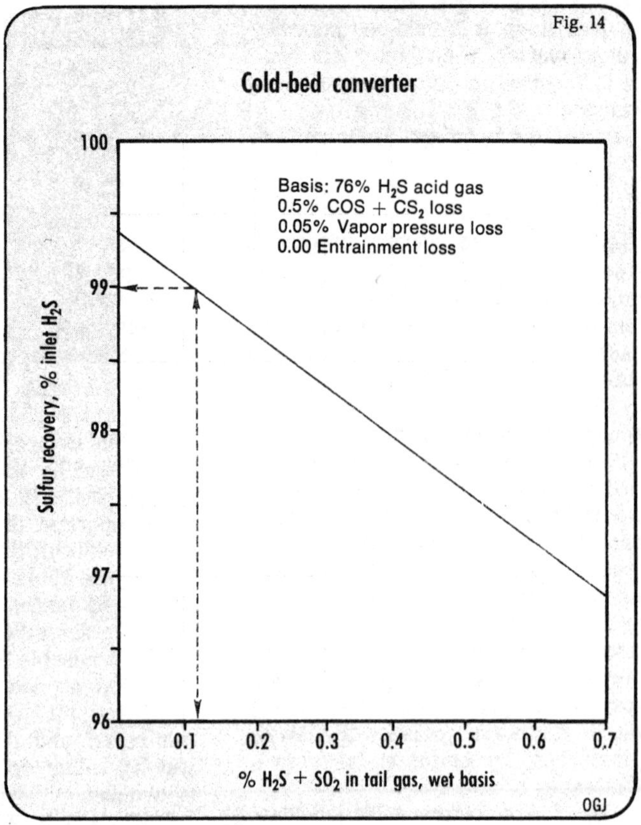

best be illustrated by the comparison in Table 11.

Recently, selective solvents such as MDEA have been used to selectively absorb essentially all of the hydrogen sulfide while absorbing only a portion of the carbon dioxide.

The technique of using selective absorption upstream of the sulfur plant to enrich hydrogen sulfide content of sulfur-plant feed gas allows the troublesome split-flow process to be avoided and the more-efficient, less-troublesome straight-through process to be used in certain new plants.

Such enrichment has the beneficial effect of increasing sulfur-recovery efficiency, reducing emissions, and reducing downtime as well as lowering sulfur-plant investment.

The effect of enrichment of hydrogen sulfide content of acid gases is illustrated by Table 12 showing typical investments and recovery as a function of acid-gas hydrogen-sulfide content for 3,739 Mcf of acid gas.

This comparison shows the potential improvement in recovery and reduction in investment which may be achieved by successful application of selective solvents for enrichment of sulfur-plant feed gas upstream of a Claus sulfur-recovery plant.

In addition to these benefits, additional benefits of selective absorption are that it will also reduce sulfur emissions and the cost of incinerator investment and fuel. The cost of tail-gas cleanup facilities is also reduced when the installation of such facilities is required.

Table 12

Effect of H_2S enrichment

Hydrogen sulfide content	16%	24%	58%	93%
Relative investment per ton of sulfur	2.06	1.67	1.15	1.00
Investment, $million	2.25	1.82	1.25	1.09
Sulfur recovery, %	93.68	94.20	95.0	95.9

Effect of internal variables. Typical losses from a straight-through two-converter sulfur plant consist of the following:

	Percent of inlet sulfur
Recovery	96.2
Claus losses, H_2S+SO_2	2.7
Organic sulfur losses, $COS+CS_2$	0.35
Vapor sulfur losses	0.25
Entrained sulfur losses	0.50

Claus losses are a function of approach to equilibrium at final Claus catalyst outlet temperatures. If catalyst is sufficient and active, and temperature is satisfactorily maintained above the sulfur dew point, sulfur yield in a typical plant may be increased about 0.5% if the temperature from the final converter is lowered 20° F.

Claus losses are also effected by insufficient catalyst, bypassing of catalyst along ends and edges of converters, fouling of catalyst with carbon and coke, and by catalyst deactivation due to sulfation. The amount of catalyst provided in converters is sufficient in most plants to allow for some deterioration of catalyst with time.

Claus losses are seriously increased if the hydrogen sulfide to sulfur dioxide ratio is allowed to vary appreciably from the optimum ratio of 2.0.

Table 13 illustrates the effect of deficiencies and excesses of air on H_2S and SO_2 losses from Claus processes. The loss in recovery is greater if the plant is air deficient than if an excess of air is present.

The table also shows that control of air within plus or minus 2% is required for good performance in a two-converter plant while control within plus or minus 1% or better is required in a three-converter plant. If the final stage is a cold-bed subdewpoint converter, even better air control is required to obtain desired performance.

Control improvement. Hydrocarbon and water content of acid-gas control is desirable in systems where high recovery in sulfur plants is required. Control of acid-gas hydrocarbon content requires automatic control of both amine and inlet-gas temperatures to the amine contractors, while water-vapor content of acid gas can usually be adequately controlled by automatic control of amine-system reflux-condenser temperature.

When gases rich in heavy hydrocarbons are processed, control of the amine and gas temperatures does not always allow heavy hydrocarbons to be held uniform even when flash tanks are used and temperatures are controlled. Some operators reduced both heavy hydrocarbon content and variations by insulating contactor and residue scrubbers to prevent heavy-hydrocarbon condensation at vessel walls.

In the 1950's, before sulfur dioxide emissions were rigidly controlled, it was generally considered satisfactory to control sulfur-plant hydrogen sulfide—sulfur dioxide in tail gas by ratio control of air to acid gas with bias manually adjusted by plant operators based on tests of hydrogen sulfide—sulfur dioxide ratios made at 4 or 8-hr intervals.

In the 1960's, as analytical instruments were improved, chromatographs capable of continuously measuring percentages of hydrogen sulfur, sulfur dioxide, carbonyl sulfide, and carbon disulfide in tail gas became available. Quality of control was appreciably improved since analyses were available about four to eight times an hour, allowing the H_2S-to-SO_2 ratio to be adjusted several times per hour instead of once or twice per shift.

A natural sequel to manual control based on frequent analyses resulted when techniques were developed to "close the loop," making adjustments to the air rate automatic for changes in acid-gas rate as well as for changes in composition.[4]

Table 13

$H_2S + SO_2$ loss, % acid-gas H_2S vs. air quantity, %

	Deficient air			Excess air			
% Air	97	98	99	100	101	102	103
Two converter	3.6	3.12	2.7	2.53	2.56	2.79	3.2
Three converter	3.1	2.14	1.32	1.05	1.20	1.54	2.1

Table 14

Effect of 10% measurement error[5]

Base value
100 tons product
105 tons inlet
5 tons loss
Recovery 95.24%

	High	Low
Method 1:		
10% error in inlet	100/115 = 86.95%	100/95 = 105.2%
Method 2:		
10% error in loss	(105−5.5)/105 = 94.76%	(105−4.5)/105 = 95.71%
10% error in inlet	(115−5.0)/115 = 95.65%	(95−5.0)/95 = 94.74%
Method 3:		
10% error in loss	100/(100+5.5) = 94.79%	100/(100+4.5) = 95.69%

The temperature to converters has sometimes been controlled automatically, but manual control of temperature is usually sufficient unless rates or compositions fluctuate frequently. It is important that the temperatures at all points in the converter be kept adequately above the sulfur dew point to prevent deactivation of catalyst due to sulfur deposition.

It is advisable to provide sufficient insulation on each converter so heat losses at the converter edges are minimized and deactivation of catalyst by sulfur deposition along edges is avoided.

Control of temperature to continuously prevent sulfur accumulation on catalyst is difficult, particularly when rates or compositions change. Therefore, it is often advisable to periodically raise catalyst temperatures substantially above normal operating levels for 1 day or longer to remove sulfur which may contribute to catalyst deactivation by vaporizing it from the catalyst beds.

Operational surveillance. Several techniques may be used to keep sulfur recovery continuously high during operation of a sulfur-recovery unit. One technique is to maintain acid-gas temperature as uniform as possible preferably as low as possible. Another is to operate the amine plant to minimize hydrocarbon content of acid gas, particularly the heavy hydrocarbons.

A third technique is to use dual-range acid-gas and air-control meters to insure precise measurement at low rates as well as high rates. Low flow shutdowns should be set at 2 roots for both acid gas and air. Acid-gas and air meters should be located just upstream of boiler entry so pressure changes will be similar for the two streams.

Also, acid gas and air leads should be self-draining, and meters should be zeroed periodically to insure metering and control accuracy.

Temperature rise across catalyst beds should be monitored to confirm that good conversion is being achieved. The final condenser outlet temperature should be kept as low as possible to reduce vapor-pressure sulfur losses. Pressure drops should be kept as low as possible by proper adjustment of all butterfly valves to reduce power load on the air blower and to improve control of air to the process.

Analytical instruments to measure H_2S and SO_2 should be serviced regularly to insure their accuracy at all times.

The temperature of all samples to UV analyzers should be kept constant at calibration value.

Concentration of H_2S and SO_2 should be continuously monitored and kept very near to the ratio of 2.0. In addition to the ratio, the absolute sum of the concentration of H_2S+SO_2 is also a valuable tool by which sulfur-recovery efficiency percentages may be closely projected since losses other than Claus H_2S and SO_2 are usually constant.

Once they have been determined, they may be added to Claus loss percentage to determine total percent losses.

Calculations may be prepared from design or test material balances. This will allow charts to be prepared which will allow the sum H_2S plus SO_2 concentration measurement to be converted directly into percentage recovery by the sulfur-plant operator.

He may then record it on plant log sheets for use of plant management. Typical charts which may be used by operators are shown in Fig. 13 and 14.

Sulfur-plant losses may also be similarly measured by surveillance of a chart recording sulfur-dioxide concentration in the incinerator stack gases. Calibration curves can also be easily prepared to allow recovery to be estimated using sulfur dioxide concentrations.

It is advisable to keep incinerator fuel readings under continuous surveillance. A decrease in fuel demand at constant plant throughput may be a signal that insufficient air is being supplied to the process, or an increase in fuel demand may indicate the opposite.

Performance testing. The recovery performance of a sulfur recovery plant can be determined by several methods. These are:

1. % Recovery = sulfur produced × 100/sulfur in inlet gas
2. % Recovery = (sulfur in inlet − tail gas losses) × 100/sulfur in inlet.
3. % Recovery = sulfur produced × 100/(sulfur produced + tail gas losses)

Measurement of inlet-gas hydrogen sulfide is difficult since it requires continuous measurement of inlet-gas volume and concentration. A continuous record of inlet concentration is not usually available and can only be obtained by frequently analyzing inlet gas for hydrogen sulfide, say at least every 4 hr.

Such analyses must be made immediately, preferably by Tutweiler titration on a spot sample taken directly from the inlet acid gas. Samples taken in cylinders for later analysis generally deteriorate in detectable hydrogen sulfide content, the greater percentage deterioration being observed in streams with low hydrogen sulfide contents.

In most sulfur-recovery plants, the tail gas is continuously analyzed for control of air to the process, or the stack-gas sulfur dioxide concentration is continuously recorded.

Sulfur production is measured by pit gauge which is usually quite accurate when measurements are carefully made and temperature corrections are properly applied to volume measurements.

It is generally advisable to measure or calculate all of the following in conducting performance tests:

- Sulfur production by pit measurement.
- Sulfur losses by outlet stream loss calculation (by computer).
- Sulfur in inlet gas.

All of these data are required to evaluate performance and to confirm the validity of the test. Sulfur in inlet gas should normally check the sum of sulfur calculated as production plus losses within 2%.

The first method of comparing production with inlet measurement is usually less accurate than other methods because the error in the numerator does not tend to be offset by the same error in the denominator, and vice versa.[5]

For example, a 10% error in measurement of inlet sulfur or losses has the effect on calculated recovery shown in Table 14.

NELSON COST INDEXES

Refinery construction (1946 basis)

Explained on p. 68 of the issue of Nov. 29, 1976

	1954	1962	1973	1976	1977	Mar. 1977	Feb. 1978	Mar. 1978
Pumps, compressors, etc.	166.5	222.5	346.9	538.6	575.0	560.3	603.3	604.4
Electrical machinery	160.0	189.5	220.2	287	301.8	296.1	314.5	316.9
Internal-comb. engines	150.5	183.4	238.3	348.3	376.9	366.7	399.3	403.2
Instruments	154.6	214.8	338.0	466.4	484.1	384.1	511.8	512.3
Heat exchangers	171.7	183.6	313.7	478.5	454.0	447.8	475.3	475.5
Misc. equip. average	160.7	198.8	291.4	423.8	438.3	431.0	460.8	462.5
Materials component	174.6	205.9	292.3	445.2	471.3	463.8	499.9	503.8
Labor component	183.3	258.8	585.2	729.4	774.1	755.5	801.9	803.6
Nelson refinery (inflation) index	179.8	237.6	468.0	615.7	653.0	638.8	681.1	683.7

Refinery operating (1956 basis)

Explained on p. 86 of the issue of Jan. 10, 1977

	1954	1962	1973	1976	1977	Mar. 1977	Feb. 1978	Mar. 1978
Fuel cost	86.5	100.9	186.8	384.5	430.5	446.8	422.2	413.4
Labor cost	90.9	93.9	102.9	145.5	149.6	146.2	176.9	174.2
Wages	88.7	123.9	221.2	314.3	346.2	344.2	386.5	385.4
Productivity	97.1	131.8	214.9	216.2	231.4	235.4	218.5	221.3
Invest., maint., etc.	92.0	121.7	193.0	256.2	263.6	258.6	273.5	274.6
Chemical costs	85.7	96.7	117.3	195.2	184.4	177.2	189.5	189.1
Nelson operating indexes								
Refinery	88.7	103.7	125.7	209.3	228.2	225.5	242.6	241.2
Process units*	88.4	103.6	168.0	267.1	287.8	290.8	296.6	293.1

*Add separate index(es) for chemicals, if any are used. +For refineries actually built (increased capacity, increased complexity). See Quarterly Costimating, July 4, 1966, p. 110.

These indexes are published in the first issue of each month. They are compiled by W. L. Nelson, technical editor and petroleum refinery consultant, Tulsa.

Charts of the indexes are published each year in a late January issue. Indexes of selected individual items of equipment and materials are also published on the Costimating page in the first issue of the months of January, April, July, and October.

References

1. Goddin, C. S., "CBA Process Ups Claus Recovery," Hydrocarbon Processing, October 1974, pp. 122-124.

2. Nobles, J. E., Palm, J. W., and Knutson, D. K., "Design and Operation of First Amoco CBA Unit," presented at Gas Conditioning Conference, University of Oklahoma, Mar. 7-9, 1977.

3. Gamson, B. W., and Elkins, R. H., "Sulfur From Hydrogen Sulfide," Chemical Engineering Progress, 49, No. 4, April 1953, pp. 203-215.

4. Saltzman, R. S., and Hunt, E. B., "A Photometric Analyzer System For Monitoring and Control of Ratio in Sulfur Plants," ISA Meeting, San Francisco, 1972.

5. Sudduth, L. F., "Evaluation of Sulfur Plant Efficiency," OGJ, Dec. 14, 1970, pp. 102-104, 109-112, and Jan. 11, 1971, p. 8.

Sulfur recovery from natural gas involves big investment

The author . . .

B. Gene Goar

B. Gene Goar is a consulting chemical engineer with the firm of Goar, Arrington & Associates Inc., Tyler, Tex. At the time of preparation and presentation of the paper from which this article was prepared, he was affiliated with Howe-Baker Engineers Inc. Goar's specialties are gas processing, gas treating, sulfur recovery, and stack-gas-cleanup technology. He is author of several articles and contributor to books on these subjects. He was previously associated with Stearns-Roger Corp., Pan American Petroleum Corp., and Struthers Scientific & International Corp. Goar is a graduate of the University of Texas in chemical engineering.

THE DECISION to build a sour natural-gas-treating facility for the purpose of sweetening the gas and recovery of elemental sulfur involves many considerations, and usually represents a significant investment.

Selection of the optimum gas-treating process and sulfur-recovery scheme may be a complex problem. Many well-established criteria are important to consider during the design phase of such a project. Subsequent operation of these facilities requires a good knowledge of potential problem areas.

There are several factors that should be considered in the selection, design, and operation of a gas-treating/sulfur-recovery facility.

Information needed. Once a sour-gas well is drilled or a sour-gas field drilling activity is completed, the producer must consider such questions as:

1. Is it more economical to lay a sour-gas line to the nearest gas-treating plant, which may have excess capacity, and pay a treating cost to have the gas sweetened?

2. Or, is it more economical to lease equipment for on-site sweetening?

3. Or, is it most feasible to build a grass-roots gas-treating plant to be owned and operated by the producer?

The answer to these questions depend upon several factors, such as (a) the volume of gas to be treated, (b) the gas composition, (c) the amount of H_2S and CO_2 to be removed, (d) the proximity of an existing plant, and (e) the estimated life of the reservoir.

If the decision is made to build a grass-roots plant, the producer will probably employ an engineering/contractor company to study the project and prepare recommendations for (a) the optimum gas-treating process to be used, (b) the most-economical overall processing scheme which is consistent with the required sales-gas specifications and the environmental regulations, and (c) the development of a definitive estimate for the erected cost of the plant.

Project study

The logical steps to be taken in conducting a project study are:

1. Define the extent and purpose of the plant.

2. Gather accurate data and infor-

mation on the specific gas to be treated.

3. Review the various processes available to industry for removal of H_2S and CO_2 from sour natural gas.

4. Decide if sufficient H_2S is present in the sour gas to justify a sulfur-recovery unit. Also, decide if a tail-gas cleanup unit is required.

5. Consider all four previous items, and eliminate those processes which are physically or chemically incompatible with the gas to be treated.

6. Make a preliminary economic comparison of the compatible processes; then select a treating process.

7. Make a detailed and objective evaluation of the overall project. This should include a definitive cost estimate, based on the processing scheme selected. An "economic justification" should be prepared. It should show an attractive payout and return-on-investment for the project.

Define plant and project. Definition of the plant/project is certainly the first order of business. Questions should be asked such as:

1. Is this plant really necessary?
2. Where shall it be located?
3. Is an extensive sour-gas-collection system required?
4. How much sales-gas line must be laid to get to the nearest pipeline?
5. Can a sales-gas contract be obtained easily?
6. What pipeline-gas specifications must be met?
7. Will the plant require operating personnel continuously, or should it be partially unattended?
8. Is electric power available, or must it be generated onsite?
9. Is adequate water available for cooling and process requirements?
10. Must air cooling be maximized?
11. How about disposal of waste streams?
12. How will the sulfur be stored and shipped?

All of these points and many more will arise during the evaluation of such a project. The earlier they can be answered or defined, the better the project will go.

Gas-stream data. One of the most important aspects of the project is the collection of accurate and meaningful data on the gas composition, the presence of contaminants, and the quantitiy of gas to be treated. Questions should be asked, such as:

1. How and when was the gas sampled for analysis?
2. Was the well(s) flowed for a while, before the sampling, to insure that water was blown from the well bore and flow line and a representative sample taken?
3. Was the flow potential measured for only one well, or were several wells tested to get an average value?
4. Was the acid-gas (H_2S and CO_2) content of the gas determined at the wellhead or in the laboratory?
5. Are O_2, COS, CS_2, mercaptans, or other trace impurities present in the gas? If so, to what extent?
6. Will corrosion inhibitors be used for down-hole corrosion control or collection-line protection? If so, what types?

The importance of obtaining an accurate gas analysis and satisfactory answers to these points cannot be overemphasized. The whole plant design and economics depend upon it.

The presence of even small amounts of impurities such as COS, CS_2, and mercaptans can have a large bearing on the selection of the type of treating process to be used. Therefore, the gas must be diligently tested for the presence of these and other impurities.

Gas-treating processes

In earlier years the choice of a treating process was much simpler, because so few processes were available. Today, the field of gas treating is so broad and the available processes are so numerous, that selection of an optimum process is a much-more-complex problem. Gas composition, available gas pressure, quantity and type of sour components to be removed, product-gas specifications, and subsequent processing of the acid-gas stream are all important factors to be considered.[3]

Several different types and categories of processes are available, such as:

1. Chemical-solvent processes such as alkanolamines (MEA, DEA, and DGA), and alkaline salt solutions (hot pot, Catacarb, Benefield).
2. Physical-solvent processes such as Sulfinol, Selexol, and Fluor solvent.
3. Direct-conversion processes such as Stretford, Thylox, Takahax, and Ferrox.
4. Dry-bed processes such as iron sponge, molecular sieves, and Haines.

All of these processes are summarized and discussed in the literature.[1 2 3 4 5 6 7] The highlights and major considerations of each process considered to be of commercial importance in the U.S. and Canada follow.

1. Chemical-solvent processes

a. Alkanolamines: (1) MEA—The monoethanolamine process has been discussed widely in the literature.[1 2 3 4 5 6 7 8 9] It is by far the most widely used process for gas treating. Advantages of the MEA process are high reactivity, low solvent cost, good chemical stability, ease of reclamation, low hydrocarbon content of the acid-gas produced, and lower plant investment when compared with many other processes.

Disadvantages of MEA are irreversible degradation by COS, CS_2, and O_2 in the gas, higher vaporization losses than any other amine-type process, ineffectiveness for removing mercaptans, nonselectivity for H_2S in the presence of CO_2, and higher utility costs than some other processes.

Unless COS, CS_2, or O_2 are present, MEA may be the best choice for treating a sour-gas stream containing low-to-medium acid-gas concentrations, with low-to-medium acid-gas partial pressures.[8] Some very general guidelines might be a sour gas containing from 4 grains H_2S/100 scf to 15 mol % total acid gas, with acid-gas partial pressures up to 100 psia.

Of course, certain factors can affect selection of the MEA process for applications outside of these guidelines.

The MEA process normally uses a solution of 15 wt % MEA in water. The solution is usually loaded with sour components in the ratio of 2.5 to 3.5 scf of acid gas (net pickup) per gallon of solution circulated. A net pickup of 3 scf/gal is typical.

(2) DEA—The diethanolamine process is the second most widely used gas-treating process, and is also discussed a great deal in the literature.[1 2 3 4 5 6 7 10 11 12] DEA's advantages with regard to MEA are its resistance to degradation from COS and CS_2, and

lower vaporization losses.

Its disadvantages are lower reactivity, higher solvent-circulation rates (conventional system), and higher solvent cost. DEA is often used to treat refinery-sour-gas streams because of the presence of COS and CS_2.

In recent years, DEA has gained much wider acceptance in Canada than in the United States for sour-natural-gas treating. In the United States, DEA is used widely in refineries but somewhat sparsely for natural-gas treating. Its popularity seems to be increasing.

There are three degrees of loadings used for DEA: (1) conventional DEA—uses solution of 20-25 wt % DEA and a net loading of 2 to 3 scf/gal; (2) high-loaded DEA—uses solution of 25-27 wt % DEA and a net loading of 5 to 5.4 scf/gal; and (3) SNPA: DEA—uses solution of 25-30 wt % DEA and a net loading of 5.5-6.2 scf/gal.

The SNPA: DEA process was developed in France and has been widely used in Canada. It is covered under Canadian patent No. 651,379 (October 1962). It is a royalty-bearing process. The "conventional" DEA process is used by many U.S. refiners and a few gas companies for natural-gas treating, where MEA cannot be used for some reason.

The high-loaded DEA process is used by a few U.S. companies and several Canadian companies for natural-gas treating, usually where the acid-gas partial pressure is about 30 psia or greater. The SNPA: DEA process has been used mostly in Canada and France, and is applied typically to gases containing 15-35 mol % total acid gases and with an acid-gas partial pressure exceeding 60 psia.[12]

With highly H_2S-sour gases, corrosion has not been reported to be a significant problem in the stronger and more heavily loaded DEA solutions.[10 12] Corrosion inhibitors have been used in some high-loaded DEA plants[10]; but, these chemicals are normally not considered necessary.

(3) DGA—The diglycolamine process was patented by the Fluor Corp. (called the Econamine process) and has been widely discussed in the literature.[1 2 3 13 14] The patent time has elapsed and now the process bears no royalty. The solution employed is 65-79 wt % DGA; and, net acid-gas loadings of 5-7 scf/gal solution are used.[3]

DGA is not irreversibly degraded by COS or CS_2. Originally, DGA offered distinct advantages over MEA and DEA (conventional) because of high solution acid-gas loadings (low circulation rates), lower utilities, no degradation due to COS and CS_2, lower vaporization losses than MEA, and lower solution freezing point (−40° F.) as compared with MEA and DEA (advantage in cold climates).

With the advent of high loaded DEA systems in the U.S. and Canada, and their wider acceptance, DGA is loosing its attractiveness over DEA in many instances. Two factors that have contributed to this situation are: (1) the 65 wt % DGA solution is more expensive than 25 wt % DEA solution, and (2) DGA solvent is available from only one supplier—Jefferson Chemical Co.

There has been a serious shortage of all amines, including DGA, in the recent past. With only one supplier, the shortage of DGA may last longer than with the other amines. MEA and DEA are manufactured by at least three large chemical companies—Jefferson, Dow, and Union Carbide.

DGA is an excellent treating process which can offer advantages over some other processes in certain applications.

(4) Adip—The Adip process is a royalty-bearing process licensed by Shell Development Co.[15] It has been used extensively for removal of acid gases from refinery gases and LP-gas liquids, mostly in western Europe and Japan. The process uses DIPA (diisopropanolamine) as the solvent.[15] It appears to be gaining popularity in the U.S. and Canada.

The Adip process can be designed to be very selective for H_2S in the presence of CO_2. Where this is a desirable situation, it can offer a distinct advantage over MEA and DEA. The Adip process is effective in removing large quantities of COS. Of course, COS does not degrade the DIPA solvent.

According to the literature,[15] the Adip process can offer dramatic utility advantages (less reflux and stripping steam) over MEA or DEA in some applications. Probably the greatest advantage Adip can offer over MEA or DEA is its selectivity for H_2S, where this is desirable.

b. Alkaline salt solutions: (1) Hot potassium carbonate—The "hot-pot" process uses an aqueous solution of 25-35 wt % K_2CO_3.[1 2 3 7] It is most applicable to large gas streams containing an appreciable amount of CO_2. If H_2S is present, with little or no CO_2, hot pot will not effectively sweeten the gas. Hot pot will remove some COS and CS_2, by hydrolysis of these compounds.

For hot pot to be attractive, the sour gas should contain 5 to 8 mol % acid gas at a contactor pressure of 300 psig or higher.[1 3 7] If H_2S is to be removed to meet normal pipeline specifications, a special design or a two-stage system may have to be used.

For certain applications, the advantages which hot pot offers over MEA or DEA are lower utilities, effective COS and CS_2 removal, lower vaporization losses, and lower plant investment due to elimination of major heat-exchange equipment.

(2) Catacarb—This is a licensed version of the hot-pot process which contains activator chemicals and corrosion inhibitors to increase the reactivity and loading of the solution. The Catacarb process has been applied most widely to CO_2 removal from hydrogen plants and ammonia plants, both in the U.S. and overseas.

(3) Benfield—The Benfield process is also a catalyzed version of the hot-pot process, containing activators to improve the treating capabilities of the solution. It is a licensed process also. It has been applied most widely to hydrogen, ammonia and SNG plants overseas.

(4) Giammarco-Vetrocoke—The G-V process is also a catalyzed version of the hot-pot process, using arsenic-containing activators to improve reactivity of the solution.[2 3] Many plants have been built worldwide; but, only one plant has been built in the U.S. to treat sour natural gas. The poisonous nature of the arsenic-bearing solution has been the process' greatest drawback.

2. Physical-solvent processes

The physical-solvent processes all use organic solvents and accomplish acid-gas removal mainly by physical absorption, rather than chemical reaction, which is directly proportional to the acid-gas partial pressure in the sour-gas stream.

These processes are most applicable to high-pressure gas streams containing appreciable quantities of sour components. The high acid-gas loadings realized account for the low solvent-circulation rates, the reduction in equipment size and cost, and the lower utilities which are characteristic of these processes.

In general, the physical-solvent processes all have two major disadvantages: (1) the solvent's affinity for absorbing heavy hydrocarbons (C_5^+), especially aromatics and olefins; and (2) the expensive nature of the solvents.

If the heavy hydrocarbons are absorbed from the feed gas and stripped from the solvent in the regenerator, a charcoal-adsorption system must be used to treat the acid-gas stream before it is processed in a Claus-type sulfur plant.[3] All of these processes are licensed by the developer and require payment of a royalty fee for their use.

A. **Sulfinol.** The Shell Sulfinol process is unique in this class of processes, in that it uses a mixture of solvents (Sulfolane and DIPA) which allows it to behave as both a chemical and physical solvent process.[15 16] Sulfinol has good affinity for sour components at low-to-medium partial pressures, and extremely high affinity for sour components at high partial pressures.

It should normally be applied to gas streams containing mostly light hydrocarbons. Sulfinol appears to have its greatest advantage when the H_2S/CO_2 ratio is 1:1 or greater, and when the acid-gas partial pressure exceeds 110 psia.[16]

The greater the H_2S/CO_2 ratio and the acid-gas partial pressure, the more likely Sulfinol will be advantageous. In general, COS, CS_2 and mercaptans can normally be removed to pipeline specifications.

A selective version of the Sulfinol process is offered by Shell, whereby essentially all of the H_2S is removed and only a predetermined amount of CO_2 is removed. In a situation where this is applicable, such as selective process is useful for improving the quality of the acid-gas feed to the sulfur plant. Also, if a sizable amount of CO_2 can be left in the sales gas, less treating "shrinkage" is experienced.

Typical Sulfinol solutions may contain 35 wt % Sulfolane, 45 wt % DIPA, and 20 wt % water. The solution composition is tailored to fit each application. The solution may be loaded (net) with 4-15 scf acid gas/gal, depending upon the application. The process has been widely accepted, worldwide, for natural-gas treating of certain types of sour-gas streams.

B. **Selexol.** The Allied Selexol process uses as a solvent dimethyl ether of polyethylene glycol (DMPG). Typically, no water is used with the solvent. This process is applicable for bulk removal of CO_2 and H_2S, primarily CO_2. Most past applications have been on streams containing large amounts of CO_2 (18-43 mol %) and less than 1 mol % H_2S.

Selexol is not particularly attractive for a condition of low acid-gas partial pressure.[3 7] It offers the advantage of selective removal of H_2S in the presence of large amounts of CO_2, where such a situation is desirable. Net loadings of 4-12 scf/gal are typical with Selexol.

Selexol suffered some serious setbacks in natural-gas treating when the air-blown regeneration scheme was used. This was due mostly to sulfur formation and resultant plugging in the regenerator. If H_2S is present in the gas to be treated, air blowing of the regenerator is no longer used, at least for the portion of the plant which removes H_2S.

Selexol has not been widely used or accepted for treating natural gas containing appreciable H_2S.

C. **Fluor Solvent** Lurgi Purisol, Lurgi Rectisol, GAF M-Pyrol, Estasolvan.[1 3 7]

None of these physical solvent processes has gained very wide acceptance in natural-gas treating in the U.S. or Canada.

3. Direct-conversion processes

The direct-conversion processes have been used most widely in Europe for removal of H_2S and recovery of sulfur from manufactured gas, coal gas, and coke-oven gases. The processes are usually best suited for sour gases containing 1 to 1,000 grains H_2S/100 scf, and with a maximum sulfur production of 10-15 long tons/day.[5 17 18]

These processes do not remove any CO_2 from the feed gas.

Most of the direct-conversion processes are based on a scheme, using oxidation — reduction reactions, involving absorption of H_2S in an alkaline solution containing oxygen carriers. The H_2S is oxidized to elemental sulfur. The oxygen is furnished by air which is normally added to the bottom of the regenerator or oxidizer vessel.

The air also acts as a flotation agent for the sulfur, which is collected at the regenerated solution interface as a froth.

The sulfur is recovered by filtration and melting.

The leading processes in this class of treating systems are Stretford, Thylox, Takahax, Ferrox, and Manchester.

Of these, only Stretford has been applied to natural-gas treating in the U.S. or Canada in recent years. Only a few Stretford plants have been built for this purpose.

One other type of direct-conversion process is represented by the Townsend and IFP processes. Both of these involve the Claus reaction in a liquid system.

Neither are of commercial importance at present.

4. Dry-bed processes

Three types of dry bed processes are represented by iron sponge, molecular sieves, and Haines. Of these, the iron sponge is of the greatest commercial significance.

The iron-sponge or iron oxide process is one of the oldest treating processes still in use today. It uses wood shavings impregnated with ferric oxide in hydrated form.[1] H_2S reacts with the ferric oxide to form ferric sulfide. Partial regeneration with air is possible, but, eventually the bed becomes plugged with sulfur and must be replaced.

The process is usually limited to treating gas streams of small to medium volumes, containing 1,000 grains H_2S/100 scf or less, and containing up to 8-10 LTPD sulfur.[1] These limitations are due to economics of bed replacement. Advantages of this process are essentially complete H_2S removal, ease of operation, and simplicity and low cost of installation.

The sulfur removed by the process must be wasted, since it cannot be recovered economically.

Amount of sulfur present. Typically, if the sour-gas stream contains sufficient H_2S to produce more than 5-10

LTPD of sulfur, a treating process should be selected which will yield an acid-gas stream that is suitable for processing in a Claus-type sulfur-recovery unit. At sulfur quantities below 5-10 LTPD, the iron-sponge process or a direct-conversion-type process should be considered, if CO_2 removal is not required.

If the CO_2 can be left in the treated gas without exceeding the sales-gas CO_2 specification, a process which is selective for H_2S should be considered. An acid-gas stream containing 60-70 mol % H_2S or greater is highly desirable for a Claus-plant feed stream. If a physical solvent process is used, the hydrocarbon content of the acid-gas stream should be considered. Excessive impurities (CO_2, hydrocarbons, water vapor, etc.) in the feed stream to a Claus plant can cause design, operational, and maintenance problems of considerable magnitude.[19] Enrichment processes for improving the quality of a poor or impure acid-gas stream are available to industry.[19]

If a decision is made to include a Claus sulfur-recovery unit, the need for a tail gas cleanup process must be considered. Depending upon the location of the plant (state and county), local pollution regulations, size of the sulfur plant, and percent overall recovery to be obtained, a tail gas cleanup process may be required.

Process, cost comparison. Eliminate the obviously incompatible gas-treating process and prepare a brief cost estimate for the remaining ones. Estimate the cost of the sulfur-recovery unit and tail gas cleanup unit, if needed. Estimate the cost of the utilities section and the sulfur-handling sections of the plant.

After reviewing the brief cost estimate prepared for the processes being considered, narrow the decision down to a single gas-treating process. Factors to consider in selecting the optimum process are plant investment, operating costs, maintenance cost, royalty payments, process reliability, chemical compatibility, climatic conditions, toxicity of the solvent, cost and availabality of the solvent, ease of operation, disposition of by-products such as sulfur, hydrocarbin liquids, etc.

Economic justification. An excellent paper was prepared on the economics of the sour-gas industry by Estep and Plus.[20] The reader is encouraged to review this paper.

Once a gas-treating process, sulfur-recovery scheme, and tail gas cleanup process selected, a detailed, definitive cost estimate for the overall project should be prepared. Then an economic justification must be prepared which should show an attractive payout and return-on-investment for the project.

Of course, the price obtained for the sales gas is a key factor. In cases where an appreciable quantity of sulfur is recovered, the price obtained for the by-product sulfur can have a dramatic effect on the economics of the project.[20]

Hypothetical gas-treating case

Assume a case based on these conditions:

1. Sour-gas flow = 50 MMscfd
Flowing conditions = 500 psig and 110° F.
2. Sour-gas composition:

Comp.	Mol%
H_2S	5.30
CO_2	1.70
N_2	1.10
C_1H_4	91.55
C_2H_6	0.35
COS	0
CS_2	0
RSH	0
	100.00

3. Acid-gas partial pressure = 36 psia
4. H_2S/CO_2 ratio = 3.1/1
5. Assume 15 wt % MEA is used, and net loading of 3 scf acid gas/gal solution. Therefore, MEA circulation = 810 gpm
6. Sulfur equivalent in sour gas = 100 LTPD. Therefore, Claus plant will be included.
7. Assume stripping steam = 1.10 lb/gal = 1.10 x 810 x 60 = 53,460 lb/hr
8. Estimated steam produced from sulfur plant = 25,730 lb/hr
9. Assume MEA-strippeer reflux drum operates at 8 psig and 105° F.
Acid-gas composition to sulfur plant:

Comp.	Mol%
H_2S	71.0
CO_2	22.8
C_1H_4	1.2
H_2O	5.0
	100.00

10. Sales gas specifications (maximum permitted:
    H_2S content: 0.25 grains H_2S/100 scf
    CO_2 content: 0.5 mol %
    Total sulfur content: 1.0 grain H_2S/100 scf
    Water content: 7 lbs H_2O/MMscf
11. Desired Claus-plant overall recovery = 95-96%, (exclusive of tail-gas treatment).
12. Claus-plant tail-gas SO_2 emissions permitted = 500 ppmv maximum. Therefore, a tail-gas cleanup unit is required.

It is assumed that several processes have been considered and the MEA process was selected. Since about 100 LTPD of sulfur is contained in the feed gas, a Claus-type sulfur-recovery unit will be included. It is further assumed that this plant will be located in an area where pollution regulations stipulate a maximum SO_2 emissions of 500 ppmv from the tail-gas stack.

Therefore, a tail-gas cleanup unit will be included. Sulfur will be shipped by tank truck in liquid form. A concrete pit for 5 days' storage (500 (LTPD) will be provided; but, a method of handling the sulfur other than in liquid form must be provided.

Design, operating considerations

The key points to be considered in the design and operation of the amine plant are:

1. Provide an adequate inlet-gas separator large enough and efficient enough to remove essentially all solids and free liquids, to keep these impurities from getting into the amine solution. In many plants, most of the iron sulfide found in the amine system enters with the sour gas. Efficient removal of this iron sulfide from the gas can prevent many operational and maintenance problems.

Of course, removal of free liquids is essential to satisfactory operation of the plant. If iron sulfide and/or liquids are known to be a problem, a special, highly-efficient separator should be used, such as (a) a "filter-sep" type, (b) a centrifugal type, or (c) a wash-oil type.

2. The inlet-gas temperature should be limited to 100°-100° F. for good acid-gas removal and low MEA losses. The temperature of the lean MEA entering the contactor should

be at least 10° F. higher than the feed gas to prevent hydrocarbon condensation and subsequent solution foaming.[21]

3. The contactor should be provided with an efficient demister pad or vane-type separator section above the top tray to reduce MEA-entertainment losses. If high MEA losses are expected, a water-wash step should be considered.

4. The bottom section of the contactor should be properly designed to prevent hydrocarbon-gas entrainment in the rich solution.

5. Normally 20 trays are used in the MEA contactor.

6. If the contactor operates at 200-300 psig or higher pressure, a solution flash tank should be included for the rich solution.

7. The final pressure reduction on the rich solution should be made downstream of the amine exchanger and just upstream of the stripper.

8. The amine/amine exchanger should heat the rich solution to 190°-200° F. before entry into the stripper. Rich amine should flow through the tubes. Linear velocities of 2-3 fps should not be exceeded through the tubes.

9. Flow velocities of 7-10 fps through piping and 15-20 fps through control valves should not be exceeded.

10. The stripper should be operated at sufficient pressure to cause the bottom temperature to be in the range of 230°-250° F., with a valid absolute maximum of 260° F.[1 21]

11. Normally, 20 trays are used in the stripper, with reflux being flowed to the top tray. Rich solution is normally added on the third tray from the top.

12. A stripping steam rate of 1.0-1.25 lb steam/gal solution is normally used for MEA systems.[6] This should result in a reflux ratio (mols H_2O/mol acid gas) of 1.0 to 3.0, which should be adequate.

13. The stripper reboiler(s) should be of a horizontal-kettle type (preferred). Ballard[21] has prepared an excellent summary of reboiler design criteria which should be considered. A brief recap of these criteria are:

a. Use a maximum steam temperature of 285° F.

b. Control the steam flow individually to each reboiler, i.e., do not attempt to control the condensate flow out. (Condensate should be on level control.)

c. Use a tube-bundle design to place the bundle on slides at least **6** in. above the bottom of the shell.

d. Use several (at least two) liquid inlets and vapor outlets on the shell.

e. Use a weir height to assure that the tubes are covered with **6-8** in. of fluid.

f. Limit the tube length to prevent condensate hammering and excessive tube vibration.

g. Consider Teflon inserts at all baffles to reduce tube grooving.

h. Use a square, rotated pitch on the reboiler and amine-exchanger tubes. Provide vapor lanes, in pattern of an "X" or "V", in the reboiler-tube bundle.

14. The proper control of the "condition" of the MEA solution can have a marked affect on solution corrosion, degradation, and foaming tendencies. The use of proper filtration techniques and a well-designed reclaimer will greatly enhance operation of the system. About 10% of the lean MEA stream should flow through a sock-type filter and then a charcoal-bed-type filter. Both types are recommended in series. There are some instances where filtering of the rich MEA may be advantageous.[22]

The use of a charcoal filter is believed to remove some heavy hydrocarbons and degradation products, as well as solid particles.[22] The importance of a clean, well-generated lean MEA solution cannot be stressed too much. Such a solution enhances operations and essentially prevents foaming and erosion/corrosion problems.

Foaming in an amine system can be caused by (1) suspended solids, (2) condensed hydrocarbons, (3) amine-degradation products, (4) corrosion inhibitors from down-hole or collection-line treatment, (5) valve grease, or (6) impurities introduced into the system by makeup water. The initial MEA solution should be mixed using steam condensate. Raw water should never be used. All makeup water added should be steam condensate.

All MEA systems should be provided with an adequate reclaimer. About 1-3% of the lean amine should be flowed to the reclaimer. Ballard[21] has given a good discussion on reclaimer design and operation in the literature.

When foaming does occur, various foam inhibitors should be tested in the laboratory for effectiveness on the specific solution before adding the inhibitor to the system. Some inhibitors are: (1) Du Pont's Ocenol, (2) Union Carbide's UCAR-470, and (3) Dow's Dow-Corning silicones (Types A & B).

15. For the most part, all equipment in an amine plant can be constructed from carbon-steel materials.[22] However, the use of stainless steels in the stripper-reboiler tubes, the overhead-condenser tubes, and certain pump parts and impellers may prove to be economically attractive. Stainless-steel float valves on carbon-steel tray decks are recommended in both the contractor and the stripper towers. It is common practice to stress relieve (1) the stripper vessel, (2) the reboiler(s) shell, (3) the amine exchanger, (4) all high-pressure piping (300# ANSI flange ratings and higher), and (5) all piping and equipment in contact with hot and/or rich amine solution. One item to remember is never to use hardened stud bolts in flanges in high-pressure sour-gas service. ASTM B7-M studs should be used.

Claus sulfur-recovery unit

The key points to be considered in the design and operation of a Claus sulfur recovery unit are:

1. The general type of Claus processing scheme must be selected. A choice must be made between the "once-through" or the "split-stream" process.[18] If the acid gas feed contains 30-40 mol % H_2S or greater, the once-through scheme should be selected, since it gives the highest overall sulfur recovery and permits maximum heat recovery at a high temperature.

In this scheme, all of the acid gas is fed to the reaction furnace, along with enough air to burn one-third of the H_2S to SO_2 and all hydrocarbons completely. Then sufficient retention time is provided to allow reaction of the SO_2 generated with the unburned H_2S to form sulfur vapor. This reaction takes place above 2,000° F., with no catalyst present. It is called the thermal-conversion step.

Up to 70% of the overall conversion of H_2S to sulfur can take place in this step.[18] The hot gases then pass through a waste-heat boiler, where they are cooled typically to about 550° F.

If a two-pass boiler is used, the gases are cooled to 1,000°-1,200° F. in the 1st pass, and on to 550° F. in the second pass. The hot gas from the first pass serves as the source for "hot-gas bypass" streams, where this method of reheating is used.

Normally, 150-300-psig steam is generated in the waste-heat boiler. In most cases, the high-pressure steam is used in a steam turbine (with 40-psig exhaust) to drive the combustion-air blower for the unit. Often times, the amount of steam generated is adequate to drive the air blower.

The low-pressure steam from the turbine exhaust is flowed to the amine-stripper reboilers, where it usually supplies 40-50% of the reboiler-steam requirement. This steam generation/utilization system makes for a very thermally efficient overall plant operation.

The acid-gas feed to the Claus plant is metered, and the flowmeter output is fed to a ratio controller, which in turn controls the amount of air added to the furnace. Analysis of the "H_2S/SO_2" ratio in the tail gas is used by the operator to adjust the ratio controller to add the proper amount of air.

The optimum H_2S/SO_2 ratio in the tail gas is 2:1, which will give the maximum sulfur conversion. A ratio either above or below 2:1 will cause a loss in conversion efficiency. If a choice must be made, it is better to run with a slight excess of air (H_2S/SO_2 ratio slightly less than 2:1), as compared with a slight deficiency of air.

For larger sulfur plants, a tail-gas analyzer, with "feedback" control of the air to the plant, is sometimes provided to optimize sulfur recovery.

2. Usually, a sulfur condenser is provided after the waste-heat boiler to condense and remove the sulfur produced by the thermal-conversion step in the reaction furnace. After the condensation step, the gas must be reheated before it flows to the first catalytic converter. The first condenser usually produces 45-psig steam and operates with a gas-outlet temperature of 340°-370° F. The gas is reheated to 450°-500° F. before entry into the first converter.

If the feed gas contains appreciable CO_2 (say more than 8-10 mol %), the first converter should be operated somewhat hotter than the subsequent converters to enhance COS and CS_2 conversion to sulfur in the 1st converter.

After the first condenser, the Claus plant consists of a series of "reheat, conversion, and condensation" steps. These steps may be repeated as many times as desired; however, experience has shown that two or three catalytic converters is usually the optimum choice.

A fourth converter cannot be justified, normally, except possibly on very large plants. If a tail-gas cleanup unit is included after the Claus plant, the use of even a third converter may be questionable. Each case must be evaluated separately. The size of the Claus plant will be a large factor.

Typically all of the condensers produce low-pressure steam (40-60 psig), with the last condenser producing 20-psig steam or serving as a boiler-feedwater preheater. It is very desirable to operate the last condenser as cool as possible to enhance final sulfur recovery.

If possible, the gas outlet from the last condenser should operate at 260°-265° F., which is safely above the sulfur-solidification point of 246 F.

3. The inlet gas to each catalytic converter is usually reheated to 400°-430° F., with the first converter inlet running 450°-500° F. There are four basic reheat schemes which may be used:

a. Hot-gas bypass.

b. In-line burners.

4. Gas-to-gas exchangers.

d. Indirect heaters, using either fuel firing or steam heating.

In general, the various reheat methods are shown in the order of increasing cost and effect of increasing the overall sulfur conversion.

The hot-gas bypass method is commonly used to reheat gas to the first converter. One of the other reheat methods is more commonly used to reheat gas to the first converter. One of the other reheat methods is more commonly used for subsequent converters. This is true, because if a hot-gas bypass reheat is used for the second or third converters, the hot-gas stream bypasses the first converter, and a loss in overall conversion occurs. In some two-converter Claus plants, hot-gas bypass streams are used for both converter reheats, because the desired overall recovery can be obtained with this scheme; and it represents a minimum investment.

4. For a well-designed Claus plant, demister pads should be included in the outlet channel of each sulfur condenser. This decreases liquid-sulfur entrainment from a condensing step into a conversion step, and thus enhances overall conversion. In addition, a separate coalescer vessel, with mesh pad, should be provided for the tail gas from the last condenser.

5. The material of construction used for a Claus sulfur plant is normally carbon steel throughout, except for (1) special stainless steeel for hot-gas bypass lines, and (2) stainless-steel demister pads in the condenser-outlet channels and coalescer vessel.

6. Several items of equipment in a Claus plant contain internal refractory linings. These normally are the reaction furnace, rear chamber of the waste-heat boiler, catalytic converters, and the incinerator stack. It is of utmost importance to control the warm-up and start-up rates of a Claus plant properly to prevent "thermal-shock" damage of the refractories. Also, controlled shutdown and cool-off rates are important.

7. The three major process-control variables in a Claus plant are: (a) H_2S/SO_2 ratio of 2:1, (b) catalytic-converter inlet temperatures, and (c) sulfur-condenser outlet temperatures. Of these, the control of the H_2S/SO_2 ratio is the most important. It is also very important to control converter-inlet temperaures sufficiently high to prevent sulfur condensation within a bed, which causes temporaary deactivation of the catalyst. However, the cooler the converter can be run, the better the Claus conversion.

8. The catalyst commonly used in Claus plants is 2/4 mesh bauxite. Typically, the beds are 3 ft deep. In a properly operated sulfur plant, with a relatively clean feed gas, the life of the catalyst beds should be 3-5 years. Periodic regenerations (carbon burn-offs) may be performed, but eventually the beds must be replaced.

New, improved sulfur-plant catalysts [23] [24] are available (such as Kaiser S-201 and Rhone-Progil CR), which are reported to offer several advantages over conventional bauxite: (a) fewer fines; (b) lower pressure drop; (c) lower bulk density; (d) greater resistance to sulfate formation; (e) less susceptibility to fouling from MEA, DEA, NH_3, etc.; and (f) better conversion of COS and CS_2 to elemental sulfur.

Certain of these new catalysts

(Kaiser S-201) have been used to good advantage in very large sulfur plants in Canada.[25] The new, improved catalysts are appreciably more expensive than bauxite; but, their advantages can outweight the additional cost in many instances.

9. Sulfur is normally produced and stored (temporarily) in liquid form. Typically a concrete pit, made of sulfate-resistant concrete, is provided for 3-5 days of liquid storage. If the liquid sulfur cannot be shipped by tank trucks or rail cars, it is usually stored in solid blocks, above ground. In situations where solid storage is required, sulfur prilling or "slanting" should be considered as alternates to blocking.

10. For the hypothetical case stated, a Claus plant using two catalytic converters, one hot-gas bypass reheat and one gas-to-gas exchanger reheat was selected. An overall sulfur recovery of 95.4% (Claus only) is expected. Liquid sulfur will be shipped by tank truck; but a small prilling tower is included for emergency use.

Selection of tail-gas cleanup process

Several processes are available for cleanup of the remaining sulfur compounds in the tail gas from a Claus plant.[1] The leading processes are:

1. Shell's SCOT process.
2. Wellman-Lord's process.
3. Parson's Beavon process.
4. SNPA/Lurgi's Sulfreen process.
5. SNPA's catalytic-oxidation process.
6. IFP's process.
7. Chiyoda's Thoroughbred 101 process.
8. Stauffer's Aquaclaus process.
9. Pritchard's Cleanair process.
10. Shell's SFGD process.
11. Westvaco's adsorption process.
12. U.S.B.M.'s Citrate process.
13. Townsend process.
14. ASR Sulfoxide process.
15. Trentham's Trencor process.

The three processes that would be considered for the hypothetical case are the SCOT, Wellman-Lord, and Beavon processes. The SCOT process was selected over the other two on the basis of (a) lowest initial investment, (b) simplified and familiar technology, (c) comparable utilities, (d) recycle of only a small amount of CO_2 along with the H_2S-recycle stream and (e) wide commercialization of the process for this type of application.

Had the Claus plant tail-gas stream contained large amounts of CO_2, the Wellman-Lord or Beavon process may have been selected, even though the initial investment may have been greater. Any of the three processes considered have the capability of yielding a treated tail-gas stream containing less than 500 ppmv SO_2.

In the SCOT process, the Claus tail gas (feed) is heated to about 570° F. by an in-line heater, mixed with a reducing gas (normally hydrogen) and flowed to a reactor where all sulfur compounds are reduced to H_2S. The gas is then cooled by heat exchange and a water-quench tower. Then, the H_2S is absorbed from the gas by a special amine which is highly selective for H_2S in the presence of CO_2.

The solvent may pickup 20-30% of the CO_2 present. The solvent is regenerated in a conventional amine regenerator (stripper) and the net overhead vapors, containing the H_2S, are recycled back to the front of the Claus plant.

The Sulfreen, SNPA catalytic-oxidation, and IFP are all viable processes; but, they will not yield a 500-ppmv SO_2-emissions level. Chiyoda's process is viable; but, it produces a gypsum by-product, which is a solids-disposal problem.

The Aquaclaus, Cleanair, SFGD, Westvaco, Citrate, Townsend, Sulfoxide, and Trencor processes are all in an early stage of development or commercialization. For this case, of these would offer any particular advantage over the SCOT process.

Approximate investment costs. The approximate investment costs for the major processing units selected for the hypothetical case are:

1. MEA gas treating plant = $3,250,000 (810 gpm MEA and 50 MMscfd gas)
2. Claus sulfur-recovery unit = $1,500,000 (100 LTPD)
3. SCOT unit = $1,200,000 (for 100 LTPD Claus unit)

These costs are only approximate, and are shown here to give some idea of the order of magnitude of costs involved in such a project. They are based on a mid-1974, U.S. Gulf Coast, erected-plant, estimated basis. The costs do not include offsites and support systems.

References

1. Maddox, Dr. R.N.: Gas and Liquid Sweetening, Second Edition, Campbell Petroleum Series, Norman, Okla. (1974).
2. Kohl, A.L., and Riesenfeld, F.C.: Gas Purification, Second Edition, Gulf Publishing Co., Houston (1974).
3. Goar, B.G.: "Today's Gas-Treating Processes - 1," OGJ, (July 12, 1971) 75-79.
4. Blake, R.J.: "How Acid-Gas Treating Processes Compare," OGJ, (Jan. 9, 1967) 105-108.
5. Riesenfeld, F.C., and Blohm, C.L.: "Acid Gas Removal Processes Compared," Hydrocarbon Processing & Petroleum Refiner (April 1962) 123-127.
6. Fitzgerald, K.J., and Richardson, J.A.: "How Gas Composition Affects Treating Process Selection," Hydrocarbon Processing, (July 1966) 125-129.
7. Swaim, Jr., C.D.: "Gas Sweetening Processes of the 1960's, "Hydrocarbon Processing March 1970) 127-130.
8. Maddox, Dr. R.N., and Burns, M.D.: "MEA Process to be Considered First," OGJ, (Aug. 21, 1967) 83-85.
9. Dingman, J.C., Allen, D.L., and Moore, T.F.: "Minimize Corrosion in MEA Units," Hydrocarbon Processing (Sept. 1966) 285-290.
10. Smith, R.F., and Younger, A.H.: "Tips on DEA Treating," Hydrocarbon Processing (July 1972) 98-100.
11. Lee, J.I., Otto, F.D., and Mather, A.E.: "Design Data for Diethanolamine Acid Gas Treating Systems," Gas Processing/Canada (March-April 1973) 26-34.
12. Dailey, L.W.: "Status of SNPA-DEA," OGJ, (May 4, 1970) 136-141.
13. Dingman, J.C., and Moore, T.F.: "Compare DGA and MEA Sweetening Methods," Hydrocarbon Processing (July 1968) 138-140.
14. Dingman, J.C.: and Moore, T.F.: "Gas Sweetening with Diglycolamine," Proceedings of the 1968 Gas Conditioning Conference, University of Oklahoma, Norman, Okla. (1968).
15. Klein, J.P.: "Development in Sulfinol and Adip Processes Increases Uses," Oil and Gas International (Sept. 1970) 109-111.
16. Goar, B.G.: "Sulfinol Process Has Several Key Advantages," OGJ, (June 30, 1969) 117-120.
17. Kohl, A.L., and Riesenfeld, F.C.: "Today's Processes for Gas Purification," Chemical Engineering (June 15, 1959) 127-178.
18. Goar, B.G.: "Today's Sulfur Recovery Processes," Hydrocarbon Processing (September 1968) 248-251.
19. Goar, B.G.: "Impure Feeds Cause Claus Plant Problems," Hydrocarbon Processing (July 1974) 129-132.
20. Estep, J.W., and Plum, E.W.: "Economics of Sour Gas Industry," Paper presented at the 19th Canadian Chemical Engineering Conference and 3rd Symposium on Catalysis, Edmonton, Alta., Oct. 19-22, 1969.
21. Ballard, D.: "How to Operate an Amine Plant," Hydrocarbon Processing (April 1966) 137-144.
22. Perry, C.H.: "Activated Carbon Filtration of Amine and Glycol Solutions," Proceedings of the 1974 Gas Conditioning Conference, University of Oklahoma, Norman, Okla. (1974).
23. Pearson, M.J.: "Recent Developments in Claus Catalysts," Gas Processing/Canada (May-June 1973) 22-36.
24. Dalla Lana, Dr. I.G.: "Chemically Modified Catalysts Found Superior," Gas Processing/Canada (Jan.-Feb. 1971) 36-43.
25. Martin, J.E.: "Plant Peaks Sulfur Recovery," Hydrocarbon Processing (April 1973) 135-137.

There are ways to smoother operation of sulfur plants

THE REALITIES of operating Claus sulfur-recovery plants have changed considerably during the past decade.

Because of more-severe air-quality regulations and economics, the old hit-or-miss operating techniques are yielding to the necessities of better plant performance.

This article offers Claus-plant operators some guides and information for achieving better performance.

But due to the diversity of plant designs and the differing situations under which various plants must operate, much of the information presented here will, of necessity, be somewhat generalized.

Since not all data will be universally applicable, the decision to employ any or all of the techniques and procedures presented must be the responsibility of the individual operator based on his knowledge of the engineering and design of the specific plant in question.

Process review. The vast majority of sulfur compounds removed from worldwide natural-gas and petroleum production are ultimately converted to elemental sulfur in the modified Claus process by the reaction:

$$2H_2S + SO_2 \leftrightarrows 3/n\ S_n + 2H_2O$$

In the typical modified Claus sulfur plant some 60% to 70% of total sulfur yield is achieved at the reaction furnace. In this first thermal step, one-third of the total plant H_2S is burned

Based on paper presented at Gas Conditioning Conference, March 1976, Norman, Okla.

The author . . .

W. S. Norman

W. S. Norman is manager of technical services for catalyst-related product lines at Kaiser Chemicals in Baton Rouge, La. He received a BS degree in geology from Louisiana State University in 1972.

with the stoichiometric amount of air at temperatures between 1,800° and 2,200° F. (1,000° - 1,200° C.) to produce SO_2 in the H_2S: SO_2 ratio of 2:1.

After passing through a condenser to remove the elemental sulfur produced at the reaction furnace, the gas stream is reheated and proceeds to a series of two or three adiabatic converters operated at temperatures just above the sulfur dew point where further reaction occurs in the presence of an alumina catalyst.

At each interstage position condensers remove sulfur from the process stream and the gases are reheated before passing to the next converter. Overall plant yield is typically in excess of 90%.

Factors affecting catalyst activity. Simply stated, the primary goals of the Claus-plant operator are maximum continuous operating time with the minimum-possible sulfur-emission level. The condition and activity of the catalyst beds are the final determinants of plant performance relative to these goals. Tailoring plant operation to the benefit of the catalyst can, therefore, go a long way toward improving plant performance.

How is this done? To understand and control the affects of plant operation on catalyst condition, the mechanisms of catalyst aging must first be understood.

All Claus catalysts in wide use today are primarily activated alumina catalysts. They are, therefore, dependent upon large numbers of catalytically active surface sites to adsorb and bring together the reactants (H_2S and SO_2) in an orientation favorable for subsequent reaction to yield elemental sulfur.

Generally, the greater the surface area of the catalyst the greater the number of active surface sites and therefore the greater the catalyst's approach to theoretically maximum activity. While a fresh catalyst will contain a comfortable excess of surface area over that necessary to assure theoretical activity, any considerable loss of surface area or any blocking of surface-active sites will cause loss of efficiency in the catalyst.

The response of an alumina catalyst, either activated bauxite or pure activated alumina, to the temperatures and contaminants typical of the modified Claus process is the slow degradation of surface area.

Since the loss of surface area is a permanent phenomenon, this slow decline can be thought of as the ideal or minimum decay curve of a Claus catalyst. Such a decay curve, however, is not typical of actual performance. Most catalyst beds experience sharp, catastrophic losses in surface area due to relatively brief excursions to temperatures in excess of 1,000° F. This usually occurs during

Fig. 1

Effect of sulfate on H_2S and COS conversion

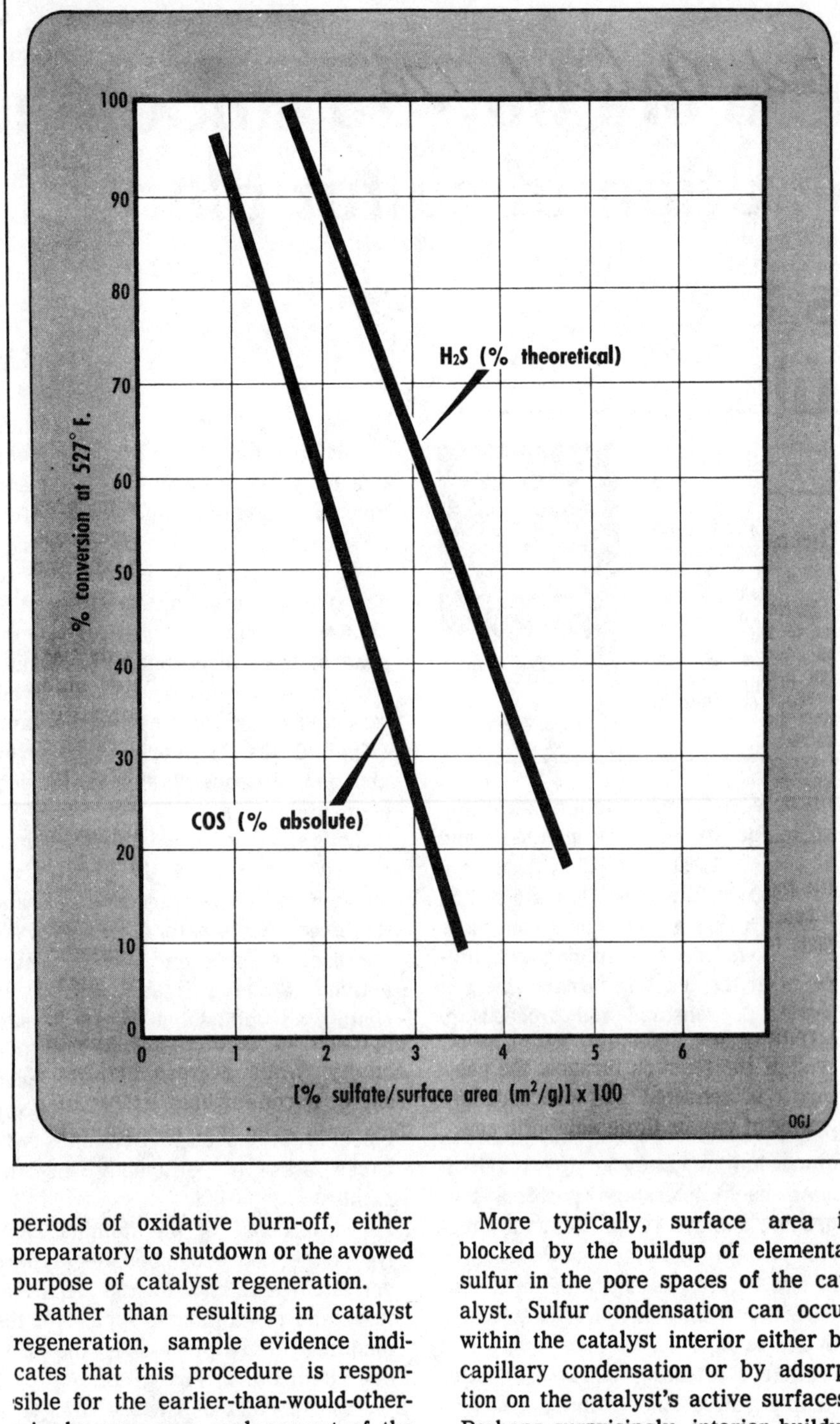

periods of oxidative burn-off, either preparatory to shutdown or the avowed purpose of catalyst regeneration.

Rather than resulting in catalyst regeneration, sample evidence indicates that this procedure is responsible for the earlier-than-would-otherwise-be-necessary replacement of the majority of Claus catalyst.

Active catalytic sites can be physically blocked by the buildup of surface contaminants which prevent the access of Claus gases to the catalyst surface. Heavy buildup may be due either to the carry-over of higher-molecular-weight hydrocarbons such as amine, or to the cracking and subsequent polymerization of lighter hydrocarbon fractions in the gas stream.

More typically, surface area is blocked by the buildup of elemental sulfur in the pore spaces of the catalyst. Sulfur condensation can occur within the catalyst interior either by capillary condensation or by adsorption on the catalyst's active surfaces. Perhaps surprisingly, interior buildup of sulfur will take place even at temperatures significantly above the sulfur dew point. Physical blocking of catalytic surfaces can result in significant, but usually reversible, losses in activity.

Individual active surface sites are blocked directly by the formation of sulfate on the catalytic surface. Although adequate surface may be present with no obstructions to prevent the free access of Claus gases to that surface, many of the individual catalytic sites become occupied by sulfate and are, therefore, unavailable to promote the necessary reactions.[2,5,7]

The surface sulfation of the catalyst may occur in several ways. Most important during normal plant operation is probably the chemisorption of SO_2 on the oxide ions or hydroxyl groups of the catalyst surface.[5,7] Any SO_3 accessible to the surface might also be expected to adsorb on an oxide site.[7]

Of major consideration and to some degree preventable is the direct oxidation of elemental sulfur previously adsorbed on the catalyst surface.

Whatever the cause, the buildup of sulfate is an equilibrium phenomenon and will continue until equilibration is reached between sulfate formation by such components as SO_2, SO_3, and O_2 and the reducing effects of H_2S. Low bed temperatures, high SO_2 and higher levels of SO_3, all favor increased catalyst sulfation.

Primary catalyst activity for a sample of purely physical catalyst (i.e., activated bauxite or pure activated alumina) uninfluenced by any physical obstruction to the catalyst interior becomes a function of the ratio of percent sulfate (by weight) to surface area (m^2/g). Total surface area is proportional to the total number of catalytic sites which might be potentially available to promote the Claus reaction.

The percent sulfate on the sample indicates the number of those sites deactivated by bound sulfate. Fig. 1 illustrates the response of H_2S and COS-conversion efficiencies to an increasing sulfate to surface-area ratio.[8] It can be seen that either an increase in sulfate level or a decrease in surface area has a detrimental effect on catalyst activity.

To summarize, the efficiency of a typical Claus catalyst is dependent on an adequate supply of active surface sites.

The total supply of surface sites available to promote the reaction is, in turn, dependent on three factors: sufficient surface area to accommodate those surface sites; free access of the Claus gases to the interior surface area; and the number of surface sites occupied by sulfate.

If we assume that the slow, long-term loss of surface area represents the ultimate and unavoidable decay of catalyst activity, then plant operation for maximum catalyst life and activity can be reduced to three maxims:

1. Prevent the catastrophic loss of surface area due to excessive temperatures.

2. Prevent the blocking of surface area due to the buildup of heavy, glossy hydrocarbons or elemental sulfur.

3. Prevent the deactivation of individual catalytic sites due to the buildup of sulfate.

Operation for increased catalyst life. To assure long-term catalyst life the primary objective must be the preservation of sufficient surface area to maintain adequate catalytic activity.

Ironically, the typical Claus-plant "regeneration" is the most-common culprit of excessive surface-area loss. Various aspects of the regeneration burn-off, its inherent problems, and suggested alternative procedures are discussed fully later. In the meantime, it is sufficient to state that the risks far outweigh any possible benefits. The use of oxidative burnoff procedures should be abandoned wherever possible.

In the discussion of those factors which block access of the Claus gases to the interior surface area of the catalyst, carbon fouling does not typically merit the importance placed on it. In most cases the carbon is the powdery or sooty variety formed by the substoichiometric combustion of light hydrocarbons.

Carbonaceous material of this type has considerable surface area in its own right and is of little or no consequence to the activity of the catalyst. Plant-performance problems may be encountered, however, should such carbon deposits reach proportions sufficient to cause excessive pressure drops or sulfur contamination.

Should the buildup of carbonaceous material be of the glossy, polymeric type derived from the carry-over of hydrocarbons, considerable catalyst deactivation can result due to the blinding off of surface area. Such deposits, however, are usually confined to the top 6 in. or so of the catalyst bed and thus, result in little or no overall loss of efficiency.

Rather than risk the unavoidable danger to both plant and catalyst posed by a burn-off, it makes better sense to simply remove and replace this top layer of catalyst at the first available opportunity.

In some instances glossy carbonaceous deposits may extend throughout the bed. Such a rare case, which calls for immediate replacement of the entire catalyst bed or burn-off of the carbon fouling, is the only good reason to conduct a regenerative burn-off. Care should be taken, however, first to strip the majority of elemental sulfur from the bed.

All Claus catalysts now in use are either activated aluminas of varying degrees of purity or activated aluminas impregnated with a promoting agent. All such aluminas are fairly strong adsorbents and, in the manner of a desiccant's removal of water vapor, will adsorb elemental sulfur on their active surface area at temperatures above the sulfur dew point.

The phenomenon of capillary condensation also occurs above the dew point and will result in the accumulation of elemental sulfur in the pore spaces of catalyst.

The combined effect of both processes is the buildup of elemental sulfur within the catalyst particle. Since the reaction gases are blocked from access to the catalyst's active surfaces, the catalyst looses activity.

The accumulation of internal sulfur typically begins in the top few inches of the catalyst bed in that region where the exothermic Claus reaction has as yet had no appreciable affect on catalyst or gas temperature. As the sulfur blockage begins to occur, the catalyst loses activity and the reaction is forced slightly deeper into the bed to the more-sulfur-free catalyst. As the reaction moves down into the bed so does the zone of temperature rise and sulfur buildup closely follows.

In this manner a zone of intragranular sulfur condensation gradually moves deeper into the bed, traveling as a wave front, and the catalyst bed progressively loses activity. Even the best run Claus units will suffer this chronic problem to some degree.

The situation is easily corrected, however, by periodic heat soaking of the catalyst beds. Simply stated, this procedure calls for temporarily increasing the bed inlet temperature of the reactor which is to be purged of sulfur. It is normally recommended that the procedure be followed on one reactor at a time beginning with the first.

Once a bed has stabilized at an inlet temperature of 25° to 50° F. (14° to 28° C.) higher than normal, 24 to 36 hr is generally sufficient to bring catalyst sulfur loading down to the lower levels in equilibrium with the elevated temperature.

After this is accomplished, the bed inlet temperature should be brought slowly back down to standard running conditions.

A significant increase in the ΔT across the catalyst bed at the lower temperature will signal a return to greater catalyst activity due to the successful removal of the elemental sulfur which had been blocking catalyst surface area.

Once precautions have been taken to minimize surface-area loss or blockage, maintaining catalyst activity becomes a task of preventing excessive catalyst sulfation. It was stated previously that the buildup of sulfate on the catalyst surface is an equilibrium phenomenon. Careful laboratory experimentation has shown that under typical Claus-stream conditions and in the absence of oxygen, sulfation in the order of 2 wt % will occur on an activated alumina catalyst.[7, 8]

It is probable that this minimum sulfation occurs by the reaction of chemisorbed SO_2 with the oxide sites of catalyst surface or through dehydroxylation of the catalyst surface.[5, 7] Whatever the method of formation, sulfate can be reduced from the catalyst by H_2S.[5, 8]

Thus unavoidable sulfation rapidly occurs and seeks an equilibrium level in response to the concentration of the Claus gases and the temperatures in the particular reactor involved. Sulfation in excess of the normally minimal equilibrium amount can cause severe catalyst deactivation by tying up a considerable proportion of the active surface sites.

The principal cause of such excessive sulfation is free oxygen at even-ppm levels. In the presence of free oxygen, sulfate formation is increased due to the oxidation of chemisorbed SO_2, and/or sulfur. The presence of small-ppm levels of SO_3 can also cause a marked increase in sulfate levels.

During normal plant operation excessive sulfate formation is usually caused by poor combustion either at the reaction furnace or at reheat burners. Faulty burner tips, poor mixing, or higher than stoichiometric air rates can all result in the breakthrough of oxygen or even minute amounts of SO_3 to the catalyst bed.

At every opportunity burner tips should be inspected and the integrity of mixing baffles should be checked. Air and fuel rates to each burner should be monitored frequently and rate indicators should be calibrated as often as possible.

Attention to such details can dramatically improve a plant's sulfation characteristics.

The use of burn-off procedures can also cause considerable changes in catalyst sulfate levels. The free oxygen present during the burn will result in extreme sulfation in the manner described previously. This is particularly true at the moderately high temperatures (750° to 1,000° F.—400° to 540° C.) experienced in the course of a well-controlled burn.

At temperatures much above this, sulfate is driven from the catalyst surface so that a mildly unsuccessful burn (1,300° to 1,500° F.—700° to 815° C.) can reduce sulfate levels to only a percent or 2 while a severe runaway above 1,600° F. (870° C.) can result in sulfate levels of only a few tenths of a percent.

The fallacy of the regenerative burn-off. Three guidelines for Claus-plant operation have now been discussed as being critical to the maintenance of good catalyst activity. The first of these, avoiding oxidative burn-off conditions, merits considerable additional discussion for several reasons. The oxidative burn-off, though commonly referred to as a regeneration, poses an extreme threat of catastrophic damage to both catalyst and plant.

Furthermore, abandoning this regeneration procedure will require considerable changes in the shutdown procedures followed by many operators.

As a method of removing carbon fouling and, hopefully, increasing catalyst activity, the burn-off is typically misapplied and, in most cases, counterproductive. The result normally falls far short of the intended goal of regeneration.

Of the many hundreds of catalyst samples taken following a burn-off and analyzed by our laboratories not one can be said to have been more than temporarily improved by the procedure.

Fig. 2

H$_2$S conversion vs. surface area and sulfate

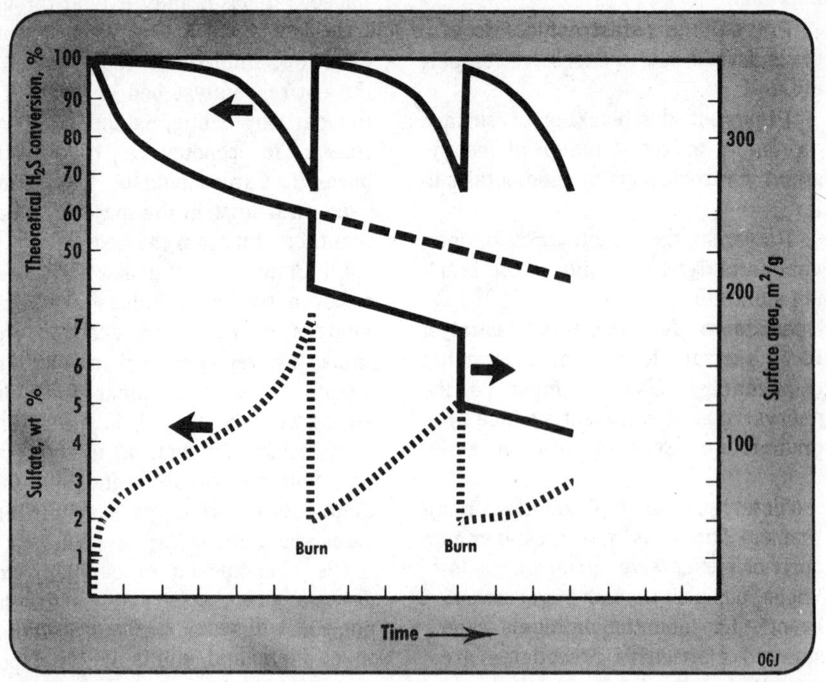

On the contrary about half of the catalyst's remaining useful surface area is lost due to high-temperature sintering.

In the event that excessive temperatures are avoided, the likely result will be heavy catalyst sulfation.

The underlying problems with the burn-off procedure are the already discussed facts that the catalyst bed normally contains a considerable quantity of sorbed sulfur and sulfur dioxide.

The burn, therefore, consists primarily of the oxidation of these two components.

The combustion of sulfur in particular can liberate vast amounts of heat very quickly and is inherently difficult to control. The thermal capacity and insulating properties of the alumina (specific heat +0.22) catalyst bed further compound the problem by:

- Retarding the response time to temperature instrumentation which would indicate a runaway condition.
- Allowing isolated pockets of extreme temperatures to develop undetected within the bed.
- By having a very slow response to efforts to quench elevated temperatures.

Under such circumstances, it is not remarkable that runaway conditions often develop, resulting in bed temperatures above 2,000°F., (1,100° C.), the complete destruction of the catalyst bed, and occasionally the slagging of vessel internals such as baffle plates and bed supports.

Even in a very-well-controlled burn in which bed exit temperatures do not exceed 900° to 1,000° F., (480° to 540° C.), there is no guarantee that high-temperature damage has not been done to the catalyst in isolated pockets. In such a case, the combustion of sorbed sulfur and sulfur dioxide will result in increased catalyst sulfation not only on the bed directly involved but also on the cooler bed downstream.

If the burn-off of a catalyst bed poses so many dangers, why is it popularly called a regeneration, and why do many operators see a short-term increase in catalyst activity?

The removal of hydrocarbon deposits and elemental sulfur from the bed provide a partial answer, but as discussed earlier, hydrocarbon deposits are not usually that great a factor and there are other methods available for the removal of sulfur.

Fig. 2 plots the percentage of theoretical H_2S conversion as a function of catalyst-surface area and sulfate content in a hypothetical plant using an activated alumina catalyst and practicing periodic regeneration of the catalyst. The normal decline of surface area with time is illustrated by the extension of the dashed line. The average effect of each burnoff (1,300° to 1,500° F.—700° to 815° C.), is a precipitious drop in surface area below the "normal" decline rate due to elevated temperatures which also drive off most of the sulfate.

The net effect is to lower the sulfate-to-surface-area ratio sufficiently to regain most of the original activity. Surface area has been lowered, though, so that activity will become critical at lower sulfate levels. With each successive regeneration the problem compounds itself and useful run times become shorter and shorter. While the short-term result of each regeneration may be a dramatic increase in activity, the long-term affect is a drastic decrease in catalyst life and more-frequent marginal plant efficiency.

It is apparent, however, that if there were some way to decrease sulfate levels without damaging surface area then there would be a true method of Claus-catalyst regeneration available. Such a procedure for catalyst rejuvenation will be introduced in this article.

Procedures. Alternative procedures for plant shutdown without the necessity of burn-off must be presented. These procedures are not universally applicable to all plants and the decision to use any or all of them must be made by the operator based on the engineering and design of the plant in question. There are several cases involved:

- Case I — A shutdown of some duration in which access must be gained to the catalyst beds. The primary objective in this case is to remove elemental sulfur and SO_2 from the bed without burning:

1. If the shutdown is planned, begin raising bed inlet temperatures on the order of 25° to 50° F. (14° to 28° C.) about 36 to 48 hr before extinguishing the acid-gas flame.
2. As the flow of acid gas to the reaction furnace ceases, begin burning fuel gas at stoichiometric or slightly substoichiometric oxygen levels. This usually requires the injection of steam into the flame both to prevent damage to reaction-furnace refractories and to prevent the formation of powdery carbon.
3. Allow fuel-gas combustion gases (oxygen free) to flow over the beds at 600° to 700° F. (315° to 370° C.).
4. Continue the fuel-gas exhaust sweep for 12 to 24 hr after the condensers run free of sulfur.
5. Begin cutting back on sweep temperature.
6. Once safely below the sulfur-combustion temperature begin cracking in small amounts of air. (Oxygen content should not exceed about 1-2% of total plant flow.) It is at this point that the greatest danger of fire exists. Bed temperatures must be watched closely for any sign of a positive ΔT. Be prepared immediately to cut back on oxygen content should this ΔT show indications of a runaway situation developing.
7. Continue cooling to ambient temperature by cracking in more and more air and cutting back on the fuel gas. Continue watching for any sign of an increasing ΔT across the beds.
8. Once at ambient temperature the beds should be free of sulfur and safe to enter. Follow accepted safety procedures.

- Case II—A shutdown of extended duration in which access to the plant interior is not required:

1. If the shutdown is planned, begin raising bed inlet temperatures on the order of 25° to 50° F. (14° to 28° C.) about 36 to 48 hr prior.
2. A short-duration fuel-gas exhaust sweep similar to that outlined above is advisable but not necessary if there is no reason to suspect unusual sulfur condensation on the beds.
3. Provide a dry, inert (nitrogen, carbon dioxide) purge sufficient to clear the plant of water vapor and Claus gases.
4. Shut in the plant and place under an inert-gas blanket. As long as no access is required, a plant can be maintained in this state indefinitely. Some means of pressure equalization may be necessary.

- Case III — A shutdown of short duration in which access to the plant interior is not required:

1. If the shutdown is planned, begin raising bed inlet temperatures on the order of 25° to 50° F. (14° to 28° C.) about 36 to 48 hr prior.
2. If possible, the plant can be simply shut in so that there is no access of air. The heat contained in the catalyst beds is sufficient to keep them at ready standby temperatures for several days. Some means of pressure equalization with ambient may be necessary.
3. Otherwise, a low-level fuel-gas exhaust sweep sufficient to maintain bed temperatures at about 400° F. will keep beds in ready standby condition.

Catalyst rejuvenation. The previous discussion outlined possible procedures to prevent the catastrophic and unnecessary loss of catalyst-surface area due to high temperatures. So long as normal surface-area decline does not take catalyst-surface area below certain minimums ($\approx$125 m^2/g for activated alumina), the relationship between surface area and sulfate content shown in Fig. 1 would indicate that the catalyst is still potentially active, providing sulfate levels are low enough.

Logic, therefore, dictates that a Claus catalyst could be rejuvenated by a procedure which lowered sulfate levels without causing any loss in surface area.

It has been stated that the buildup of sulfate occurs as an equilibrium phenomenon, which, of course, implies that sulfate formation is reversable under Claus-plant conditions. The mechanism involved is the reduction of sulfates by H_2S to form elemental sulfur and water vapor. As catalyst temperature is raised above normal operating conditions in the presence of H_2S, sulfate levels are reduced below previous equilibrium levels. Increased levels of H_2S further decrease sulfate levels.

The rate and degree of sulfate reduction are dependent upon temperature and H_2S partial pressure and upon the original temperature of sulfate formation,[5] but, in general, increase with increasing H_2S content and catalyst temperature. A practical maximum temperature should be considered to be about 750° F.

This knowledge can be applied to provide a method for Claus-catalyst rejuvenation during on-stream operation. The procedure, which must be applied with discretion based on the individual characteristics of the plant in question, is as follows:

1. Raise bed inlet temperatures sufficiently to achieve 650° to 700° F. (340° to 370° C.) in the first bed and at least 50° F. (28° C.) above normal in the second and third catalyst beds. It should be noted that while the degree to which the procedure is effective is dependent upon the amount of temperature increase realized, any significant increase, even if

only 20° F. (11° C.), is of some benefit in the second and third converters.

2. Allow 12 to 24 hr for the removal of condensed sulfur from the catalyst beds.

3. Go off ratio at the reaction furnace to an $H_2S:SO_2$ ratio of at least 2.5:1 (ratios as high as 10:1 have been reported as successful but are not necessarily advised) and maintain at least that ratio through the entire plant for 24 to 36 hr. Emission regulations or tail-gas unit limitations may require a rate cutback during this period of off-ratio operation.

4. Return to normal ratio conditions followed by a return to normal temperature conditions.

5. Increased catalyst activity will be noted by increased ΔT's and greater overall sulfur recovery.

The rejuvenation procedure, in one variation or another, is rapidly gaining favor in North America. It is an apparently effective procedure which can be applied with relatively little inconvenience to overall plant operations.

While further refinement will undoubtedly occur, results are already impressive. One West Coast refiner, for example, despite being unable to reach the full recommended temperatures, and only being able to achieve a 2.4:1 ratio, realized a 50% reduction in sulfur emissions from the Claus plant after following the procedure.

Catalyst maintenance. During the course of this article two catalyst-maintenance procedures have been discussed: (1) periodic heat soaking to desorb elemental sulfur; and (2) on-stream rejuvenation for sulfate reduction.

It is recommended that these procedures be incorporated into a regularly scheduled program of catalyst maintenance. The interval between heat soakings is best determined by setting bed inlet temperatures some 15° to 20° F. (8° to 11° C.) above the sulfur dew point immediately following an initial heat soak or rejuvenation.

If a record is made of the ΔT across each bed at this time, any appreciable drop in ΔT (although dependent on the requirements faced by an individual plant, roughly 10°-15° F. in the first bed, 5° - 10° F. in the second, and 3° - 5° F. in the third) will signal the necessity of the next heat soak.

If the interval is less than 2 weeks, the bed inlet temperature is probably too low. If the interval is greater than 1 month the bed inlet temperature is probably too high for maximum equilibrium efficiency.

Typically, then, such a program should consist of a heat soak applied at an interval of between 2 weeks and 1 month with the rejuvenation procedure substituted for every third or fourth heat soak depending on the requirements of the specific plant.

To be more than nominally successful, no program of catalyst maintenance can function in the absence of sound plant operation and the attention to catalyst condition discussed earlier.

Such preventive maintenance however, coupled with awareness of the three maxims for catalyst-oriented plant operation, and a regularly scheduled program of catalyst maintenance will assure peak catalyst activity throughout a significantly improved catalyst lifetime.

Additional benefits should accrue due to smoother plant operation with less ups and downs and consistently high sulfur-recovery yields.

Acknowledgment

The author gratefully acknowledges the support provided by his colleagues at Kaiser Aluminum & Chemical Corp., in particular that of M. J. Pearson and his associates at the Kaiser Center for Technology, whose theoretical work provides the foundation for much of the information presented here.

References

1. Barry, C. B., "Reduce Claus Sulfur Emission," Hyrocarbon Processing, April 1972, p. 102.
2. Dalla Lana, Dr. I. G., "Some Aspects of Claus Process Catalytic Technology," Gas Processing Canada, March-April 1973, p. 20.
3. Fischer, H., "Burner/Firebox Design Improves Sulfur Recovery," Hydrocarbon Processing, October 1974, p. 125.
4. Goar, B. G., "Feed Gas Impurities Can Create Claus Plant Problems," Proceedings Gas Conditioning Conference, University of Oklahoma, 1974.
5. Graulier, M., and Papee, D., "Optimization of a Claus Unit," Paper presented to Canadian Gas Processors Association, Calgary, Alta., March 1974.
6. Parnell, D. C., "Claus Sulfur Recovery Unit Startups," Chemical Engineering Process, August 1973, p. 89.
7. Pearson, M. J., "Recent Developments in Claus Catalysts," Hydrocarbon Processing, February 1973, p. 81.
8. Pearson, M. J., Unpublished communications.

Modified sulfur-recovery process meets air-quality regulations

ROBERT D. WRIGHT*
JERRY W. STRANGE
El Paso Natural Gas Co.
El Paso

IN MID-1974, the New Mexico Environmental Improvement Agency (EIA) adopted regulations that required installation of a sulfur-recovery plant at the El Paso Natural Gas Jal 4 gas-processing complex in Southeast New Mexico. The regulations required 88% minimum sulfur recovery.

Bids were solicited as a first step in achieving sulfur-emission compliance with New Mexico air-quality control regulations. A bid by Dresser Engineering Co. was accepted for design and construction of a sulfur-recovery plant.

The process selected was an Amoco-licensed modified Claus process to recover sulfur from the low H_2S acid gas from the existing DEG-MEA natural-gas-treating plant.

Formal design of the sulfur plant was started in late 1974. Construction was started about 10 months later. With introduction of acid gas from the treating plant, start-up operations were completed within 18 months after awarding the contract.

Results. Start-up and initial operation of the Jal 4 sulfur plant were generally good and were completed with relatively few serious problems. The basic plant operation has been quite stable with minimum unscheduled shutdowns.

Performance and operating experience of the Dowtherm preheat system have been quite good. Problems en-

*Now affiliated with Exxon Production Research Co., Houston.

Paper titled "Innovative Process for Claus Sulfur Recovery," presented at 27th Annual Gas Conditioning Conference, University of Oklahoma, Norman, Okla., Mar. 7-9, 1977.

Fig. 1

Process flow sheet—Jal 4 sulfur plant

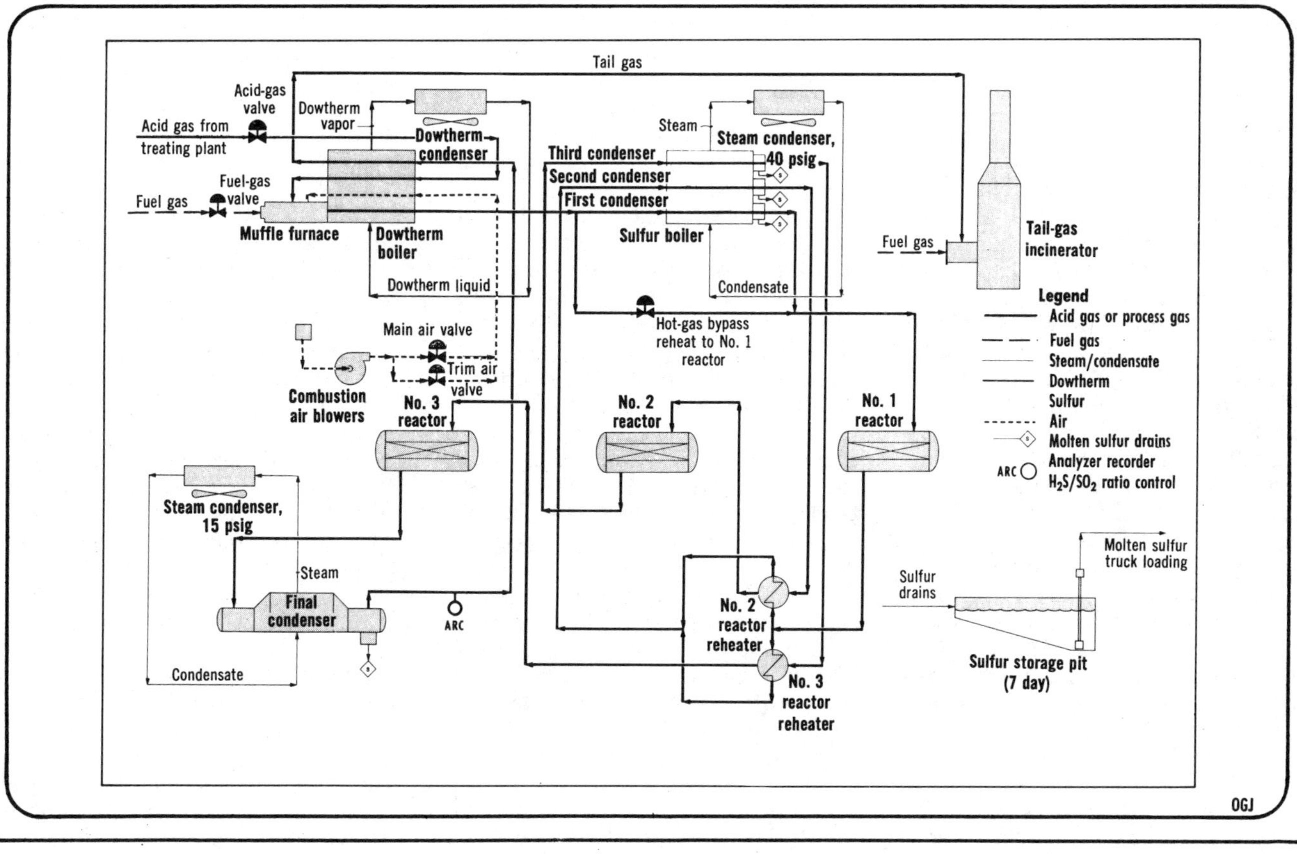

OVERVIEW of El Paso's Jal 4 sulfur plant in Southeast New Mexico (Fig. 2).

countered were primarily due to significant changes to operating conditions of the Claus plant.

Following start-up, plant performance tests were conducted to determine optimum plant operating conditions, and to identify several needed design changes. As a result of the test program, changes were made to the design of the acid-gas burner and the Dowtherm condenser. Plant operating conditions were also defined to minimize formation of carbon sulfides in the furnace and to maximize hydrolysis of carbon sulfides to H_2S in the No. 1 reactor.

Plant performance tests have successfully demonstrated capability of the plant to comply with New Mexico environmental regulations. A 60-hr performance test, using improved operating conditions developed from earlier tests, showed an 89.9% sulfur recovery which was well within the minimum required 88% recovery.

The process. The sulfur-recovery process incorporates several unique features. The sulfur plant is designed for 4.51 MMscfd of acid-gas feed at 14.4% H_2S and a design sulfur production of 25.3 short tons/day.

Acid-gas feed to the sulfur plant originates from an existing DEG-MEA treating plant which treats sour natural gas downstream of a lean-oil absorption gasoline plant. The sulfur-recovery process uses a Dowtherm heat-recovery system.

Heat-recovery exchangers, located in the vapor space of the Dowtherm boiler, preheat the acid-gas feed and combustion air to 625° F. by condensing Dowtherm from the vaporizer section. A similar exchanger reheats the tail gas to 625° F. by condensing Dowtherm.

The preheated acid gas and part of the preheated air are introduced to the furnace through an acid-gas burner. A small amount of fuel gas is combined with the balance of the air for combustion in a separate fuel-gas burner.

Combined combustion gases from the acid-gas and fuel-gas burners exit the furnace at 2,150° F. and are cooled to 1,000° F. in the Dowtherm vaporizer where Dowtherm is vaporized to recover waste heat from the furnace effluent stream.

An air-cooled condenser cools and condenses excess Dowtherm vapor which is returned to the Dowtherm vaporizer. Fig. 1 shows the simplified process flow sheet. Fig. 2 shows the Jal 4 sulfur plant.

Part of the hot process stream from the Dowtherm vaporizer is cooled in the first pass of the sulfur boiler by generating 40-psig steam on the shell side of the boiler. The cooling step condenses and removes part of the elemental sulfur from the process stream.

The balance of the hot stream from the vaporizer is bypassed around the sulfur boiler through a hot-gas bypass valve to reheat the total process stream before the first catalytic reactor. The following primary reactions occur in the first catalytic reactor:

$$2H_2S + SO_2 \rightarrow (3/e)S_e + 2H_2O \quad (1)$$

$$COS + H_2O \rightarrow CO_2 + H_2S \quad (2)$$

$$CS_2 + 2H_2O \rightarrow CO_2 + 2H_2S \quad (3)$$

where e = 2, 6, and 8.

The process stream enters the first reactor at 600°-650° F. and exits at 700°-750° F. The high operating temperatures in the first reactor promote hydrolysis reactions 2 and 3. These convert much of the carbon sulfide present to H_2S for Claus conversion and recovery of sulfur in the second and third reactors.

The No. 1 reactor outlet stream is cooled by preheating the process streams to the second and third reactors by gas-gas exchange. The process stream is then cooled in the second pass of the sulfur boiler where additional sulfur is condensed and removed from the process by generating 40-psig steam.

Gas exiting the second condenser is then reheated to 400°-440° F. in the No. 2 reactor reheater and introduced to the second reactor where additional sulfur is produced primarily by reaction 1. The process stream exits the second reactor and is cooled in the third pass of the sulfur boiler where additional sulfur is condensed and removed from the process.

Gas exiting the third pass of the sulfur boiler is reheated to 370°-400° F. in the No. 3 reactor reheater and introduced to the third reactor where final sulfur conversion occurs primarily by reaction 1. The stream exits the third reactor and is cooled to 250° F. in the final condenser which generates 18-psig steam on the shell side. A small amount of sulfur is condensed and removed from the outlet of the final condenser.

Tail gas exits the final condenser at 250° F., and is reheated to 625° F. in the heat-recovery section of the Dowtherm boiler. Reheated tail gas is then introduced to the incinerator, combined with air and fuel gas, and combusted to convert all sulfur compounds in the tail gas to SO_2 before release from the incinerator stack.

Excess 40-psig and 18-psig steam from the sulfur boiler and final condenser is cooled and condensed in two steam coils, combined into a single air-cooled condenser unit. Steam condensate is returned to the sulfur boiler and final condenser by gravity flow.

Product liquid elemental sulfur is collected by gravity flow into a 7-day concrete storage pit. A vertical centrifugal sump pump delivers product sulfur into tank trucks for shipping. A sulfur overflow area is provided for emergency use.

Unique process features

Several features of the Jal 4 sulfur-recovery process are of particular interest due to the very lean acid gas (12-13% actual H_2S concentration) and high hydrocarbon content (4-5%) of the acid-gas feed. High sulfur-recovery levels from lean acid gases have been a difficult application for Claus sulfur plants due to the extremely low acid-gas heating value, difficulties that result from low flame temperature, unstable combustion, and dilution of process gas.

Dowtherm heat-recovery system. The Dowtherm heat-recovery system consists of five basic components:

1. Dowtherm waste-heat-recovery vaporizer.
2. Acid-gas preheat exchanger.
3. Air-preheat exchanger.
4. Tail-gas-reheat exchanger.
5. Air-cooled excess Dowtherm condenser.

The system uses Dowtherm A as a heat-transfer medium. The Dowtherm vaporizer cools the furnace effluent stream from 2,150° F. to 1,000° F. by vaporizing Dowtherm. Dowtherm vapor from the vaporizer is partially condensed in the preheat and reheat exchangers and simultaneously heats the acid gas, combustion air, and the tail gas.

Any excess Dowtherm vapor from the preheat and reheat exchangers is condensed in an air-cooled exchanger. Condensed Dowtherm is returned from the condenser to the vaporizer by gravity flow.

The Dowtherm system permits preheating the acid gas and combustion air to much higher temperatures than are possible with a steam waste-heat-recovery system. For this application, the higher air and acid-gas preheat temperatures enable selection of the "straight-through" Claus process which has many operating advantages compared with conventional processes for handling lean acid-gas streams.

Reheat methods. The Jal 4 sulfur plant combines two methods to reheat the process stream before catalytic reaction.

The first reactor feed is reheated by a hot-gas bypass valve. It bypasses part of the hot (1,000° F.) Dowtherm vaporizer outlet gas around the first pass of the sulfur boiler on temperature control. The hot-gas bypass reheat to the first reactor provides the most-economical method of reheating the process stream and does not significantly reduce overall sulfur recovery since the hot gas is not bypassed around the reactors.

The process streams to the second and third reactors are reheated by gas-to-gas exchange with the hot gas exiting the first reactor. In both cases, reheat temperature is controlled by control valves which bypass part of the cold reactor feed stream around the reheat exchanger on temperature control.

Gas-gas exchange provides maximum sulfur recovery which was required for the very lean process stream. Use of gas-gas exchange has also proven an excellent reheat method from the standpoint of simplicity of controls for start-up and operation.

Start-up and operating experience

After initial sulfur-plant start-up, several design modifications and changes to certain key operating conditions were required to comply with sulfur emission regulations.

Start-up and operation. Start-up progressed with relatively few serious problems. Before lighting the muffle furnace, the Dowtherm system and both steam boilers were chemically cleaned rather than conducting a conventional boil-out procedure. The process train was heated by fuel-gas combustion for initial drying of refractory.

Fuel-gas heating continued until all condenser temperatures were above the sulfur solidification point (240° F.) and all reactor temperatures were near design levels. Following completion of heat-up and drying, the plant was shut down and restarted with acid-gas feed to the furnace.

Start-up procedure. The initial heat-up and dry-out of the sulfur plant were conducted with fuel gas and large amounts of excess air. Furnace temperature was carefully controlled by varying excess air during refractory drying and subsequent heat-up of the plant.

As soon as plant heat-up was completed, the plant was shut down and restarted with acid-gas feed. During the final start-up, air supplied to the muffle furnace was controlled to maintain the stoichiometric amount required for fuel-gas combustion plus an amount required for combustion of the hydrocarbons and one-third of the H_2S in the acid gas.

A Du Pont tail-gas photometric H_2S/SO_2 analyzer was put into service shortly after start-up and the trim airflow controlled to maintain an approximate 2-to-1 mole ratio of H_2S to SO_2 in the tail gas. Plant operation was allowed to stabilize with the muffle-furnace temperature and the reactor inlet temperatures set near design conditions.

Dowtherm system performance. During initial plant heat-up, several problems developed with the Dowtherm system. However, subsequent operation of the Dowtherm system has been generally satisfactory.

Leaks appeared at several hand holes and through valve packings. Asbestos valve packing in all the Dowtherm-system valves was replaced with aluminum foil packing selected for high-temperature Dowtherm service.

Through the heat-up, periodic boiler-level adjustments were made to allow for the 43% volumetric expansion of Dowtherm during heating from ambient temperature to 650+° F. This was done by pressuring Dowtherm from the boiler to the Dowtherm storage tank.

It was found that, due to the low surface tension of Dowtherm, it is difficult to locate potential leaks by hydrostatic pressure testing of the Dowtherm system.

Following start-up, process gas leaving the Dowtherm boiler was about 150° F. below design level. This low vaporizer-outlet temperature caused the Dowtherm condenser to be severely loaded with additional Dowtherm vapor.

The original design of the Dowtherm condenser fan motors and louvers permitted exposure of the motors to internal hot-air recirculation resulting in damage to the

R. D. Wright

J. W. Strange

The authors . . .

Robert D. Wright at the time of preparation of the article was a senior process engineer in the Process Engineering Group for El Paso Natural Gas Co., El Paso. He is now affiliated with Exxon Production Research Co., Houston. He was previously a process engineer with Continental Oil Co. in the Process Engineering Department at Ponca City, Okla., and at the Billings, Mont., refinery. He holds a BS degree in chemical engineering from Tulsa University, Tulsa, and an ME degree in chemical engineering from Oklahoma University, Norman.

Jerry W. Strange is a senior process engineer with El Paso Natural Gas Co. in the Process Engineering Group, El Paso. He was previously with Dow Chemical Co. in operations at the Freeport, Tex., chemical plant. Strange holds BS and MS degrees in chemical engineering from New Mexico State University, Las Cruces.

motors. The condenser was modified by moving the louvers from above the condenser coils to the air space between the bottom of the cooling coils and the fans.

This modification effectively prevented exposure of fan motors to internal hot-air recirculation. The Dowtherm condenser is presently unable to provide full Dowtherm condensing duty due to higher-than-design muffle-furnace firebox temperatures. Another air-cooled condenser will be located parallel to the present Dowtherm condenser to provide additional condensing duty.

Sulfur color/recovery problems. Following start-up, initial sulfur production from condenser No. 1 was black, causing off-color sulfur product. It was later determined that black sulfur was caused by an acid-gas combustion problem in the furnace.

One-day performance tests conducted shortly after start-up showed a sulfur recovery of 75-80%. This was well below the minimum required 88% recovery. Performance tests were then initiated to identify conditions that influenced sulfur-recovery efficiency of the sulfur plant.

Tail-gas-analyzer inlet-air-control system. Tail-gas analysis is performed by a Du Pont ultraviolet photometric analyzer. Initially, analyzer windows became dirty very rapidly causing inaccurate measurement of tail-gas H_2S and SO_2 concentrations. The dirty analyzer windows were caused partly by products of poor combustion in the muffle furnace and partly by oil contamination of the instrument air used to purge the analyzer windows.

Modifications to the acid-gas burner design, change of muffle-furnace operating temperature, and repairs to the instrument-air compressor combined to eliminate this problem.

Muffle-furnace temperature-sensor problems. Three thermocouples were used for muffle-furnace temperature control, high-temperature alarm, and high-temperature shutdown, respectively. The thermocouples were originally located on the side of the furnace wall adjacent to the acid-gas burner.

It was determined that the thermocouples were located too close to the burners, causing widely varying temperature measurements. The thermowells were moved to the top of the furnace circular cross section and about 10 ft downstream of the acid-gas burner. The new locations produced consistent, reliable temperature measurements.

Acid-gas combustion problem. It was eventually determined that modifications to the acid-gas burner were needed. The burner design was modified to promote better mixing of the acid gas with combustion air, thus improving combustion. This burner modification resulted in immediate improvement of plant operation as indicated by substantially improved sulfur color, increased sulfur recovery, and reduced carbon sulfides in the tail gas.

Performance and compliance testing

Sulfur-recovery performance tests were conducted on the Jal 4 sulfur plant following the start-up and initial operating period.

Investigation of operating conditions. Sulfur-recovery efficiency was influenced by a number of plant operating conditions. Some important parameters were muffle-furnace temperature, inlet temperature to the three reactors, volumetric H_2S/SO_2 ratio in the tail gas, and outlet temperature from the final condenser.

Variations of muffle-furnace temperatures in conjunction with COS/CS_2 chromatographic analyses revealed that increasing furnace temperature increased concentration of COS and decreased concentration of CS_2, producing a net decrease in total sulfur as carbon sulfides. The muffle-furnace temperature was, therefore, increased to a practical limit of 2,150° F.

The No. 1 reactor inlet temperature was increased by steps until a large increase in temperature rise across the bed indicated conversion of carbon sulfides to H_2S. This evidence was confirmed by monitoring the concentration of COS and CS_2 entering and leaving reactor No. 1.

Maintaining the maximum temperature rise required that a certain minimum reactor inlet temperature be maintained to hydrolyze the carbon sulfides. The No. 1 reactor inlet temperature was then maintained slightly above the minimum.

The inlet temperatures to the second and third reactors were adjusted to maximize the Claus reaction while maintaining bed temperatures slightly above the sulfur dew point. Reactor No. 2 inlet temperature was set about 30° F. above design and reactor No. 3 inlet temperature about 7° F. above design.

Variation of the H_2S/SO_2 mole ratio in the tail gas between 4.0-to-1 and 0.5-to-1 indicated that the best operating ratio is the standard 2-to-1. Zero suppression allowance in the tail-gas-analyzer calibration (for elemental sulfur and COS content) provided improved accuracy of the measured H_2S and SO_2 concentrations. END

World's largest processing complex handles Orenburg gas

W. L. SCHEIRMAN, J. F. Pritchard & Co., Kansas City

A GIANT natural-gas processing complex is operating in the Orenburg gas field in the Ural area of the southeast part of the Russian Federal Soviet Republic. The Orenburg field is one of the world's major gas fields, although only fourth in size in the Soviet Union. Three larger fields are in western Siberia.

A sour-gas-condensate field, the Orenburg field was discovered in the 1960's. Reserves are more than 70 x 12^{12} cu ft.[1] The field measures 128 km long and 20 km wide.[2] It lies along the Ural River in the vicinity of the city of Orenburg in the Orenburg province of European Russia.

The central gas-treating complex includes three sour gas treating plants. In discussing this system, emphasis will be on the second of the three plants, designed and engineered by J. F. Pritchard Co. and COCEI. Financing of materials and equipment of foreign supply was done in France.

Fig. 1

Orenburg 2 central gas-treating plant

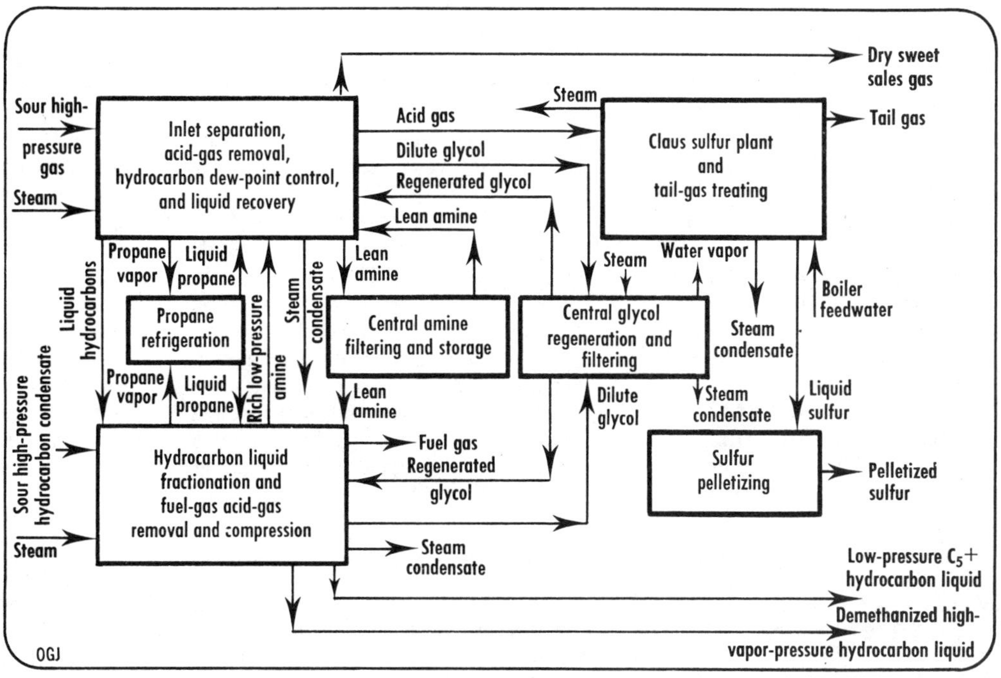

Table 1

Product capacities of Orenburg gas plants*

Product	Measurement unit	Plant number 1	2	3
Commercial gas	10^9 m³/year	14.6	14.6	14.6
Sulfur	1,000 metric tons/year	432.0	432.0	432.0
Light hydrocarbons	1,000 metric tons/year	267.0	267.0	267.0
Stable condensate	1,000 metric tons/year	937.0	543.0	66.0

*From Ref. 3.

Table 2

Design inlet gas composition-Orenburg Plant No. 2

Component	Mol % range
CH_4	82.26
C_2H_6	5.20-5.32
C_3H_8	1.80-2.16
iC_4H_{10}	0.70-0.84
nC_4H_{10}	0.34-0.41
iC_5H_{12}	0.30-0.36
nC_5H_{12}	0.30-0.36
C_6+	0.20-0.24
N_2	4.84-3.87
CO_2	1.40
H_2S	2.58-2.70
H_2O	0.10

The first of the central gas-processing plants for the sour gas-condensate field was put into operation in 1974.[3] Before construction of the central gas-treating complex, a number of gas-oil separation plants (GOSP) was built within the borders of the gas field.

Ultimately, there will be 10 of these plants of 2.5 and 5 x 10^9 m³/year (280 and 560 MMscfd) capacity.[3] These GOSP's separate the sour condensate, dehydrate the sour gas, and lower the hydrocarbon dewpoint of the gas to condition it for transmission to the central complex.

Separation. A conventional low-temperature separation flow scheme is used. Inlet aqueous-phase and high-pressure condensate liquids are each separated at wellhead flowing pressure. The high-pressure separated gas is then cooled in a cold gas-to-inlet-gas heat exchanger.

Wellhead flow-line pressure is reduced to central plant feeder-line pressure through a control valve using the Joule-Thomson effect to refrigerate the gas stream. This condenses additional hydrocarbon liquids and removes the water vapor.

Liquids are separated in a three-phase gas-liquid separator downstream of the Joule-Thomson valve. Cold gas is warmed as it cools the incoming high-pressure gas.

Several of these low-temperature separators in each GOSP supply a 720 mm (28.3 in.) feeder line at a pressure of 66 kg/cm² (939 psig). The gas flows to the central gas-treating complex about 29-35 km away.[3]

Hydrate inhibition. Methanol is used as a hydrate inhibitor in the wellheads and their flow lines to the GOSP's, in the gas cooler, and in the Joule-Thomson valve. At first, the GOSP's used methanol injection only in the high-pressure part of the system, with glycol injection ahead of the Joule-Thomson valve.

The methanol-water phase was separated at inlet pressure and the glycol-water phase at low-temperature separator pressure. Methanol vapor in the warm high-pressure gas stream was absorbed in the glycol and lost in the glycol-regeneration system.

The glycol-injection system was converted to methanol injection and more economical operation resulted with the methanol being separated at low-temperature separator pressure. Now only methanol injection is used. Methanol is stripped from the water and recirculated. It is distributed through a central high-pressure pumping system in each GOSP to the wellheads and the inlet gas coolers.

Corrosion. Some corrosion problems were met, and a corrosion-inhibitor program was put into effect. Corrosion inhibitor No. 1 (Table 3) is dissolved in the methanol going to the wellheads and the low-temperature separator.

This methanol, along with condensed water, is separated from the gas in the GOSP separators. To protect the feeder lines to the central gas-treating complex, every few months a large dose of methanol containing corrosion inhibitor No. 2 (Table 3) is pigged through the feeder line to a pig trap at the gas treating complex inlet. Part of the methanol solution is recovered at the pig traps.

Generally, no corrosion problems are now encountered in the vessels or pipelines, although there are still some problems with trim in those valves having high-pressure drops.

The unstable sour hydrocarbon condensate from the inlet and cold separators of the GOSP's flows through a separate 377 mm (14.8 in.) pipeline to the main gas-treating complex at a pressure of 60 kg/cm² (853 psig).[3]

Central gas-treating complex

The central gas-treating complex consists of three plants in parallel. Each one is designed for a gas capacity of 15.1x10^9 m³/year (normal flow of 1,683 MMscfd), but with some excess gas capacity. The plants are designed to handle varying amounts of separate inlet unstable condensate. Products are shown in Table 1 for each plant.

The first of these three plants was put into operation in 1974.[2] The third plant is now under construction.[4] The plants vary in design to incorporate different technology.

Second plant design. Design of the second plant was started in 1972. Construction started in 1973 and plant start-up took place on schedule in the autumn of 1975. Subsequent discussion will refer to this plant No. 2. Table 2 shows the design gas composition for Plant No. 2. It was necessary to use a nominal design gas composition for the central gas-treating plants since the gas composition varies somewhat in the Orenburg field.[3] The hydrogen sulfide content varies from 1.3 to 4.6%.[2]

Gas Plant No. 2 includes the following units:

1. Gas treating, including inlet liquid separation, carbon dioxide, and H_2S removal; and refrigeration for hydrocarbon dewpoint control and light hydrocarbon liquid recovery.
2. Claus sulfur recovery.
3. Sulfur plant tail gas treatment and incinerator.
4. Hydrocarbon liquid H_2S removal and fractionation, including acid gas removal from the low-pressure gas, which is used in the fuel systems.
5. Propane refrigeration for unit 1 above.
6. Central glycol solution regeneration, filtering, storage, and pumping.
7. Central amine solution storage, filtering, and pumping.
8. Sulfur pelletizing.
9. Utilities and other off-site facilities.

Based on paper titled "Sour Gas Treating at the World's Largest Natural Gas Processing Plant," presented at 27th Annual Gas Conditioning Conference, University of Oklahoma, Norman, Okla., March 1977.

Table 3

Inhibitors used in gas system

Inhibitor No.	Function	Where used	Type
1	Corrosion inhibitor	Wellheads, well-stream flow lines, field gas plants.	Persistent organic filming compound diluted in methanol.
2	Corrosion inhibitor	Separated sour-gas-gathering pipelines.	Complex mixture of tars and higher pyridine bases which are by-products in the manufacture of 2-methyl-5-ethyl pyridine. It is the still residue after distillation of the oil layer formed in the above synthesis. It can be produced as a condensation product of paraldehyde and ammonia.* Diluted in methanol for use.
3	Corrosion inhibitor	Diethanolamine system in central gas-treating plant.	Mixture of long-chain amine derivatives and light amines, with antifoaming tendencies in amine solutions.
4	Foam inhibitor	Antifoam in the diethanolamine system.	30% silicone emulsion, diluted.

*From Ref. 13

Any inlet liquids are separated and routed to the hydrocarbon fractionation facilities. Acid gas is removed from the natural gas by countercurrent contact with diethanolamine (DEA) solution in the amine contactors. Sweet gas flows to the hydrocarbon dewpoint-control system. There, it is cooled, first by gas-gas heat exchange, and then by propane refrigeration.

Condensed liquid hydrocarbons and glycol solution are separated in the low-temperature separator, producing a gas with water and hydrocarbon dewpoint adequate for the pipeline. As the cold gas exits, refrigeration is recovered from it in the gas-gas heat exchanger. Ethylene glycol solution in the gas-treating system is regenerated with steam heat. Acid gas from the DEA regenerators goes to the Claus sulfur-recovery units.

Acid-gas feed to the Claus units has a favorable H_2S/CO_2 ratio for good Claus-plant operation. Part of the H_2S is converted to sulfur dioxide (SO_2) in the furnace. Conversion of SO_2 and the remaining H_2S to sulfur and water vapor is accomplished in the furnace and the Claus reactors, recovering liquid sulfur.

A tail-gas treatment plant is being installed, but until it is in operation, tail gas is incinerated to convert all remaining sulfur compounds to SO_2.

Sour hydrocarbon condensate from the gas field separately enters the hydrocarbon-liquid fractionation unit for inlet separation. In a stabilizer, these liquids and recovered liquid from the gas-treating units of the plant are fractionated for removal of lighter hydrocarbons and acid gas, producing a stabilized C_5+ condensate product.

The sour low-pressure-gas streams are treated with DEA, recompressed, and refrigerated for light-hydrocarbon liquid recovery. The product gas goes to the fuel-gas system. The condensed light hydrocarbon liquids are then partially de-ethanized to produce a broad-fractionation high-pressure hydrocarbon liquid product.

A central propane-compression-refrigeration unit supplies liquid propane to the gas chillers in the hydrocarbon dewpoint control portion of the main gas-treating units and the light hydrocarbon liquids recovery section of the hydrocarbon liquid fractionation system.

The dilute glycol solution streams are separated from the hydrocarbon condensate in each refrigeration section employing glycol injection. This glycol is routed to a central regeneration system for removal of excess water, cooling, filtering, and temporary storage. Central glycol pumps send the glycol back to the gas-cooling systems. Provision is made for antifoam injection.

Each treating unit has its own DEA regeneration. Regenerated DEA is collected into a central surge tank from which a sidestream is filtered through both carbon and precoat filters. Central booster pumps return the amine to each individual amine delivery pump system.

Facilities are provided for injecting antifoam, corrosion inhibitor, and soda ash into the DEA aystem.

Recovered liquid sulfur from both gas plants 1 and 2, as well as from gas plant 3 when it comes into production, go to a central sulfur-pelletizing facility.

A central steam power plant serves the whole gas-plant complex. Steam condensate from steam users is returned to the power plant.

Operating experience

Start-up. The start-up of Gas Plant No. 2 in the autumn of 1975 was accomplished on schedule without serious problems. The plant has been in continuous operation ever since. Outlet gas meets pipeline specifications of 5.7 mg H_2S/nm^3 (0.22 grain/100 scf) and a hydrocarbon dewpoint of $-15°$ C. ($+5°$ F.) at normal flow rate.

The various trains of the plant were operated up to capacity. Actual inlet concentration of H_2S has been lower than design so less than design percentage of DEA in the amine system is generally maintained.

No significant operating problems have been experienced with the design of the plant, but some difficulties occurred due to foaming in the amine contactors. The investigation of this problem is of interest.

Foaming problem. Some DEA was carried over in the glycol-injection system. This created a problem in glycol-liquid hydrocarbon separation due to increased viscosity. An aqueous solution of 7.6 wt % DEA and 62.7 wt % ethylene glycol at 20° C. had a viscosity of 10.15 cp vs. an expected viscosity of about 6.8 cp for a 70.3 wt % glycol aqueous solution at 20° C.

No purging of the glycol system was done, as makeup glycol gradually reduced the concentration of DEA to a reasonable level. An investigation of the DEA foaming was made in the spring of 1976. Study of operating conditions revealed the bad foaming condition had occurred after the large dosage of corrosion inhibitor to the plant inlet lines by injection of the methanol-corrosion inhibitor liquid slugs at the field gas plants.

At the time of the investigation, the DEA solution was not foamy, so it was necessary to create a foamy solution. The corrosion inhibitor (Table 3) being used in the central plant complex is an antifoaming type, so it was not suspected to be a problem. Its use has been discussed by Heisler and Weiss, (their N-3).[5]

It was found that addition of inhibitors No. 1 or No. 2 used in the gas field either singly or mixed together into pure DEA solution or by making a concentration of 40 ppm of them in the regenerated DEA from the plant caused considerable foaming. Foam-testing apparatus used was as described in Ref. 7. It consisted of 200 ml of liquid in a 1,000 ml graduate with an air flow of 4,000 cc/min through a vertical surface gas disperser.

An antifoam (Table 3) is being used when required in the DEA systems. But the foaming that occurred in the spring of 1976 was severe enough that this antifoam could not control it immediately.

Antifoam injection had been continued during part of the time following the period when foaming was experienced and the period of the investigation. It was noted that plant operating DEA solution sampled at a later date than the first test sample was more resistant to foaming caused by addition of corrosion inhibitors No. 1 and/or No. 2. About twice as much of the inhibitors was required to produce foaming similar to that in the earlier sample.

Several antifoams were laboratory tested to try to find a more efficient one, but the silicone already in use in the DEA system was the most effective.

From the laboratory tests and operating records it was concluded that the inhibitors from the inlet gas lines were most likely the cause of foaming. There is no practical method of analyzing the DEA solution to determine the concentration of the corrosion inhibitors present.

Corrosion inhibitor No. 1 presumably would be separated in the GOSP separators. But calculations indicated that, if there is as much as 14 liters/MM nM^3 carry-over (0.1 gal/MMscf) of aqueous phase liquid from the cold-gas separators in these field gas plants, the amount of inhibitor added to the feeder pipelines to the central plant would be equal to 40 ppm in the DEA systems in 60 days.

It is assumed that eventually some of this quantity concentrates in the DEA system. A malfunction of a field gas-plant-separator level controller could increase this amount going to the gas line.

In the case of corrosion inhibitor No. 2, the amount left in the inlet pipelines from the unrecovered portion of the periodic methanol-corrosion inhibitor dosage amounts to 4,000 ppm of corrosion inhibitor in the plant DEA system per treatment. It must be assumed that some small portion of this carries through the inlet separators into the DEA systems.

On June 16, 1976, an incident occurred which fairly well confirmed the conclusions made from the tests and calculations. Foaming started in the DEA contactors which are equipped with differential-pressure recorders. Antifoam injection was commenced at once and the system was under control in 2½ hr.

The DEA content of the glycol solution before and after the incident indicated no carry-over of DEA into the cold-gas system. Normally, no liquid is collected in the central-plant inlet separators, But, in this instance, there was some liquid accumulation. Analysis of this liquid revealed the presence of corrosion inhibitor No. 1 from the field gas plant.

Investigations of other corrosion inhibitors was started to find a suitable nonfoaming type, but results of their use are not yet available.

Gas outlet system

Natural gas from the Orenburg complex is distributed via the 48-in. (1,220 mm) Orenburg-Ostrogzhsk line, to the northwest, and other systems into the national pipeline grid. Now under construction is 1,420 mm (55.9 in.) gas pipeline 2,700 km (nearly 1,700 miles) long from Orenburg west to Uzhgorod on the Czechoslovakian border for sending gas to eastern Europe.

This huge pipeline will have 22 compressor stations, employing a total of 123 gas turbine-compressor sets. Seventeen of the stations will be equipped with American compressors. The largest single centrifugal compressor order ever awarded has been given to the American compressor manufacturer for 88 of the compressors with 10,000-kw drivers.

An Italian manufacturer will supply compressors for the other five stations. Seventy-seven of the 88 are designed to handle up to 620 MMscfd each at 50-kg/cm² (710-psig) suction and 74.5-kg/cm² (1,059-psig) discharge pressure. The other 11 compressors for the head station will be rated at 425 MMscfd each with an inlet pressure of 45.5 kg/cm² (647 psig) and discharge pressure of 75.4 kg/cm² (1,073 psig.).[9, 10, 14]

Construction of the 1,420-mm pipeline has been divided into five sections, with construction crews from Hungary, Czechoslovakia, Poland, East Germany, and Bulgaria each responsible for a section. Romania is furnishing money for material purchases.[10, 11] American automatic welding equipment is being used on this pipeline project.[12]

Acknowledgment

Acknowledgment is made to the Ministry of Gas, Moscow; COCEI, Rueil-Malmaison, France; and J. F. Pritchard & Co. for permission to publish this paper.

References

1. Rigassi, D. A., "U.S.S.R. Growth Rate Slows, but Still is Impressive," World Oil, V. 183, No. 3, Aug. 15, 1976, p. 126.
2. Sulphur (London), No. 111, p. 16, March/April 1974.
3. Zaitsev, Y., Korshunov, E., Konovalov, V., Karpov, A., and Shatalov, A., "Complex Utilization of Raw Resources of Large Gas Fields with Compound Gas," Paper IGN/F5-76, 13th World Gas Conference, London, June 7-11, 1976.
4. Sulphur (London), No. 121, p. 16, November/December 1975.
5. Heisler, L., and Weiss, H., "Experience with an Austrian Gas Plant," Hydrocarbon Processing, V54, No. 5, May 1975, p. 157.
6. Shapovalova, L., Scheirman, W. L., and Chernomyrdina, N., "Ispitanie Ingibitorov Korrozii na Vspenivanie Amina" ("Test of Corrosion Inhibitors on Amine Foaming") Central Plant Laboratory, Orenburg Gas Treating Plant, Orenburg, Russia (Russian text), May 28, 1976.
7. Dow Chemical Co., "Gas Conditioning Fact Book," 1962, p. 339.
8. "Massive Compressor Contract for Russian Pipeline," Petroleum Times (London) V. 80, No. 2032, July 23, 1976, p. 11.
9. Canadian Petroleum, V. 17, No. 8, Aug. 1976, p. 11.
10. Ebel, R. E., "Pipe Shortage Slows Soviet Construction," Pipeline Industry, V. 41, No. 5, November 1974, p. 34.
11. "Orenburgskii Gaz," Molodaya Gvardiya, Moscow (Russian text), 1976.
12. Quarles, B., "Automatic Welding is used Successfully on Orenburg Line," Pipeline News, V. 48, No. 6, November/December 1976, p. 28.
13. Perry, R. B., and Lyle, F. F., Jr. "Soviet Corrosion Inhibitors," Materials Performance, V. 15, August 1976, p. 38.
14. V. de Biasi, "USSR Pipeline Plans Dwarf Past Projects," Gas Turbine World, Vol. 6, No. 3, July 1976, p. 18.

The author . . .

W. L. Scheirman

William L. Scheirman is a senior process engineer with J. F. Pritchard & Co. Joining the firm in 1969, his responsibilities include process design of chemical plants, natural-gasoline plants, natural-gas absorption and adsorption, dehydration, and desulfurization plants, sulfur-recovery plants, crude-oil central production systems, and other projects. Scheirman holds eight patents. He obtained a BS degree in mechanical engineering from Oklahoma State University in 1943 and has done graduate study there in chemical engineering. He has been responsible for process design work for plants in the United States, Canada, U.S.S.R., Austria, Algeria, and Japan. A registered professional engineer, Scheirman has been affiliated with Rhodes Technology Corp.; Black, Sivalls & Bryson; Seattle Gas Co.; Douglas Aircraft Co.; and John Deere Tractor Co.

Taking H_2S from liquid sulfur—why and how

FRANK W. KING
Ralph M. Parsons Co. Ltd.
Calgary, Alta.

AN important new process is proving itself for eliminating the hazards formerly associated with shipment of sulfur in liquid form.

The process is also improving the safety and quality of sulfur-slating operations.

Soc. Nationale des Petroles d'Aquitaine (SNPA) developed the process in France after several years of study into the solubility characteristics of H_2S in sulfur. The investigation covered the risk of explosion from the presence of H_2S dissolved in sulfur and then liberated during handling and transport.

Till now, only a relatively small portion of sulfur has been shipped as a liquid. The market for liquid sulfur, however, is growing. A number of producers in Canada and elsewhere are now moving into the liquid-sulfur market. For these producers, removal of H_2S from the liquid sulfur, to some safe level, has become a vital matter.

Moreover, H_2S content is important to sulfur-slating operations. In these operations, usually undertaken to meet changing market and environmental needs, concern has mounted for the safety of the men working around the slating machine. Substantial quantities of H_2S can evolve from the liquid sulfur in this region.

Besides, the presence of H_2S in liquid sulfur has an adverse effect on sulfur-slate strength.

For the sulfur-slate producers, the removal of H_2S from sulfur, to some safe level, has also become an important industrial technique.

Sources of H_2S in sulfur. Most of the sulfur in a Claus plant is produced in the reaction furnace and first converter. At this point in the process, both H_2S concentration and temperature are high. Both conditions appear to favor H_2S "involvement" with liquid sulfur.

In one investigation, it was found that sulfur produced from the reaction furnace contained 500-700-ppm H_2S; the first-condenser sulfur 180-280 ppm; the second-condenser sulfur 70-110 ppm; the third-condenser sulfur 10-15 ppm; and the fourth-condenser coalescer sulfur 5-10 ppm H_2S.'

The "apparent solubility" of H_2S in sulfur at the front end of the Claus plant is much higher than expected for the temperatures involved. It is evident, then, that some anomaly must exist in the H_2S/sulfur solubility relationship.

H_2S solubility anomaly. Normally in a gas-liquid system, the absorption rate of the gas is lower at higher temperatures. H_2S and sulfur are different. In the Claus temperature range, solubility of H_2S **increases** as the temperature increases. But, curiously, when the sulfur is cooled the H_2S does not evolve as quickly as expected.

The work of many, including Dr. Hyne and his associates at Alberta Sulphur Research Ltd., has provided evidence that the H_2S solubility anomaly exists because of the reaction of H_2S with sulfur to form polysulfates.[3] The polysulfide formation reactions are quite favorable at Claus temperatures so H_2S appears to "dissolve" in sulfur at an unusually high rate. But the polysulfides don't release the H_2S nearly as willingly as the temperature is reduced.

This explains why sulfur-plant operators have experienced very little problem with H_2S release in the sulfur-plant area and yet liquid-sulfur shipments can evolve relatively high and dangerous concentrations of H_2S several days after being produced.

The true solubility of H_2S in liquid sulfur follows normal expectations. That is, solubility decreases as temperature increases. But the total or "apparent solubility" of H_2S has two elements:

- True solubility element.
- Polysulfide formation element.

Published data[4] show that at front-end Claus plant conditions, the true solubility element may be less than one-third of the total "apparent solubility of H_2S in sulfur.

Sulfur-transport tests. After the Sulphur Queen disaster of 1963 when the liquid-sulfur-carrying ship went down without explanation, transportation tests[1] were carried out on railway and road tankers. Aim was to monitor the development of temperature and H_2S content of the liquid sulfur and of the vapor phase under the tank dome.

To minimize the hazards associated with early tests, the rail cars were loaded 50% full with coalescer sulfur containing only 5-10 ppm. The first test was 65 hr long. During the first 36 hr the car was left sitting still and the sulfur cooled from 290° to 237° F. H_2S in the gas above the liquid rose from 172 ppm to 799 ppm, illustrating the desorption due to the cooling effect.

For the next 12 hr the rail car was shunted at low speeds and the temperature remained constant. H_2S content in the gas rose from 799 ppm to 2,800 ppm, illustrating the desorption effect due to agitation.

During the next 17 hr the temper-

This is an edited version of a paper presented to the Canadian Gas Processors Association, Calgary, Nov. 30, 1973.

ature stayed the same and the car was again left sitting still. H_2S content of the gas dropped from 2,800 ppm to 2,300 ppm, illustrating the true solubility effect of H_2S in equilibrium with liquid sulfur at lower than Claus temperatures.

The H_2S content of the sulfur had dropped from 8 ppm to 4 ppm from the beginning to the end of the test.

It was concluded that liberation of the H_2S from liquid sulfur takes place in two ways:

- Through a drop in temperature.
- Through physical agitation.

The rail-tanker tests continued in a similar fashion except that sulfur containing higher concentrations of H_2S was used. The tests were designed to find the H_2S concentration at which the lower ignition limit (about 36,000-ppm H_2S in air) would be reached.

In one test, liquid sulfur from the condenser containing about 100 ppm was used. The lower explosive was reached within 10 min with very little cooling and no movement of the car.

After 40 hr, H_2S content of the gas space had risen from 38,130 ppm to 47,760 ppm and the liquid-sulfur H_2S content had dropped from 100 ppm to 63 ppm.

As experience with the tests increased more precise data on the liberation of H_2S were obtained without taking serious risks.

Finally, the tank cars were filled normally, that is, with an empty space of only 1-2%. In later testing, road-tank trunks with full loads of liquid sulfur and varying H_2S concentrations were also used. In each case, the tests indicated that, if tankers were filled with liquid sulfur containing more than 15 ppm, the H_2S concentration in the gas space would reach the lower ignition limits after only 7 or 8 hr of transport. At this time the H_2S content in the liquid would be reduced to 8 ppm.

Degasification

Based on its transportation tests, SNPA concluded that 15 ppm was the upper dangerous limit for H_2S in liquid sulfur. And 5 ppm was decided upon as a commercial specification. Some reports indicate that current U.S. regulations limit the H_2S content in liquid-sulfur shipments to 10 ppm.

Development of the processes. Following the transportation tests, SNPA embarked on a program to develop a degasification process which could produce liquid sulfur with 5-ppm H_2S or less. After an additional 8 months of concentrated research and testing, the original SNPA process for degassing large commercial quantities of sulfur was ready.

The process takes advantage of the two principal degasification mechanisms studied:

- Cooling.
- Agitation.

The method found most effective was to pump the liquid sulfur through jets and against an obstacle. But further testing had discovered that a third important mechanism improved degasification remarkably. It was found that chemical injection of ammonia catalyzed the release of H_2S tied up as polysulfides. Recall that nearly two-thirds of the total dissolved H_2S is tied up as polysulfides.

Basis for the SNPA process, for which patents have been issued, is to recirculate liquid sulfur through huge spray nozzles and into a receiving pit or tank. Small quantities of ammonia are injected upstream of the pump to catalyze the decomposition reaction and shorten the cycle time required to reach 5 ppm.

The original SNPA process is a batch process, so a separate pit or tank is required for degassing. Naturally, during degassing, the H_2S content of the gas above the pit increases as the treatment proceeds. It is, therefore, essential, if the pit is flushed with air, not to exceed the lower explosive limit in the pit. The H_2S evolved during degasification operations can be evacuated using an eductor drawing on the pit and discharging to the tail-gas line upstream of the incinerator. Or the draft available from the annulus in tail stacks can be used. The eductor system is considered more suitable for Alberta operations.

The SNPA degasification process has been operating successcfully in France for about 5 years. More than 2,000 long tons/day are being treated to a final H_2S level of less than 5 ppm.

Modified versions of the process will soon be operating in two major Alberta sulfur plants.

Texas Gulf Sulfur Co. has been operating another version of a degasification process at its Whitecourt plant since 1969 with reports of final treated sulfur containing less than 50 ppm of H_2S.

Measurement, control. Complete understanding of the H_2S-sulfur solubility relationship has been hampered by the lack of reliable plant-measurement techniques. Several interesting analytical techniques are available for laboratory-study work, but these methods are not always practical for plant operations. SNPA has developed a semiautomatic analyzer to measure traces of H_2S in liquid sulfur. The analyzer uses an inert gas carrier to bubble through a liquid sulfur sample for as long as it takes to remove all measurable H_2S. The H_2S removed is collected and measured with a lead acetate strip. The analyzer is available in Canada through Aquitaine Co. of Canada Ltd.

A second analyzer (usually an in-line ultraviolet or infrared analyzer and/or a chromatograph) is used to monitor the H_2S concentration in the vapor phase of the degassing operation.

If the H_2S concentration in the vapor phase exceeds 25,000 ppm, the liquid-sulfur circulation pump shuts off automatically.

New developments. As demand for slated and liquid sulfur grows, it is likely that H_2S degasification will become a common sulfur-plant processing step.

The basic design parameters for the original SNPA degasification process still remain valid. Recent attention to the process, however, has brought rapid evolution of process improvements—some still now under commercial test.

The most encouraging recent design development has been the adaptation of the SNPA batch system to operate as a continuous process. This design will soon be tested on a commercial plant using a relatively minor modification to existing run-down lines and storage pit.

The cost vs. performance effect of this new design is expected to be very favorable for existing plants because it will eliminate the need to construct a separate batch-degassing pit.

References

1. J. Solinhac, September 1967.
2. W. J. Rennie, B. Andreassen, D. Dunay, J. B. Hyne, Quarterly Bulletin, Alberta Sulphur Research Ltd., Vol. VII, No. 3.
3. E. Muller, A. Parthasarathy, H. J. Langer, Quarterly Bulletin, Alberta Sulphur Research Ltd., Vol. V, No. 4.
4. T. K. Wiewiorowski, F. J. Touro, August 1965, The Sulphur-Hydrogen Sulfide System Journal of Physical Chemistry 70 234, 1966.

Part 3
Tail-Gas Cleanup

Performance compliance stands as major sulfur-plant problem

Arthur R. Laengrich
William L. Cameron
Ortloff Corp.
Midland, Tex.

COMPLIANCE with air-pollution regulations has recently emerged as one of the most significant problems for the owner of a sulfur-recovery plant. This is especially true for a sulfur-recovery unit associated with a gas-treating plant.

Adapted from paper, "Sulfur Recovery Plants—Solving the Performance Compliance Problem," presented at Gas Processors Association Panhandle Plains Regional Meeting, Amarillo, Tex., Oct. 13, 1977.

For a refinery sulfur plant, costs of compliance can be absorbed more easily since the sulfur-plant cleanup costs are relatively small compared to the total cost of the facility. Technical expertise in operations, optimization, and cleanup-unit design is also readily available, sometimes within the plant itself.

For gas processor's sulfur plants, these advantages are not usually available. Costs of modifications or additions are much more significant, often exceeding the cost of the original installation. For a number of the available tail-gas-cleanup processes, the technology will be unfamiliar, and operation will require more attention.

The critical issue is compliance. If the sulfur plant is found to be in violation of the regulations, it must be brought into compliance or it could be forced to shut down.

When it goes down, so does the gas-treating plant and the sales gas and liquid lines.

So just exactly what are the regulations and their effect on the individual sulfur plant? If a plant is out of compliance, what can be done to bring it into compliance? And how much will it cost?

This two-part article discusses regu-

Fig. 1

Recovery required by TACB Regulation II

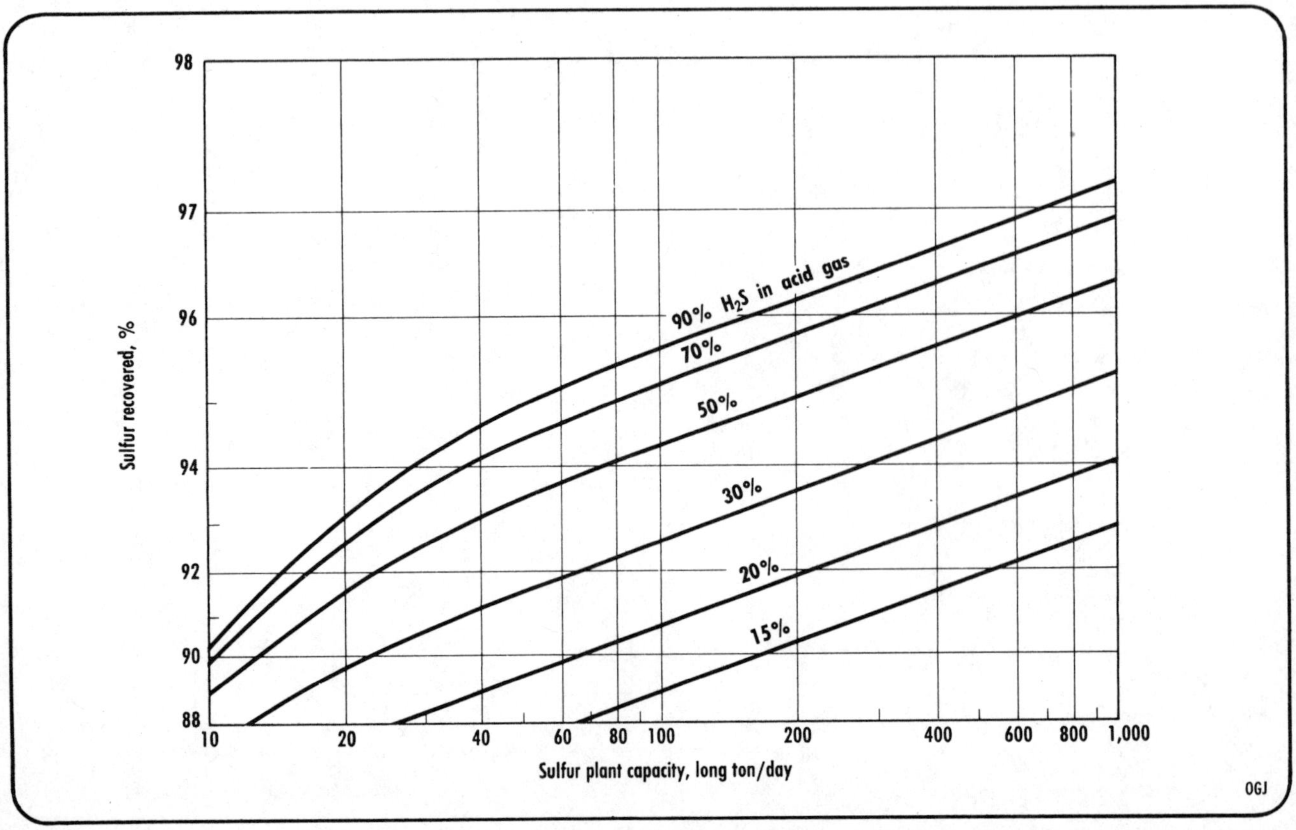

lations, and processes and techniques that can help operators comply with requirements.

The conclusion of this two-part article will appear next week.

Performance requirements

The performance requirements imposed on sulfur-recovery units included in natural-gas-treating plants by the different states vary. This discussion will be restricted to selected regulations that apply to facilities located in Texas.

Existing sulfur-recovery plants. The current regulations of the Texas Air Control Board (TACB) establish a maximum permissible rate of emission of sulfur dioxide from existing sulfur-recovery plants. This is covered in TACB Regulation II.

The upper limit is based on one of two formulas, depending on plant size. For a total stack-effluent rate of up to 4,000 scfm, the maximum SO_2 emission rate is set by the equation:

$$SO_2 \text{ (lb/hr)} = 123.4 + 0.91Q$$

where Q is the stack-gas rate, scfm. For larger plants, the following equation is used:

$$SO_2 \text{ (lb/hr)} = 0.614Q^{0.842}$$

Fig. 1 graphically presents the approximate recovery levels required by TACB Regulation II for sulfur-recovery plants as a function of size and H_2S content in the feed gas. As these curves show, these regulations are more lenient in overall recovery requirements for small plants and for plants with low H_2S content feeds. This realistically reflects the difficulties encountered in achieving high sulfur recovery from a low H_2S content feed stream.

To use Fig. 1 to estimate the sulfur-recovery performance required for an existing facility, locate the plant design capacity on the horizontal scale. Go vertically until the constant capacity line intersects the correct H_2S content for the sulfur-plant feed. Then move to the left to determine the approximate recovery level required.

Fig. 2 displays typical recovery capability curves for well-engineered Claus sulfur-recovery units of different process design configurations. Both straight-through and split-flow process configurations were evaluated for: (1) two-stage plants with bypass reheat, (2) maximum recovery two-stage plants with indirect reheat, and (3) maximum recovery three-stage plants with indirect reheat.

Fig. 3 combines these two sets of data, as well as recovery capability data for more energy-optimized two and three-stage configurations. It can be used to help determine whether one of the Claus designs alone can satisfy the present TACB requirements for existing plants.

Use of this figure is basically the same as for Fig. 1. From the intersection of the capacity and H_2S composition lines, move to the right along the composition curve. The key in the upper portion of the figure is used to identify the process configuration of the first plant-recovery curve intersected. This basic design should be able to satisfy the requirements of Regulation II.

For example, an existing 100-long-ton/day (Lt/d) sulfur-recovery plant operating with a feed gas containing 50% H_2S will probably require a three-stage process design to comply with TACB Regulation II.

The graph indicates a well-designed two-stage plant would closely approach Regulation II requirements. Therefore, an existing plant of this type might be maintained in compliance with optimized and closely controlled operation.

Had the facility been a 250-Lt/d unit, a two-stage configuration would be unlikely to satisfy the regulation while a well-designed three-stage plant almost certainly would be capable of compliance.

New sulfur-recovery plants. Note that TACB Regulation II applies to existing sulfur-recovery plants only. Any new sulfur-recovery plant should expect somewhat more stringent requirements, with the provisions of Regulation II offering an absolute minimum achievement target.

Fig. 2

Maximum recovery levels from typical Claus plants

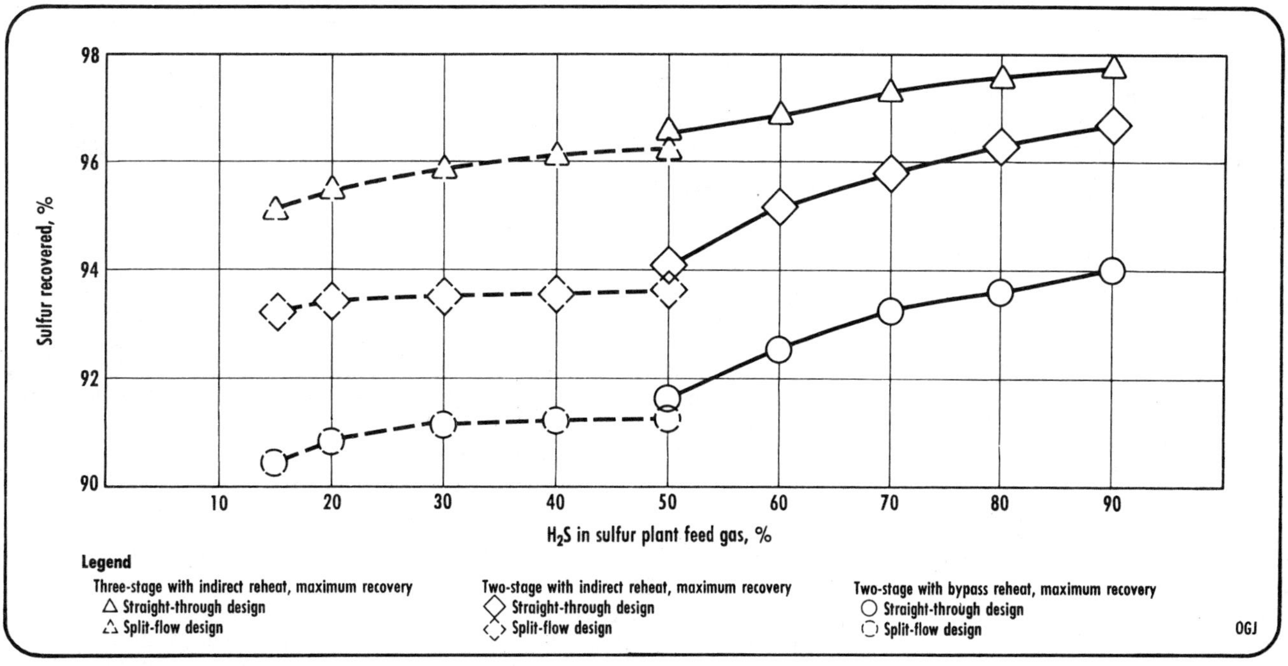

Fig. 3

Calculated plant requirements vs. TACB Regulation II

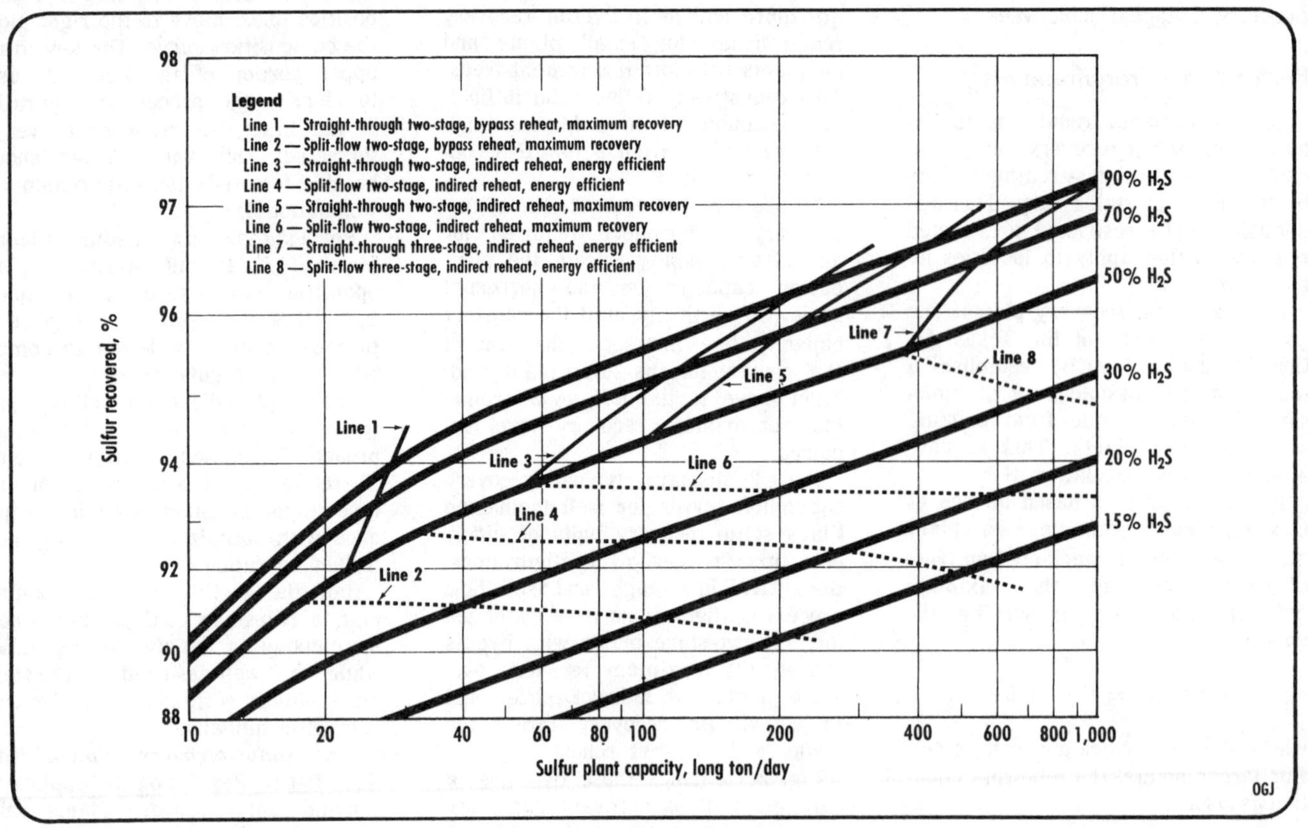

Regulation VI outlines the general requirements for new plants. The most significant of these are that it: (1) "will not prevent the maintenance or attainment of any applicable ambient air quality standard", (2) "will not cause significant deterioration of existing ambient air quality", (3) "will utilize the best available control technology," and (4) "will meet . . . any applicable new source performance standards".[1]* The "best available control technology" requirement will be evaluated considering technical and economic factors.

The permissible sulfur-dioxide-emission level for each new facility is evaluated in light of the existing SO_2 ambient air concentration, present and potential industrial developments, population density, and other factors. These same factors should be considered in finalizing a design to be presented in an application for a construction permit.

In addition to the existing state regulations, new federal regulations are expected soon. Preliminary drafts of tentative standards were made available early last year. These preliminary standards limited SO_2 emissions to 250 ppm and forbid any gas containing more than 160 ppm H_2S to be combusted directly to the atmosphere.

A recent conversation with EPA personnel indicated that these preliminary standards had been reviewed and were being reconsidered, with a new study being conducted.

A second set of tentative regulations is expected to be available early in 1978.

One subject the EPA will be studying is the use of sliding scales for either small facilities or low H_2S concentration plant feed gases. The original tentative regulations simply granted an exemption for all sulfur-recovery facilities handling less than 2 Lt/d.

This blanket exemption may be revised or replaced with a formula for maximum emissions related to plant size, or H_2S content, or both. Such a formula would lessen the burden on small installations and recognize the difficulties of achieving high recoveries from low H_2S content gases.

Existing plant compliance options

The owner of an existing sulfur-recovery plant experiencing a compliance problem is interested in a problem solving procedure that costs the minimum. As the following various options are discussed, it will be generally safe to assume they become more costly in terms of investment expenditure in ascending order.

Improving operations and maintenance. Catalyst performance and service life are quite important factors to consider and evaluate. Catalyst degradation is usually due to surface sulfation, or to contamination by certain compounds in the sulfur-plant feed stream.

Design modifications are sometimes required to eliminate catalyst degradation. But quite frequently, changes in plant operating procedures and selected conditions can be effective.

For instance, maintaining certain temperatures and pressures in the gas-treating unit at acceptable levels and under close control can significantly reduce hydrocarbon content in the sulfur recovery unit feed stream.

Operating the sulfur-recovery plant to always maintain a reducing atmosphere at all points upstream of and in the catalyst beds during normal operation, and during startup and shutdown periods, can significantly reduce catalyst sulfation rate.[2]

Maintaining a plot of temperature rise across each catalyst bed vs. time, which is a direct measure of conver-

*Numbers designate references listed at the end of Part 2.

sion in the catalyst bed, can provide an effective guide for maintenance planning from the standpoint of predicting optimum catalyst change-out dates.

Many existing sulfur-recovery plants have been given very little attention in the past with regard to optimizing process conditions and controls to obtain maximum performance. If the basic plant design is good, actual performance of the plant can be forced to approach quite closely the calculated sulfur recovery performance capability of its particular design configuration by closely controlling near the optimum values.

The single most important variable, of course, is the air feed rate. Some sulfur plant tail gas analyzers have now been developed to a point where they can truly be classified as reliable devices for continuously monitoring and controlling the process air feed rate.

If the sample systems required by these analyzers are carefully designed and installed, performance of the analyzers can even be acceptable in closed-loop control.

The closed-loop control system can be simple and straightforward. A complicated optimizing system is not necessary to obtain 90% or more of the potential benefits available from a continuous-monitoring closed-loop control system.

A noteworthy improvement in an existing plant's overall sulfur recovery performance, in the range of 1.0 to 2.0%, can frequently be attained just by optimizing the temperature profile throughout the unit. In general, lower temperatures force the reactions to proceed further in the desirable direction and allow more sulfur to be removed from the process stream at each condensation and separation point.

Where conversion of more refractory compounds such as carbonyl sulfide and carbon disulfide is a serious consideration in the first catalyst bed, the lowest possible temperature may not be the optimum operating point, however.

Also, selection of a particular catalyst that will give improved conversion of these compounds at a given temperature level should be considered.

When an existing plant has sufficient reactor feed reheat capacity, a valuable increase in overall performance can usually be obtained by reducing the sulfur condenser outlet temperatures to the minimum acceptable operating level.

This change in operating conditions usually reduces the overall energy efficiency of the process in that the steam produced from the sulfur condensers drops in pressure to a level that makes it unacceptable for most process uses.

However, this and the changes mentioned earlier are frequently quite attractive from an economic standpoint if they accomplish the needed improvement in plant performance. Therefore, if the information obtained from Figs. 1 and 2 indicates a particular plant should be capable of complying with Regulation II requirements, consider and try the items discussed above before looking seriously at modifications that would require a major capital expenditure.

Design modifications. Valid incentives usually exist for upgrading an existing sulfur-recovery plant rather than replacing it entirely. The capital requirement will usually be lower.

If the modification project is initiated before compliance problems force action, and if the basis for the modification does not include an increase in design throughput, a construction permit from the Texas Air Control Board and other agencies may not be required.

For these reasons, it is often desirable to upgrade an existing plant by performing various modifications before a contest regarding compliance capability occurs.

A large number of potential modifications exist that can bring meaningful improvements in the performance capability of an existing sulfur-recovery plant. Some potential modifications are feasible and practical for an existing plant; some are not.

Some existing sulfur-recovery plants were designed as split-flow plants; i.e., a portion of the feed gas is bypassed around the combustion or thermal reaction chamber for reasons other than valid process limitations.

These plants can sometimes be easily converted to straight-through process-configuration plants. This conversion will frequently cause an improvement in overall performance capability that greatly exceeds the increment obtained from Fig. 2, since the basic conversion has a net effect of removing duty from existing reactors and sulfur condensers.

Addition of a third catalytic conversion stage with its associated sulfur condenser/separator will typically give a minimum of 2.0% improvement in overall sulfur-recovery performance of an existing two-stage plant. Frequently, the actual improvement will far exceed the calculated potential improvement due to design deficiencies that exist in the existing equipment.

This particular modification, when applicable, is almost always possible and practical to implement.

The method of reheating the process stream prior to routing it to each catalyst bed has an important bearing upon performance capability of an existing plant. When an antiquated method of reheat is used, such as hot-gas bypass or in-line burners, a conversion to an indirect method of reheat can be justifiable.

Referring again to Fig. 2, converting a hot-gas-bypass reheated plant to an indirect reheated plant can easily improve overall performance capability by more than 2.0%. If high pressure steam, 350 psig or higher, is available at the location, conversion to indirect reheat is usually simple and modest in cost. The conversion is certainly feasible even when this condition does not exist.

One of the more noticeable and desirable benefits of this conversion is the improvement in typical control system simplicity and controllability. The typical plant becomes much easier to operate consistently at near optimum conditions. Therefore, the average performance improvement obtained can again far exceed the calculated minimum increment.

In some situations, the optimum modification for an existing sulfur-recovery plant can involve total replacement of a large segment of the plant. The combination of age deterioration, original design deficien-

The authors . . .

Arthur R. Laengrich

William L. Cameron

Arthur R. Laengrich is vice president of the Ortloff Corp., Midland, Tex. His area of responsibility includes consulting services, process engineering, conceptual plant design, and plant start-up. A graduate of the School of Chemical Engineering at Oklahoma State University, Laengrich is a registered professional engineer and has been intimately and continuously involved in the design and operation of sulfur-recovery systems since 1964. He has performed or directly supervised the process and detailed design of more than 30 sulfur-recovery systems, including both Claus sulfur-recovery and tail-gas-cleanup units.

William L. Cameron is a process engineer with the Ortloff Corp., Midland, Tex. His previous assignments have included design and specification of gas treating, sulfur recovery, cryogenic-liquids recovery, and condensate-stabilization facilities. Cameron received a masters degree in chemical engineering from Rice University in 1975.

cies, and similar factors can sometimes cause the modifications discussed previously to be completely impractical.

When this case exists, it can be entirely practical to replace the main process portion of the plant with a new system that incorporates current "best available technology."

Several equipment items from the existing installation, including the inlet gas separator, process air blowers, tail gas incinerator, sulfur rundown tank, and sulfur storage and loading facilities can remain in their present service.

The result is a modified existing sulfur-recovery plant which incorporates current "best available technology" in the process section, which determines overall performance capability, for a cost significantly less than the cost of a total new installation.

Also, it is entirely possible to accomplish this type of modification without interrupting the operation of the existing sulfur-recovery plant prior to final tie-in time if adequate adjacent plot space is available.

The ultimate upgrading "modification" of an existing Claus sulfur-recovery plant is that of converting it to a low-temperature Claus process plant using such technology as the Amoco CBA process. The CBA process is capable of attaining an overall sulfur recovery of approximately 99%.

Referring to Fig. 1, TACB Regulation II compliance requires higher than 96% overall recovery performance from many existing plants larger than about 150 Lt/d capacity. Unless circumstances are ideal, it may be difficult to maintain the average performance level of a Claus plant above 96%.

This consideration plus other special requirements sometimes imposed upon an existing facility could make conversion to the CBA process a highly attractive alternate to adding a tail gas cleanup process to the end of the Claus plant.

Although the CBA process plant is slightly more complex to operate and somewhat less energy efficient than a standard Claus sulfur recovery plant, it is a far better choice than any add-on tail gas cleanup process facility with respect to both of these important considerations.

If the existing Claus plant is in good condition and has good design quality incorporated into it, conversion to the CBA process to attain a higher recovery performance can be accomplished by adding a minimum number of additional items while requiring a minimum amount of downtime.

Part 2, the conclusion, will appear next week. END

Tail-gas cleanup addition may solve sulfur-plant compliance problem

ARTHUR R. LAENGRICH
WILLIAM L. CAMERON
Ortloff Corp.
Midland, Tex.

AS mentioned in Part 1 (OGJ, Mar. 20), compliance with air pollution regulations is one of the most significant problems facing owners or operators of sulfur-recovery plants, particularly those associated with gas treating plants.

Since performance requirements vary somewhat by states, this discussion is restricted to selected regulations that apply to facilities in Texas.

Continuing with the existing-plant compliance options discussed in Part 1, the last resort for solving a performance compliance problem is addition of a tail gas cleanup system. These should be considered last because they are expensive. They are expensive in terms of investment, energy consumption, or both, depending upon the process selected.

Quite a large number of processes have been studied for application to Claus sulfur-recovery plant tail gas cleanup service. A recent process screening study performed for a client revealed more than 70 processes advertised or discussed in the literature for possible application in this service.

The following processes briefly discussed appear noteworthy at present from the standpoint of commercialization and applicability to tail gas cleanup service. These processes seem to logically group under two categories: one to attain "intermediate" overall sulfur-recovery levels up to about 99%, and the other to attain "stringent" overall sulfur recovery up to about 99.9% or higher.

Hopefully, none of the existing gas treating and processing plant sulfur-recovery units located in Texas will ever have to be upgraded to the maximum capability of these processes. However, if that requirement should occur, some commercially proven processes are available.

Adapted from paper, "Sulfur Recovery Plants—Solving the Performance Compliance Problem," presented at Gas Processors Association Panhandle Plains Regional Meeting, Amarillo, Tex., Oct. 13, 1977.

Table 1

Summary of system capabilities

Recovery system	Overall recovery capability, %
Two-stage Claus with bypass reheat	90-94
Two-stage Claus with indirect reheat	93-96.8
Three-stage Claus with indirect reheat	95-97.8
Two-stage Claus (96% recovery) with one CBA stage in tail-gas-cleanup unit	Up to approx. 99[3]
Two-stage Claus (96% recovery) with two CBA stages in tail-gas-cleanup unit	Up to approx. 99.5[3]
Two-stage Claus (95.5% recovery) with Sulfreen process tail-gas-cleanup unit	Up to approx. 99[4]
Two-stage Claus (95% recovery) with IFP process tail-gas-cleanup unit	98.1-99.4[5]
Claus plant (94% recovery) with SCOT process tail-gas-cleanup unit	99.9+[7]
Claus plant with Beavon process tail-gas-cleanup unit	99.9+[9]
Claus plant with Wellman-Lord process tail-gas-cleanup unit	99.9+[10]

The negative net effect is just one of converting the Claus sulfur recovery plant from a moderate cost energy efficient pollution control device to a high-investment, energy-inefficient pollution control complex. The overall cost per additional unit of emissions reduction above the 99% recovery level is extremely high.

Meeting intermediate cleanup requirements

This first class is made up of tail gas cleanup processes which allow the Claus reaction itself to take place under more favorable conditions. These processes are often referred to as extended Claus processes. All are claimed to offer an overall sulfur-recovery performance of approximately 99%, including sulfur recovered in the Claus unit.

Three of the most important commercially tested processes are Amoco's Cold Bed Adsorption process, the SNPA/Lurgi Sulfreen process, and the IFP/process.

Cold Bed Adsorption (CBA). An extended Claus process to decrease sulfur-plant emissions has been developed by Amoco Production Co., the Amoco Cold Bed Adsorption, or CBA, process. Here, the Claus reaction takes place in the gas phase at a temperature below the sulfur dewpoint.

Amoco has built a CBA unit for its East Crossfield plant, a 1,500 Lt/d sulfur plant near Calgary, Alta.

In the CBA process, gas from the main Claus plant final condenser is fed to an adsorption reactor operating about 260° to 300° F., which contains conventional Claus catalyst. The low temperature favors equilibrium conversion. The condensation of sulfur vapor on the bed removes the produced sulfur, shifting the equilibrium toward even higher conversion. Gas from the reactor is then incinerated.

Catalyst activity gradually declines as sulfur condenses on the bed. While one reactor is on adsorption cycle, a second reactor is being regenerated. Hot gas from the first Claus reactor vaporizes the condensed sulfur and reactivates the catalyst. The gas is cooled and sulfur vapor condensed in a sulfur condenser. Then the gas is returned to the Claus cycle just downstream of the first Claus sulfur condenser.

The CBA process uses the same construction and materials proven in the Claus plant. Since an internal stream is used for bed reactivation, only two reactors, a sulfur condenser, and a blower are required to add this process to an existing two-stage plant.

Energy consumption is low, and operation is similar to a conventional sulfur plant. The CBA process is quite suitable for a retrofit situation as it requires little plot space and only minor modification to an existing Claus plant.

Capital cost to convert an existing unit to use CBA is estimated to be approximately half that of the Claus plant. A new sulfur-recovery unit using the CBA process is estimated to cost approximately 1.5 times as much as a standard three-stage Claus unit.

The CBA process is reported to achieve up to 99% overall recovery with one CBA stage. For this level of recovery, the incinerated tail gas SO_2 content should be reduced to about 1,500 ppmv.

Levels of COS and CS_2 are not reduced by this process. Careful design and operation of the Claus plant is required to minimize the amount of these compounds in the Claus tail gas.

Based on the Claus reaction, the CBA process also requires good control of the stoichiometric H_2S/SO_2 ratio for maximum sulfur recovery.[3]

Sulfreen process. Similar to the CBA process, the Sulfreen process was developed in Europe by Soc. Nationale des Petroles d'Aquitaine (SNPA) and Lurgi Gesellschaft G.m.b.H. Like the previous system, Sulfreen uses a vapor-phase extended Claus reaction carried out below the sulfur dewpoint.

The Sulfreen process has been used in a number of sulfur plants in France and Canada. The design has undergone significant development since the first unit was built in 1968. Initially using using six reactors filled with a special activated charcoal, the process now can be designed with two reactors using conventional Claus catalyst beds.

The Sulfreen process operates in the same temperature range as the CBA design with the produced sulfur being deposited on the alumina catalyst bed. In the two-reactor case, one reactor is in service while the other is being regenerated.

The primary difference between the CBA and Sulfreen process is in the design of the regeneration system. While the CBA system is an open cycle with one additional sulfur condenser required, the Sulfreen design uses a closed regeneration loop containing a sulfur condenser and a regeneration gas heater, usually an indirect-fired unit with stainless steel tubes.

Recovery levels are comparable with the CBA process. As with the CBA design, the Sulfreen process operates much like a Claus stage and is influenced by the same factors. Carbon-steel construction is used except in the regeneration-gas heater where stainless steel is used for the tubes. The facility is compact and can easily be added to an existing sulfur plant.

The Sulfreen process uses fuel gas which the CBA design does not, but produces more low-pressure steam. Costs, both capital and operating, are also comparable for both of these processes.[4]

With a closed regeneration loop and gas heater, the Sulfreen process seems slightly more complicated than the CBA process. For a Claus unit, the simplicity of operation for either of these vapor-phase systems makes operator familiarization more simple.

Since these processes are extended Claus reactions, they require careful operation of the parent sulfur plant to achieve maximum recovery.

IFP process. A treating process used for Claus plant tail-gas cleanup has been developed by the Institut Francais de Petrole (IFP). It is basically an extension of the Claus reaction in a liquid-phase medium and is capable of reducing the SO_2 content of the incinerated tail gas to as low as 1,000 ppmv.

Tail gas from the Claus unit is contacted by the IFP solvent, a proprietary catalyst in a high-boiling-point glycol, and H_2S and SO_2 are absorbed. The Claus reaction then converts these compounds to sulfur.

The entire process takes place above the melting point of sulfur. As sulfur has low solubility in the solvent, the liquid sulfur accumulates in the bottom of the tower and is withdrawn. Treated gas leaves the top of the packed contactor and is incinerated. Solvent is circulated back to the top of the tower.

The IFP solvent is reported to have good thermal and chemical stability and low volatility, reducing solvent losses. Recovered sulfur is of high quality. The process itself appears straightforward, requiring only the contactor, a solvent circulating pump, a solvent heater for start-up, and solvent storage and makeup facilities.

In normal operation, no fuel or steam is required other than condensate for makeup. Stainless steel is used only for column internals and the solvent circulating pump. The IFP solvent is also relatively inexpensive, keeping initial and operating costs down for this process.

Retrofit is not complicated as the IFP installation requires little plot space and does not recycle any gas to the Claus plant feed.

The IFP process encounters many of the same difficulties present in the CBA and Sulfreen processes. Air control to maintain the correct H_2S/SO_2 ratio and proper operation of the Claus plant to minimize COS and CS_2 in the Claus tail gas are required for optimum performance. Operation of an IFP installation is quite different from the parent Claus unit, presenting a new set of operating problems.[5, 6]

Meeting stringent cleanup requirements

The second class of tail gas cleanup processes includes those capable of achieving overall sulfur recoveries in the range of 99.5-99.9%. This recovery level corresponds to about 300 ppmv or less total sulfur in the stack gas.

Recoveries at this level are now required for refinery sulfur plants. They were called for in the preliminary drafts of the tentative New Source Performance Standard proposed for sour gas treating sulfur-recovery facilities.

While this standard is being reconsidered, it is possible that this performance level will be required of larger plants. Three processes for stringent clean up which have been commercially applied are the Shell Claus Off-Gas Treating (SCOT) process, the Beavon process, and the Wellman-Lord process.

SCOT process. The Shell Claus Off-Gas Treating (SCOT) process was announced in 1972. It consists of a reduction stage followed by a concentration stage which provides an H_2S-rich stream to be recycled to the Claus plant.

The concentration process is similar to the amine gas-sweetening processes which are commonly used in gas processing. This makes the SCOT process considerably easier to operate. Flexibility and overall process reliability are reported to be good.

In the SCOT process, the Claus plant tail gas is reduced over a cobalt-molybdenum catalyst supported on alumina at about 575° F.

The tail gas contains some H_2 and CO. Additional reducing gas may be supplied by a small furnace equipped with a special burner or from some existing source.

Catalyst life is reported to be good and studies have shown that presulfiding of the catalyst is not critical.

Controlled burn-off followed by resulfiding is said to restore catalyst to its original activity.

The reduction step converts essentially all sulfur-containing compounds to H_2S. The hot gas is then cooled, and water is condensed in a direct-contact cooling tower. The cooled gas is then countercurrently scrubbed by an alkanolamine solution or other appropriate solvent.

Design and operation of the absorption column is aimed at achieving essentially complete removal of H_2S while coabsorbing only a fraction of the CO_2 present. Solvent selection also is used to minimize CO_2 absorption.

A conventional stripper is used to strip the acid gases from the solvent. These gases are recycled to the Claus plant inlet. The remaining tail gas, normally containing 200-300 ppmv H_2S, is then incinerated.

The recycled gas has little effect on Claus-plant operations as long as CO_2 content is held at a low level. As the CO_2 content of the SCOT feed gas increases, the amount of CO_2 recycled becomes a significant diluent. Therefore, the process may not be a good selection for direct treating of the tail gas from a Claus plant that processes a feed gas with a high CO_2, low H_2S content.

The SCOT unit can be designed to operate from 20% of design feed rate up to full rate. Changes in the feed will have only a small effect on overall sulfur recovery. The unit also reduces the sensitivity of the overall sulfur-recovery facility to variations in the air supply rate.

This process can also be designed for minimum pressure drop to make it more suitable for add-on installation. Proven and familiar equipment is used in each step of this process.

Carbon-steel construction can be used for almost all equipment in the SCOT unit. Capital costs for an add-on unit are about equal to the cost of the Claus plant. For a new facility with the SCOT design optimized for the best possible fit, the cost can be as low as 75-85% of the Claus unit.[7, 8]

Beavon process. The Beavon process, developed by Ralph M. Parsons Co. and Union Oil of California, has been used in several Claus tail gas cleanup units with good results. A number of these units are in operation with the longest experience in Claus cleanup applications.

This process is reported to be able to achieve 100 ppmv or less total sulfur in tail gas before incineration.

In this process, tail gas from the Claus unit is first treated by a reducing gas over a cobalt-molybdenum catalyst to convert all sulfur-containing species to H_2S.

The Claus gas itself usually contains a significant portion of the reducing agents required. Additional reducing gas is supplied by an auxiliary burner which also is used to maintain a temperature of 650°-700° F.

Residual concentrations of COS, CS_2, and CH_3SH from the reactor are low. The reduced gases are then cooled in a direct-contact condenser to about 300°-375° F. and contacted by a sodium carbonate-bicarbonate solution at a pH above 7. This solution cools the gas and also scrubs out any SO_2 which might have passed through the catalyst bed without being reduced.

The cooled gas then enters a Stretford absorber where it is contacted by a sodium carbonate-sodium bicarbonate solution containing sodium vanadate and an oxidation catalyst. The H_2S in the feed gas is absorbed and oxidized to sulfur by the vanadium. A reaction tank provides additional holding time for this reaction. Air is then used to oxidize the vanadium back to the pentavalent state.

The recovered sulfur forms a froth at the top of the oxidizer which is skimmed off, filtered, washed and dried, and melted. Molten sulfur flows into the sulfur tank. Reactivated solvent flows through a cooling tower to control solvent temperature between 90° to 105° F. and maintain the water balance. A solution heater is also provided to evaporate water as needed.

Beavon process metallurgy is carbon steel except the direct-contact condenser which is stainless steel. Some sections of the Stretford unit are coated with plastic to avoid corrosion by deposited sulfur.

No tail gas incinerator is required during normal operations, but some provision must be made for sour gas disposal during absorber cleanouts which occur about every 6 months, reduction catalyst changeouts, and any mechanical failure.

Operating experience has located several trouble spots in this process. Absorber plugging is reported to have been encountered as well as plugging in the sulfur froth lines. The reducing catalyst life is reported to be about 2 years.

Since no streams are recycled to the Claus unit, this process is suitable for add-on installation to any type sulfur-recovery plant if adequate plot space is available. Total investment for a Beavon tail gas cleanup unit is reported to be approximately equal to that of the parent Claus plant.[8, 9]

Wellman-Lord process. A different approach was taken in the development of the Wellman-Lord process. Several Wellman-Lord facilities have been installed in the U.S. and Japan. These plants have reported the process's ability to achieve emissions levels at about 200 ppmv SO_2 or less.

In the Wellman-Lord processing scheme, the Claus plant tail gas is incinerated then cooled to about 140° F. and fed to the absorber. It is contacted by an aqueous solution of sodium sulfite, which reacts with the SO_2 to form the bisulfite.

Steam is used to drive off the SO_2 and much of the solution water in the evaporator/crystallizer. Sodium sulfite crystals precipitate here, forming a slurry.

Gas from the evaporator/crystallizer is cooled to recover most of the vaporized water which is used to dissolve the crystals. The SO_2 gas is recycled to the front end of the Claus unit.

The solvent is also treated with sodium hydroxide to complete the regeneration process, reacting with any remaining bisulfite to form sodium sulfite and water.

As the Wellman-Lord process removes sulfur after the tail gas is incinerated, H_2S/SO_2 ratio control is not critical to design tail gas treating. Tankage can also be designed into the facility to allow design operation for up to 3 days while the regeneration cycle is down.

This process is well-suited to high CO_2 streams as it does not recycle CO_2 with the SO_2. The chemistry of the Wellman-Lord process requires exotic metallurgy, significantly increasing capital costs. These cost penalties are particularly severe for smaller installations.

For a 100 Lt/d sulfur-plant tail gas treating unit, costs are estimated to be between 130 and 150% of the parent Claus unit costs. The process chemistry and equipment are not familiar to many plant personnel, compounding operating and training difficulties.[8, 10]

New sulfur-recovery unit compliance options

It is safe to assume that a sulfur-recovery plant to be installed as a part of any new facility will have a more stringent performance requirement placed upon it than that required to satisfy the TACB Regulation II. The exact requirement will depend primarily upon location and size.

If the project is located in West Texas 30 or 40 miles from the nearest population center, the requirement imposed may be quite similar to that set by Regulation II. However, if the project is located in East Texas where future sulfur-dioxide emissions from lignite-burning power plants are predicted to approach the national secondary ambient air quality standard limits, and the size of plant is significant, the recovery requirement can be anticipated to be within the 99.8% overall recovery range.

The following is an over-simplified guide for use in arriving at the new sulfur-recovery system process configuration:

1. Try a "Best Current Technology" Claus sulfur-recovery unit first.
2. Consider a "Low Temperature Process" Claus sulfur recovery unit second.
3. Choose a Claus sulfur-recovery unit plus a tail gas-cleanup unit system as a last resort!

Table 1 summarizes the recovery capability of several sulfur-recovery systems ranging from the simple to the elaborate. This information should help in reaching a preliminary determination of the system required to satisfy a particular need.

Of course, a detailed engineering study should be performed to confirm the selections for a specific case.

References

1. "Control of Air Pollution Permits for New Construction or Modification." Texas Air Control Board Regulation VI, Draft dated Jan. 4, 1977.
2. Laengrich, A. R., et al., "Sulfur Recovery Units Designed to be More Reliable." Paper presented at 1977 annual meeting of National Petroleum Refiners Association, San Francisco, Calif., Mar. 29, 1977.
3. Goddin, C. S., et al., "Amoco CBA Process for Improving Claus Plant Recovery." Paper presented at 53rd annual convention of GPA, Denver, Colo., Mar. 25-27, 1974.
4. Cameron, L. C., and Grancher, P., "Sulfreen Process: Status of Development." Paper presented at 53rd annual convention of GPA, Denver, Colo., Mar. 25-27, 1974.
5. Andrews, J. W., et al., "Control of Claus Plant Sulfur Compound Emissions." Journal of Air Pollution Control Association, Vol. 26, No. 7, July 1976, pp. 664-667.
6. Barthel, Y., et al., "The Technology of Claus Unit Tail Gas Cleanup with the IFP Process." Paper presented at 53rd annual convention of GPA, Denver, Colo., Mar. 25-27, 1974.
7. Naber, J. E., et al., "The Shell Claus Off-Gas Treating Process." Paper presented at 74th National Meeting of American Institute of Chemical Engineers, New Orleans, La., Mar. 11-15, 1973.
8. Goar, B. G., "Today's Claus Tail Gas Cleanup Processes." Paper presented at 25th Annual Gas Conditioning Conference, University of Oklahoma, Norman, Mar. 3-5, 1975.
9. Kouzel, B., et al., "Treat Low Sulfur Gases with Beavon Sulfur Removal Process and the Improved Stretford Process." Paper presented at 27th Annual Gas Conditioning Conference, University of Oklahoma, Norman, Mar. 7-8, 1977.
10. Potter, B. H., and Earl, C. B., "The Wellman-Lord SO_2 Recovery Process." Paper presented at 23rd Annual Gas Conditioning Conference, University of Oklahoma, Norman, 1973.

Tougher air-quality standards face sulfur-recovery plants

W. D. SAULMON, Shell Oil Co., New Orleans

IT NOW appears that National New Source Performance Standards covering emissions from sour-gas treating and sulfur recovery facilities will be issued this year. These standards, in most cases, will have a significant impact on new plant design, operations, and economics.

Air-pollution regulations have had their greatest effect on the natural-gas processing industry in the area of sour-gas treating and sulfur recovery. Air-emission regulations for sulfur-recovery plants have been established at the state level. Although they can vary considerably between states, they have all grown to a more stringent plateau in recent years.

The new standards will probably be considerably tougher than existing state regulations.

Federal Clean Air Act. The Federal Clean Air Act was amended in 1970. It provides the basis for all air-emis-

Fig. 1

Capital cost comparison

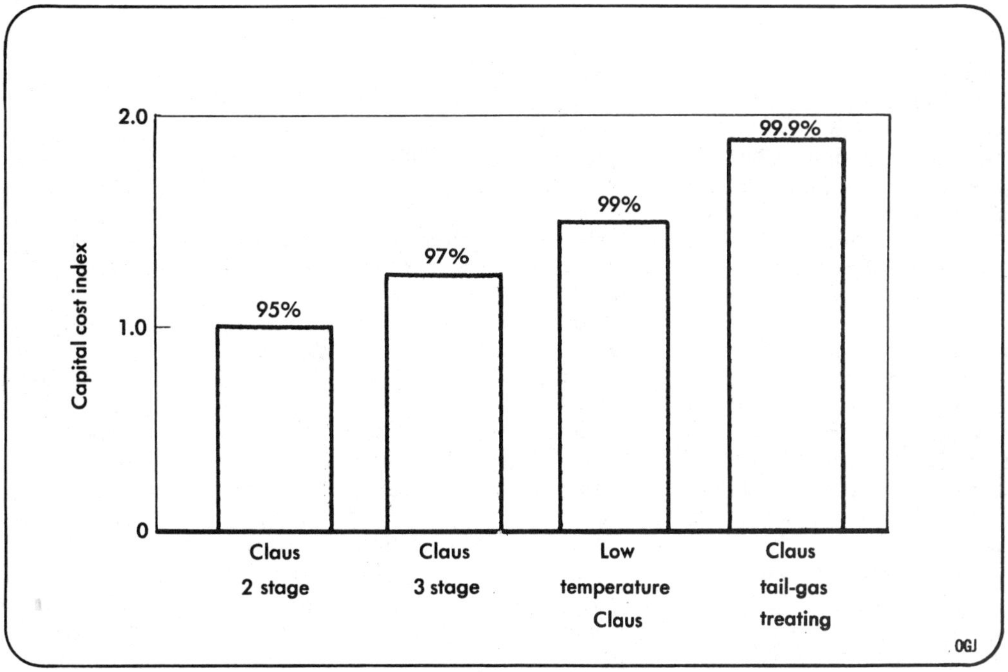

Fig. 2

Profitability comparison

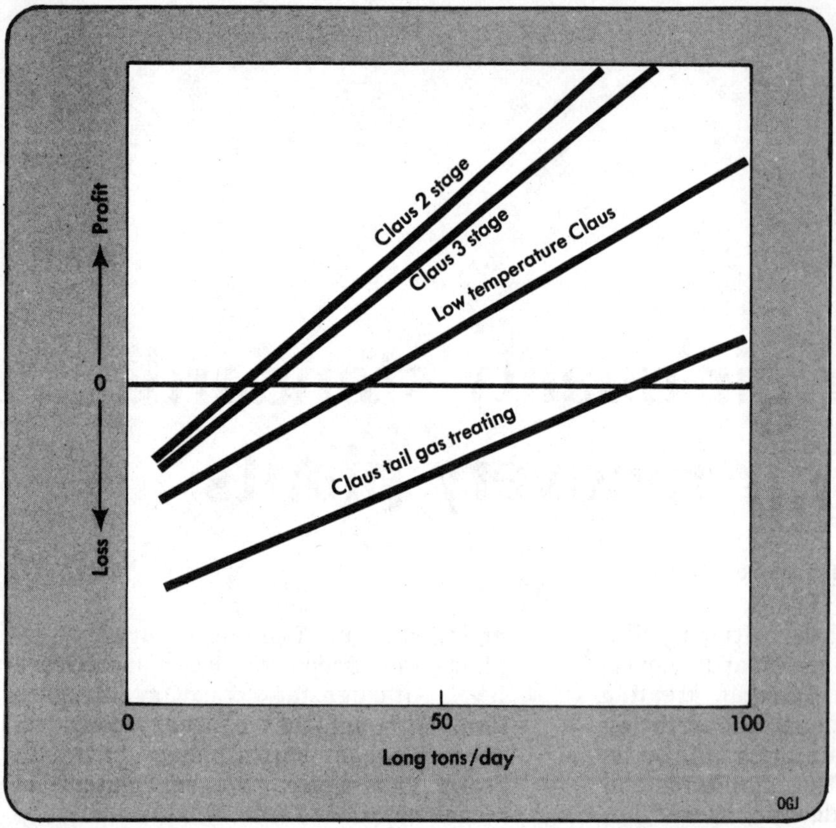

Table 1

Air regulations status

Federal
Ambient Air Standards—established for SO_2.
New Source Performance Standards—expected during 1977 for sulfur-recovery plants.
Significant Deterioration Standard—established for SO_2.
States
Ambient Air Standards—SO_2 standard may be equal to or more stringent than national, and may have H_2S standard.
Emission Limits—usually require incineration of H_2S; usually set SO_2 maximum after incineration, requiring sulfur recovery if exceeded, and may set minimum sulfur-recovery-plant performance level.

sion regulations now in effect in the U.S. EPA regulations under the act have grown more complex and broader in scope during the past 6 years. Significant provisions of the act are: (1) establish ambient air standards, (2) provide new source-performance standards, (3) require state implementation plans, and (4) include development of the significant deterioration standard.

Maximum Ambient Air Quality Standards have been established which define the maximum allowable ground-level concentrations for six major air pollutants. These include sulfur dioxide, particulate matter, carbon monoxide, photochemical oxidants, hydrocarbons, and nitrogen dioxide.

Primary standards are defined as those necessary to protect the public health. Secondary standards are defined on a broader basis of protecting the public welfare. Secondary standards are more stringent and are seen as a longer-term goal to be achieved.

New-source performance standards have been issued for some categories of plants to insure that facilities constructed or modified after a given date incorporate the best system of pollutant-emission reduction (considering cost) that has been adequately demonstrated. These standards directly limit the amount of emissions to the atmosphere, but do not necessarily achieve any specific air-quality level.

Adapted from paper titled "Air Quality Regulations for Sulfur Recovery plants—Where are We?", presented at Gas Processors Association 56th Annual Convention, Dallas, Mar. 21-23, 1977.

Under the Clean Air Act, states have the primary responsibility to prevent and control air pollution at its source. But if a state does not assume this responsibility, EPA will take it over. States are required to submit implementation plans for EPA approval which will provide for achieving and enforcing national ambient standards in the various air-control regions in each state.

Implementation plans were to achieve primary standards by mid-1975 and secondary standards in a reasonable time thereafter. The states may adopt their own standards if they are as stringent as national standards.

During 1972, EPA, acting on court order, disapproved all state implementation plans because they did not provide for prevention of significant deterioration of existing air quality. The courts ruled that the Clean Air Act required EPA to prevent air quality in relatively clean-air areas from significantly deteriorating—even if the ambient standards were being met.

Subsequently, EPA issued regulations establishing three classes of allowable incremental deterioration in air quality for total suspended particulates and sulfur dioxide.

These range from Class 1 in which practically any worsening of air quality would not be allowed, to Class 3 in which deterioration up to the ambient air standards would be allowed. National legislation is currently pending on the significant deterioration provisions which may change the regulations.

Other provisions of the Clean Air Act can affect plant operations. EPA has broad authority over Federal enforcement, inspections, monitoring, and right of entry. Fines for intentional violation of regulations range from $10,000 to $50,000 per day and from 6 months to 2 years imprisonment. EPA can require a plant operator to establish and maintain records, install certain monitoring equipment, and conduct tests as deemed necessary.

Current regulations. Sulfur-recovery plants associated with sour-gas-treating facilities convert the H_2S removed from the sour-gas stream to elemental sulfur. When the amount of H_2S removed from the sour-gas stream is too small to justify a sulfur-recovery plant, the H_2S is combusted to the less noxious SO_2.

This is done by flaring or incineration and venting. But state regulations usually limit the amount of SO_2 which can be emitted. This may require installation of a sulfur-recovery plant where it would not be economically justified.

Specific emission regulations for sulfur-recovery plants are currently applied at the state level. Direct interface with national regulations is usually confined to diffusion modeling to show compliance with ambient standards and significant deterioration standards. Table 1 summarizes the general status of air regulations for sour-gas treating and sulfur recovery.

Federal ambient-air standards have established the maximum ground level concentration for SO_2. The primary standard is 0.03 ppm annual arithmetic mean and 0.14 ppm maximum 24-hr average. The secondary standard for SO_2 is 0.5 ppm maximum 3-hr average, not to be exceeded more than once per year.

While Federal New Source Performance Standards (NSPS) for sulfur-recovery plants associated with sour-natural-gas-treating facilities are expected to be issued this year, they have already been issued for refinery sulfur-recovery plants.

Federal Significant Deterioration Standards have been issued for SO_2, although, as mentioned previously, pending legislation may revamp the regulations. The SO_2 standards are shown in Table 2. These are allowable increases in SO_2 concentrations over a base-line air-quality concentration.

The EPA has initially designated all areas of the U.S. as Class 2. Reclassification of an area can be done by state action involving public hearings and EPA approval. No plant construction or modification which increases emissions of SO_2 can begin after June 1, 1975, until a preconstruction review by the EPA has determined that the level of emissions will not exceed the allowable air-quality increments shown in Table 2.

The significant deterioration provisions also list sulfur-recovery plants among those facilities requiring any new or modified plant to meet the best-available controlled technology (BACT) for emissions reduction. Since NSPS have not been issued for sulfur-recovery plants, BACT is decided on a case-by-case basis.

State regulations usually adopt the Federal ambient-air standards for SO_2 and set their own emission limits for H_2S and SO_2. These emission limits establish when a sulfur-recovery plant will be required.

Many states also have specific emissions standards for sulfur-recovery plants, while some use a case-by-case approach based on meeting ambient standards. About 90% of the sour-gas-treating sulfur-recovery-plant capacity in the U.S. is located in Texas, Mississippi, Alabama, and Florida. Regulations pertaining to sour-gas treating and sulfur-recovery operations in these states are summarized in Table 3.

Table 2

SO_2 deterioration increments, $\mu g/m^3$

	Class 1	Class 2	Class 3
Annual arithmetic mean	2 (0.00075 ppm)	15 (0.0056 ppm)	(up to ambient standard)
24-hr maximum	6	100	
3-hr maximum	25	700	

Table 3

State sulfur-recovery-plant air standards

(ALABAMA, FLORIDA, MISSISSIPPI, TEXAS)

Alabama

Ambient air-quality standards
- SO_2—same as national.

General emission limits
- SO_2—no limit for streams containing up to 10 long tons/day of sulfur (rural counties).
- H_2S—emissions over 0.1 gr/scf must be combusted to maintain ground-level concentration of less than 0.02 ppm.

Sulfur-plant emission limits
- All plants in rural counties:

Sulfur, long tons/day	SO_2 emission limit
10-50	560 lb/hr
50-100	0.1 lb SO_2/lb S processed (95% recovery)
over 100	0.08 lb SO_2/lb S processed (96% recovery)

Increased SO_2 emission limits are allowed for acid-gas streams containing less than 60% H_2S. More stringent limits are set for urban counties.

Florida

Ambient air-quality standards
- SO_2—0.02 ppm annual arithmetic mean, 0.1 ppm maximum 24 hr, 0.5 ppm maximum 3 hr.

General emission limits
- H_2S—no direct flaring of H_2S allowed.

Sulfur-plant emission limits
- Existing plants (before July 1975)—0.08 lb SO_2/lb S processed (96% recovery).
- New plants (after July 1975)—0.004 lb SO_2/lb S processed (99.8% recovery).

Mississippi

Ambient all-quality standards
- SO_2—same as national

General emission limits
- SO_2—2,000 ppm
- H_2S—emissions over 16 ppm H_2S must be incinerated or otherwise processed.

Sulfur-plant emission limits
- All plants—0.12 lb SO_2/lb S processed (94.0% recovery).

Texas

Ambient air-quality standards
- SO_2—0.28 to 0.4 ppm average over 30 min.
- H_2S—0.08 to 0.12 ppm average over 30 min.

General emission limits
- SO_2—set as necessary to meet ambient air standards.
- H_2S—meet ambient standards by direct measurement or specified calculations.

Sulfur-plant emissions limit
- Existing plants—Up to 4,000 scfm, lb/hr SO_2 = 123.4 + 0.091 Q; over 4,000 scfm, lb/hr SO_2 = 0.614 $Q^{0.8042}$, where Q = stack effluent rate, scfm (must also meet standard effective stack heights).
- New plants—case-by-case basis consistent with ambient standards, NSPS and BACT.

Compliance with current standards. The Claus process is the most proven and widely used sulfur-recovery process in sour-gas treating. Claus plants in operation today range in capacity from 5 to over 1,200 long tons/d. The process as currently employed uses two or three stages of catalytic conversion.

Under the best operating conditions, a two-stage plant is capable of 95% sulfur-recovery efficiency, and a three-stage plant is capable of 97% recovery efficiency.

Many variables can affect these ideal percent efficiencies, including percent H_2S in acid gas, catalyst activity, and contaminants in acid gas. Unreacted H_2S and other sulfur compounds leaving in the tail gas from a Claus plant are usually burned to SO_2 in an incinerator and vented through a stack.

An efficient, well-maintained Claus plant is capable of meeting national ambient-air standards by a wide mar-

Table 4

Emission control cost

Facility	\$/long ton SO_2 removed
Coal burning installations	172-259
Oil desulfurization	156-356
Clause two-stage @ 10 long tons/day ..	13
Claus three-stage @ 10 long tons/day .	*116
Low-temperature Claus @ 10 long tons/day ..	*442
Claus tail-gas treating @ 10 long tons/day .	*2864

*Incremental unit cost over previous recovery level.

Table 5

Tentative NSPS for sour-gas treating and sulfur-recovery facilities

1. No gas containing over 160 ppm H_2S can be combusted to atmosphere.
2. Sulfur-recovery plant emissions shall not exceed 250 ppm of SO_2 and 300 ppm of COS, CS_2, and H_2S. (Approximately equivalent to 99.9% sulfur-recovery efficiency).
3. Gas-treating facilities handling less than 2 LT/D of sulfur are exempt from standards.

gin under normal circumstances. Continuous ambient-air monitoring and dispersion modeling have been used to verify compliance.

As listed in table 3 for state regulations, a three-stage Claus plant can meet all emission limits with the exception of Florida's new plant requirement of 99.8% recovery efficiency. This level of sulfur recovery requires a Claus plant with special tail-gas-treating facilities.

Sulfur-recovery NSPS. In the Oct. 4, 1976, Federal Register, EPA proposed standards of performance for new petroleum - refinery sulfur - recovery plants. These standards limit SO_2 emissions to 250 ppm in the tail gas discharged to the atmosphere.

However, since some emission-control systems installed to meet this SO_2 limit would emit H_2S, COS, and CS_2, the standards also limit emissions to 10 ppm H_2S and 300 ppm H_2S, COS, and CS_2.

These emission limits correspond to 99.9% sulfur-recovery efficiency.

This level of sulfur recovery will require a Claus plant with tail-gas-treating facilities or a new, highly efficient, sulfur-recovery process. Capital costs for Claus tail-gas treating would approximately double the base Claus-plant cost.

The author . . .

William D. Saulmon is project-engineering-section leader for Shell Oil Co. in the Onshore Production Division at New Orleans, La. Previous Shell assignments have included operations, design, construction, and economic studies of gas-processing facilities in Texas, Louisiana, and Mississippi. Saulmon is a 1965 chemical engineering graduate of Tulsa University and is a registered professional engineer.

W. D. Saulmon

Obviously, tail-gas treating cannot be economically justified since it would only recover the last 3 to 6% of sulfur not recovered in the Claus unit. These standards will be applicable to any refinery sulfur plant built after Oct. 4, 1976, or to any previously existing plant modified after then.

The economic impact estimated by EPA to comply with these regulations includes an investment of \$110 million between 1976 and 1980, increased annual operating costs of \$16 million/year by 1980, and a resulting petroleum-products price increase of 0.1 to 1.0%.

EPA concluded this was a reasonable impact on the refining industry.

Natural-gas sulfur-recovery plants were not covered under the standard since EPA felt the economic and environmental impact could differ significantly between refineries and gas plants.

During 1976, EPA developed background information to support NSPS for natural gas treating sulfur recovery plants. Various levels of emission control were studied. Basically, these can be summarized in three categories:

1. A sulfur recovery level of 97%, which could generally be achieved by three-stage Claus plants.

2. 99% sulfur recovery, which is achievable by processes which extend the Claus reactions at lower temperatures.

3. 99.9% sulfur recovery, which would require Claus tail-gas treating. This option would be equivalent to that proposed for refinery sulfur plants.

These control levels were considered for sulfur-recovery plants on sour-gas streams containing over 2 long tons/day of sulfur.

The cost of meeting these control levels was determined for representative model plants. Makeup of the sour-gas treating and sulfur-recovery segment of the industry was characterized and industry growth projections were made.

This information was then used to determine the economic impact of controls on the industry.

Sulfur-recovery plant capital cost is a function of: (1) percent sulfur recovery, (2) sulfur capacity, and (3) percent H_2S in the acid-gas feed. A generalized capital cost comparison between the different per cent sulfur-recovery levels is shown in Figure 1, which is based on EPA published data.

Sulfur-recovery plant capability to operate at a profit depends on revenue from sulfur sales, direct operating costs, and amortization of investment at an acceptable rate of return.

A qualitative comparison of profitability for the various control levels is shown in Fig. 2, which is based on EPA data for plants up to 100 long tons/day capacity. At \$40/long ton, the minimum break-even capacity for two or three-stage Claus plants is about 20 long tons/day. Low temperature Claus and Claus tail-gas treating plants would pay out at about 40 and 80 long tons/day, respectively.

For reference, Table 4 shows emission control unit costs for SO_2 removal for several industries and the cost for the various control levels considered for sulfur-recovery plants based on EPA data.

Some of EPA's preliminary findings were transmitted during 1976 to various interested and involved segments of the petroleum industry, including an ad hoc task group that was set up by API.

This group basically suggested that the cost effectiveness of tail-gas treating over a three-stage Claus plant was questionable at best in terms of benefit to the environment. The group also did not feel that the full economic impact of NSPS had been adequately defined.

As mentioned, the Clean Air Act provides for new-source performance standards to reflect the best system of emission reduction, considering costs, that has been adequately demonstrated.

EPA advises that judicial decisions have established that, in determining costs, it was not necessary to provide a cost-benefit analysis and that EPA was not required to justify standards that required different levels of control for different industries.

The EPA's primary concern is whether or not a new sulfur-recovery plant would be built after new-source performance standards were enacted. In other words, EPA is saying that the question is one of affordability rather than cost effectiveness.

The EPA presented a preliminary draft of tentative standards to members of the National Air Pollution Control Techniques Advisory Committee (Napctac) at an open meeting on Mar. 2, 1977. A summary of these tentative standards is shown in Table 5. If the EPA still considers these standards to be necessary and reasonable, they will be published in the Federal Register.

Formal comments will be duly considered, after which final standards would be issued; however, the effective date of the regulations will be the date of the first publication in the Federal Register.

Assuming no unusual delay in these steps, new source performance standards for natural-gas-treating sulfur-recovery plants could become effective during this year. Interested parties may receive preliminary drafts of the standards and comment on them by contacting the United States Environmental Protection Agency, Office of Air Programs, Research Triangle Park, N.C. 27711.

Although these standards primarily apply to new plants built after date of publication in the Federal Register, existing sources which are modified may also become subject to NSPS.

A physical or operational change to an existing sulfur-recovery plant which results in an increase in sulfur emissions may be considered as a modification requiring that facility to comply with NSPS.

Physical changes resulting from routine maintenance, and throughput increases not requiring capital expenditures would not be considered modifications.

However, existing sources which are reconstructed may be considered as a new source regardless of whether emissions increase or not. END

Processes clean up tail gas

THE trend to reduce emissions of air pollutants to the atmosphere and to increase the severity of air-pollution regulations has created the need for a new group of processes designed to clean up the tail gas from Claus sulfur-recovery units. Numerous tail-gas cleanup (TGCU) processes are available. These TGCU processes represent a wide spectrum of process technologies and capabilities of reducing SO_2 emissions to the atmosphere.

For most potential applications, a detailed review of all commercially viable processes should be made to determine the best or optimum process scheme to be used for reducing SO_2 emissions from a new or existing Claus plant.

The preceding comments were made by B. Gene Goar, Goar, Arrington & Associates Inc., Tyler, Tex., the concluding panelist at the panel session on H_2S removal methods at the recent Gas Processors Association Convention in New Orleans. Goar described numerous current Claus tail-gas cleanup processes, essentially as follows:

An awakened public awareness to the potential threat of air pollution has increased attention on all industrial processes which can contribute to the worsening environmental condition. The Claus process has been used in industry for over 75 years to recover elemental sulfur from hydrogen-sulfide-bearing streams.

Until recently, most Claus plants were considered economically attractive and significant in reducing SO_2 emissions to the atmosphere. But since all opinions are only relative, the role of any process must be considered as "good or bad" in view of prevailing public opinion. Now the popular viewpoint in industry appears to be that all Claus plants are potential major offenders of air-pollution regulations.

At present, the supply of sulfur is plentiful, and its price is somewhat depressed. Many operating companies consider their Claus plants as only a "necessary evil." They realize very little profit from the plant, if any, and they are under frequent scrutiny and criticism from the local air-quality control board.

Most operators are having to face the need to add a Claus tail-gas cleanup process to reduce SO_2 emissions and comply with existing or suggested regulations.

Fig. 15

Claus tail-gas cleanup processes

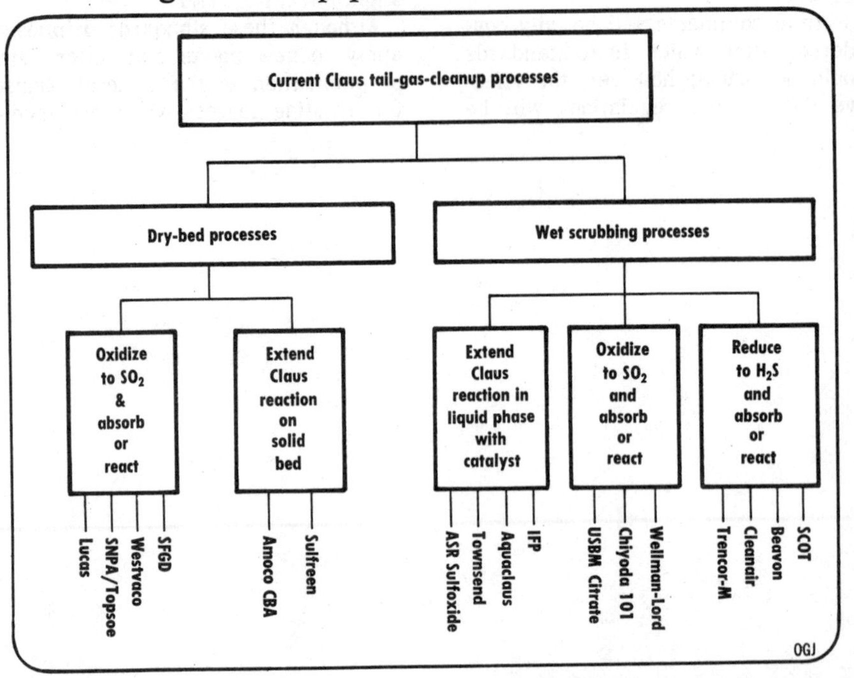

A properly designed and operated Claus sulfur plant will yield a sulfur recovery of only 93-97%, even with the use of an optimum reheat scheme and up to four catalytic converters. The tail gas from a properly designed and operated sulfur plant may contain from 8,000-28,000 ppmv of sulfurous compounds (H_2S, SO_2, COS, CS_2, S vapor).

The typical concentration of such compounds in a Claus-plant tail-gas stream is 10,000-15,000 ppmv. The federal EPA has set standards for Claus sulfur plants located in refineries which limit the concentration of SO_2 in the stack gas to 250 ppmv on a basis of zero excess air and no moisture.[1]* It is believed that EPA may extend these standards to sulfur plants located in field processing plants in the fairly near future.

Certain municipalities (Los Angeles and New York) have set a rigid standard limit for SO_2 emissions to the atmosphere of 500 ppmv, or lower. It is anticipated that more cities and states will enact standards of a maximum of 500 ppmv SO_2.

To satisfy these emission standards, the overall conversion and recovery efficiency of a Claus unit must be in the range of 99.5-99.9%. It is virtually impossible to attain such efficiencies by conventional means, regardless of how many catalytic converters are used.

To comply with the more stringent emission regulations, "tail-gas cleanup" processes have been developed. These processes are commonly called "add-on" processes, since they are added to an existing Claus plant.

There are over 10 commercial Claus tail-gas cleanup processes available. They vary widely in processing technology and capability to reduce emissions.

As would be expected, the more complex and expensive processes yield the lowest emissions.

The potential user must select the most economical process which will enable his Claus plant to comply with the air-quality regulations enforced in the area where the plant is lo-

*Such numbers designate references to be listed at the end of Part 5, Sept. 11.

cated. If a new Claus plant is being considered, the operating company should endeavor to select the optimum "Claus/cleanup" processing scheme that will result in a reliable plant, built for the minimum investment, which can be operated in compliance with emission regulations.

The available processes can be divided into two broad categories: (1) wet scrubbing processes, and (2) dry bed processes. These categories may be subdivided further as shown in Fig. 1 as (1) wet-reduction to H_2S with subsequent absorption or reaction, (2) wet-oxidation to SO_2 with subsequent absorption or reaction, (3) wet-extension of Claus reaction in liquid phase with catalyst present, (4) dry-extension of Claus reaction on a solid bed, and (5) dry-oxidation to SO_2 with subsequent adsorption or reaction.

Wet scrubbing processes

The first set of wet scrubbing processes discussed involve reduction of all sulfurous compounds to H_2S and subsequent absorption or reaction:

Shell SCOT process. The Shell Claus offgas treating (SCOT) process, announced to industry in late 1972, was developed by the Royal Dutch/Shell Laboratories in The Netherlands and is licensed in the U.S. by Shell Development Co., Houston.

The first two commercial installations were small, skid-mounted units installed for Champlin Petroleum and Douglas Oil in California.[2] There are 17 SCOT units in operation in the U.S. and Canada.

Numerous other SCOT plants (about 15) are in some stage of design, construction, or commissioning at present on a worldwide basis. The sizes of Claus plants using SCOT units vary widely from as small as 10 long tons per day (lt/d) (Douglas Oil) to 2,100 lt/d (Shell in Canada).

The process appears to be very reliable as long as the parent Claus plant is operated reasonably close to the proper conditions. If the H_2S/SO_2 ratio in the tail gas from the Claus plant is grossly "off," the SCOT unit may have operating difficulties.

However, this should not be considered an unreasonable requirement of the SCOT process. After all, if the need for a SCOT unit exists, certainly the need to operate the Claus plant properly must coexist. The SCOT process is quite sensitive to tail-gas composition changes.

If an operating company is willing to invest in a SCOT unit, it certainly should be willing to invest in adequate instrumentation, analyzers, and operator attendance to make sure the Claus/SCOT combination system has a reasonable chance of accomplishing the goal set when the SCOT unit was installed, i.e., to reduce SO_2 emissions to the atmosphere.

The SCOT process is reported to have several benefits, such as: (1) easy adaptation to an existing Claus plant, (2) use of familiar process technology and equipment, (3) easy and flexible operation, (4) elimination of secondary air and water pollution, and (5) high degree of sulfur removal over a wide range of operating conditions.[3] It is considered to be one of the most flexible processes available.

The SCOT process occurs in essentially three stages: heating and reduction; cooling and quenching; and H_2S absorption, stripping, and recycle. In the first stage, the Claus tail gas is heated to about 570° F. and reacted with hydrogen (or a reducing gas) over a cobalt molybdenum catalyst.

All sulfurous compounds (SO_2, S, COS, CS_2, etc.) are reduced to H_2S. The hot gas from these highly exothermic reactions is cooled in a waste-heat boiler and finally quenched in a water-quench tower to about ambient temperature.

The final stage involves the selective absorption of H_2S in a selective-type alkanolamine solution. The vent gas from the SCOT absorber containing about 200-500 ppmv of H_2S is normally incinerated before discharging to the atmosphere.

The rich amine is stripped in a conventional manner, and the H_2S-rich stream is recycled back to the front of the Claus plant. As a general rule, the solvent will coabsorb about 20-30% of the CO_2 present in the tail gas.[4]

When considering the use of the SCOT process with a new Claus plant, it is believed that in most instances the optimum recovery level for the Claus plant alone should be about 95%. Lower Claus plant recoveries are acceptable to this scheme, but higher Claus recoveries probably cannot be justified unless the plant is very large or some unusual circumstances exist.

For most common applications with a rich-acid-gas Claus plant feed (80% H_2S or more), only two catalytic Claus converters are justified when a SCOT unit is planned. A third Claus converter is appropriate in some unusual cases, especially when a relatively lean acid-gas feed is experienced.

The capital investment for an add-on SCOT unit (existing Claus plant) equals about 100% of the cost of the parent Claus plant.[3] For the integrated SCOT unit (new facility with SCOT optimized), the capital investment equals about 75-80% of the cost of the Claus plant.[2,3]

For a 100-lt/d Claus plant, a SCOT add-on unit would cost about $1,000,-000, installed (1973 basis),[3] or $3,500,-000 installed (1978 estimated basis).

Nearly all mechanical equipment in a SCOT plant can be fabricated from carbon steel. The reduction catalyst is relatively cheap and is estimated to have a normal 3-year life span.[2]

Operating costs for a SCOT unit on a 100-lt/d Claus plant was estimated at about $720/stream day[3] (1973 basis), or about $1,000/stream day (1978 estimated basis).

Parsons Beavon process. Ralph M. Parsons Co. and Union Oil Co. of California have developed the Beavon sulfur-removal process (BSRP) for treating Claus tail-gas streams. The process employs three distinct steps: hydrogenation of sulfurous compounds to H_2S in a catalytic converter, cooling of the converter effluent gases, and conversion of the H_2S in the tail gas from the cooler to elemental sulfur by use of the Stretford process (developed in England by the North Western Gas Board of Manchester, England).

To prove the hydrogenation-conversion step, Parsons built and operated a small pilot plant (equivalent to processing the tail gas from a 1-lt/d sulfur plant) in southwestern Canada. The unit was operated for several months, and performed very well.[5]

The test unit was moved to the Los Angeles refinery of Union Oil Co. of California, and a small Stretford plant was added. The integrated unit was operated for some time and served to prove the technical feasibility of the Beavon process.

Union Oil Co. of California installed two Beavon units for Claus plant tail-gas cleanup at its Los Angeles refinery for a cost of about $6 million (1971 basis). Each unit combines a 100-lt/d sulfur plant with a Beavon process plant.

The Beavon process is licensed to industry by Union Oil Co. of California. In excess of 20 Beavon units is in commercial operation. Several other Beavon units are in some stage of design or construction.

The typical tail gas from a properly designed and operated Claus unit will contain CO_2, N_2, and H_2O vapor, plus sulfur compounds in concentrations shown in Table 15.[6]

The concentration of contaminants is obviously too high to meet new regulations. Parsons developed their Beavon tail-gas cleanup process to satisfy the overall efficiencies required. The Beavon process is easily capable of surpassing the goal of 250-

Table 15

Claus tail gas

Component	Claus sulfur-plant tail-gas composition range, ppmv (dry basis)
H_2S	5,000-12,000
SO_2	2,500- 6,000
CS_2	300- 5,000
COS	300- 5,000
S_8 (vapor)	100- 200

Table 16

Beavon plant

Component	Sulfur-plant tail-gas	Gas from Beavon process reactor	Gas from Stretford absorber
H_2S, ppmv	30,000	40,000	5
SO_2, ppmv	4,000	<20	<20
COS, ppm	1,400	50	50
CS_2, ppmv	2,500	20	20
CH_3SH, ppmv	<5	<5	<5

ppmv SO_2 in the final clean tail gas from the process. Typical results from the Beavon demonstration plant at Union's refinery are shown in Table 16.[6]

It is apparent that the final clean tail-gas sulfur-compounds content from the Beavon unit was well below 250 ppmv. But, the Stretford process does not remove COS, CS_2, or CH_3SH from the gas.

In the Beavon process, the tail gas from the Claus unit is first heated by mixing with "fuel gas and air" combustion products, and then fed to the catalytic converter which contains a cobalt molybdate catalyst, which is both rugged and cheap.[7]

Partial combustion of fuel gas in an in-line burner supplies hydrogen required for reduction of all sulfur compounds to H_2S and simultaneously raises the tail-gas stream temperature to the level required for the hydrogenation reactions.

The converter effluent gases then enter a direct-contact condenser where most of the H_2O vapor is condensed and the tail gas is cooled. The purge water produced from the condenser is good quality, having only a small amount of dissolved H_2S. H_2S removal in a small "sour-water stripper" makes it suitable for cooling-tower makeup water.[6]

A recirculation-water loop with a shell-and-tube cooler maintains a flow of water into the top of the direct-contactor condenser and out of the bottom. A small bleed or purge stream of water goes to further processing.

The cooled gas then enters the Stretford absorber where H_2S is removed almost quantitatively by the circulating Stretford solution. The treated gas from the absorber is odorless and may be vented directly to the atmosphere or incinerated before discharge.[6]

The Stretford process portion of the Beavon process uses an aqueous solution of sodium carbonate which reacts with the H_2S to form NaHS, as:[7]

$$H_2S + Na_2CO_3 \rightarrow NaHS + NaHCO_3$$

The hydrosulfide is oxidized to sulfur by sodium vanadate also in the solution.

$$NaHS + NaHCO_3 + 2NaVO_3 \rightarrow S + Na_2V_2O_5 + Na_2CO_3 + H_2O$$

Subsequently, the vanadium is oxidized back to the pentavalent state by blowing with air, with anthraquinone disulfonic acid (ADA) or sodium anthraquinone disulfonate acting as an oxidation catalyst:

$$Na_2V_2O_5 + \tfrac{1}{2}O_2 \xrightarrow{ADA} 2NaVO_3$$

The Stretford process takes place essentially in five steps: H_2S is absorbed by the alkali, the $NaVO_3$ is reduced by a sulfur shift, elemental sulfur is liberated by dissolved oxygen in the solution, the $NaVO_3$ is reoxidized in the presence of ADA, and the alkaline solution is oxygenated by air blowing, which also floats the sulfur out of solution.[8]

The spent absorbent solution flows from the absorber to the regenerator (oxidizer) where air is sparged or stirred into the tower or tank. Tiny particles of sulfur collect as a froth at the top of the oxidizer tower, are skimmed off, and go down a duct to a filter or centrifuge for removal of the filtrate from the sulfur cake. The clear (regenerated) absorbent solution and the filtrate from the filter are recycled back to the absorber.[7]

The sulfur cake may be washed, dried, melted, and added to the product from the parent Claus unit. The solid sulfur produced from the Stretford process is usually melted in an autoclave.

The Japanese Takahax[7] process is essentially the same as the Stretford process, except for the chemicals used. Takahax uses an absorbent solution of sodium carbonate; 1,4-naphthoquinone; and 2-sulfonate sodium.[8] The Takahax process could be used in conjunction with the Beavon process in lieu of the Stretford process, if desired.

With the exception of the hottest section of the direct-contact condenser, which requires stainless steel, the entire Beavon process plant including the Stretford plant may be constructed of carbon steel. Certain sections of the Stretford unit are usually coated with plastic to prevent corrosion by deposited sulfur.

Utility requirements for a Beavon process plant treating the tail gas from a typical 100-lt/d Claus sulfur plant are shown in Table 17:[32]

Table 17

Utility need

(FOR A BEAVON PLANT)

Utility	Requirement
Power (operating), kw	300
Boiler feedwater makeup, gpm	7
Fuel gas, MMBTU/hr	5.2

In general, the total investment, including royalty in a Beavon process plant for tail-gas treating, will approximately equal that of the parent Claus sulfur plant.[6] It is felt that the Beavon process will meet or exceed current as well as anticipated future SO_2 air-pollution standards now under consideration in all major industrial areas. Normally, the Beavon process should be considered for larger Claus plants (100 lt/d or larger).

Pritchard Cleanair process. J. F. Pritchard & Co., Kansas City, Mo., is developing a process for the treatment of tail-gas streams from Claus plants. The process, named Cleanair, is a joint effort of Pritchard and Texas Gulf Inc.

The process is being designed to provide an overall removal efficiency of 99.9% of the sulfur compounds fed to the Claus unit. The Pritchard process bears a similarity to Parsons' Beavon, in that final cleanup is accomplished in a Stretford unit. Cleanair is designed around three separate process steps.

The first step is the reduction of the COS and CS_2 compounds from about 12,000 ppmv to 150 ppmv by operating the Claus reactors at elevated temperatures.

The second step is designed to remove SO_2 and elemental sulfur in the tail gas. This is accomplished by aqueous scrubbing in a tower which quenches the offgas from 270° F. to 120° F.

The third and final step of Cleanair is a Stretford unit which is a proven design for the removal of H_2S from acid-gas streams. Pritchard projects that effluent gas from the Stretford tower will contain less than 250 ppm

Table 19

Wellman-Lord plants

Item	200-lt/d size 1972 basis[12]	200-lt/d size 1978 est. basis	1,500-lt/d size 1972 basis[12]	1,500-lt/d size 1978 est. basis
Capital cost W-L system w/waste-heat boiler	$1,550,000	$2,575,000	$3,440,000	$5,700,000
Operating cost, W-L system, $/yr	$ 428,000	$ 650,000	$ 908,400	$1,360,000

erated, then cooled in a waste-heat boiler and by an adiabatic water-quench system to 140° F., and finally fed to the W-L absorber for SO_2 removal. The product SO_2 released from the absorbent in the evaporator-crystallizer is cooled and recycled to the front end of the Claus unit.[13]

In the Toa Nenryo plant, the W-L absorber is composed of a prescrubber section (one sieve tray) and two SO_2 absorption stages (two sieve trays for each stage). Also, this plant uses a two-stage glass-lined evaporator-crystallizer vessel. Ceramic booster compressors are used to recycle the wet product SO_2 back to the front end of the Claus unit at about 10-in. water-differential pressure.

The Toa Nenryo W-L unit operation includes the use of a proprietary chemical additive for reduction of sulfate formation in the absorber. Indications are that the oxidation rate (to sulfate) is reduced to about one-half, with consequent savings in caustic makeup to the W-L unit.[13]

Table 19 shows economics of the W-L process for Claus-unit tail-gas treatment.

For a Claus plant of 100-lt/d size, a W-L unit for cleanup of the tail gas would cost about 130-250% of the cost of the parent Claus unit. For smaller Claus units, a companion W-L unit would cost 200% or more of the parent Claus-plant investment.

Since the W-L process uses considerable exotic metallurgy, the W-L plants are relatively expensive. It is probably not feasible to consider the W-L process for small Claus plants (50 lt/d and smaller).

Many gas-plant and refinery operators are not very receptive to the W-L process because of the unfamiliarity of the process chemistry and chemical-plant type equipment employed. They prefer a plant that uses more familiar technology and equipment. As general guidelines, for the W-L to be attractive for treatment of Claus tail-gas streams, one or more of the following conditions should exist:

- A single-train large Claus plant should be considered (200-500-lt/d size).
- Multiple-train Claus units (100-150-lt/d size each) should exist at a single location. Several W-L absorbers are used, but only one central regeneration system.
- The Claus tail gas contains appreciable CO_2 (50-70 vol % or greater). If the tail gas from a Claus plant contains appreciable CO_2, the W-L process may be a good choice since it does not create a CO_2 recycle problem.

In some W-L installations, chemical storage tanks are used to store both the lean and rich solutions (for 2-3 days). Then, if the regeneration section is down for a brief maintenance period, the absorber section(s) can be operated to allow operation of the Claus plant(s) without exceeding emission limitations.

Chiyoda Thoroughbred 101 process. Chiyoda Chemical Engineering & Construction Co. Ltd. of Yokohama, Japan, has developed a process for SO_2 removal called the Chiyoda "Thoroughbred 101" flue-gas-desulfurization process. It was tested and proved during 150 days of operation of a pilot plant in Japan which handled 1,000 normal cu m/hr of flue gas.[14]

Eight commercial plants using the process have been installed in Japan, three of which are treating tail gas from Claus plants, and others are planned. One small Chiyoda plant is under construction in Florida.[15]

The Chiyoda process has three basic stages: absorption, oxidation, and crystallization. For a Claus-plant tail-gas application, the tail gas is incinerated initially and then cooled. After scrubbing with water, the flue gas goes to the absorber where a circulating stream of dilute sulfuric acid absorbs the SO_2.

The treated gas leaves the absorber and passes on to the stack. The acid bottoms from the absorber flow to the oxidizer where air is blown into the tower. The oxidizing catalyst is an inexpensive nonpoisonous compound that is soluble in the acid.

Part of the acid goes from the oxidizer back to the absorber while the rest goes to a crystallizer. Limestone is mixed with the acid solution in the crystallizer to form gypsum crystals and the bottoms go to a centrifuge for separation of the crystals.

The mother liquor and wash water are recycled to the absorber, and the gypsum crystals may be dried and stored.[14] There are no waste-effluent streams from the process. Commercially usable gypsum is the by-product produced.

Chiyoda claims that the process is nonclogging, inexpensive to install and operate, and has high process flexibility. It claims SO_2 removal of 97% for certain applications.[14]

For a Claus plant, an emission level of 500-ppmv SO_2 can be achieved if the incinerated tail gas does not contain greater than about 15,000-ppmv SO_2.

Since the Chiyoda process produces gypsum as a by-product, it is not considered to be of great commercial significance in the U.S. There appears to be a better market for gypsum in Japan than in the U.S. or Canada.

Also, since the process involves handling sulfuric acid (exotic metallurgy) and operation of chemical-plant type equipment, it will probably never gain wide acceptance in the refining and gas-processing industries in the U.S. or Canada.

USBM Citrate process. The U.S. Bureau of Mines has been working on development of its "Citrate" or "buffered SO_2-H_2S" process for a number of years. The process is still in a development stage, with two pilot plants having been built and tested extensively. The process is reported to be capable of removing 95-99% of the SO_2 present in a tail-gas stream.[16]

In this process scheme, the Claus tail gas is first incinerated and cooled by conventional means. Then the gas flows to an absorption tower where SO_2 is absorbed in an aqueous solution of citric acid and other carboxylates. The loaded or rich solution flows to a stirred-reactor vessel where H_2S is added to precipitate elemental sulfur.

The sulfur is concentrated by oil flotation and is ultimately melted and drawn off from the system as a liquid. The H_2S required for the reaction step is taken from the feed stream to the Claus plant.[16]

The USBM made a cost estimate for a Citrate-process plant to treat the tail gas from an 85-lt/d Claus plant. The estimated capital cost was about $2,500,000 (1973 basis),[16] or $4,000,000 (1978 estimated basis), which represents about 250% of the cost of the parent Claus plant.

In its present state of development, the Citrate process is considered to be very expensive. No commercial plants using this process have been built to date. END

More Claus cleanup processes explained

THE fifth and concluding panelist on H_2S removal methods at the recent panel session held during the Gas Processors Convention in New Orleans went into a number of processes available to gas processors to meet air-pollution regulations in cleaning up the tail gas from Claus sulfur-recovery units.

The panelist was B. Gene Goar, Goar, Arrington & Associates Inc., Tyler, Tex. In Part 4 of the panel discussion (OGJ, Aug. 28), Goar went into two sets of wet scrubbing processes for Claus tail-gas cleanup. In this final installment, he goes into a third set of wet scrubbing processes and two sets of dry-bed processes.

Table 20

IFP process results

(Basis: 100-lt/d Claus plant)

	Case			
	A*	B	C	D
Claus recovery, %	95.0	95.0	95.0	95.0
IFP unit recovery, %	90.0	80.0	70.0	60.0
Overall recovery (Claus + IFP), %	99.4	99.0	98.6	98.1
Relative SO_2 emissions to atmosphere, ppmv	2,000	4,000	6,000	8,000
Battery-limits investment for IFP unit, $1,000 (1975 basis)	900	580	500	420
IFP solvent & catalyst consumption, $/hr	1.67	1.65	1.57	1.54
Utilities:				
Condensate (lb/hr)	370	320	280	250
Power (kw-hr/hr)	21	17	16	15

*Case A investment represents about 65% of the cost of the parent Claus plant. A 100-lt/d Claus plant would cost about $3,500,000 (1978 estimated basis).

Wet scrubbing processes

The third set of wet scrubbing processes involves an extension of the Claus reaction in the liquid phase with a catalyst present:

IFP process. Institut Francais du Pétrole (IFP), of Rueil-Malmaison, France, has developed a process for treatment of tail gases from Claus sulfur plants and flue gases from boilers. The process was thoroughly piloted in France, and IFP now has over 25 commercial plants (Claus tail gas) in operation or under construction. Sixteen units are now in operation.[17]

The IFP process is quite different from the processes described thus far. In the IFP process, the Claus unit tail gas, at about 260° F., enters a vertical, packed-tower reactor vessel. The gas is contacted countercurrently by a proprietary catalyst/solvent liquid solution.

The solution is recirculated around the reactor (from bottom to top), and catalyst/solvent makeup and steam condensate is added to the solution entering the reactor.

The liquid temperature in the reactor is maintained at 260°-280° F. by the steam condensate injected into the solution entering the reactor. The catalyst/solvent solution has good thermal and chemical stability.

The solvent is reported to be a polyethylene glycol (molecular weight—about 400) containing a metal-salt catalyst.[17, 29] No liquid foaming problems have been encountered. The treated tail gas leaves the top of the reactor, at about 260° F., after incurring a total pressure drop through the system of less than 1 psi. The gas may be incinerated and flowed to a stack.

Liquid sulfur, at about 270° F., is withdrawn from the bottom of the reactor vessel. The sulfur purity is reported to be 99.7% pure.

The basic reaction of the IFP process is the same as in the Claus unit:

$$2\,H_2S + SO_2 \xrightarrow{\text{Liquid catalyst}} 3S + H_2O$$

The H_2O and SO_2 dissolve in the solvent and, in the presence of the catalyst, react to form elemental sulfur. Sulfur is only slightly soluble in the solution. The IFP process does not reduce COS and CS_2 levels in the feed gas to it.

While appreciable quantities of these components are formed in the Claus furnace, IFP recommends two steps to reduce the COS and CS_2 levels in the Claus-unit tail gas.[17, 18] First, operate the first Claus converter at a hotter temperature than usual, which enhances the conversion of COS and CS_2 back to elemental sulfur vapor. Second, replace the bauxite catalyst normally used in Claus converters with a more-sophisticated catalyst which tends to promote COS and CS_2 conversion.

By using these recommendations and installing an IFP plant, IFP states that an overall sulfur recovery of higher than 99.3% can be achieved.[17, 18] This would probably correspond to an SO_2 concentration in the clean tail gas from the IFP unit of about 1,500-2,000 ppmv.

Steam condensate is injected into the reactor to be vaporized and remove the exothermic heat of H_2S and SO_2 reacting to form sulfur. Since there is no buildup of water in the system, corrosion is essentially nil, and the unit can be fabricated of carbon steel.

IFP points out that a new "Claus/IFP" system should be optimized to comply with emission regulations for the minimum investment.[17] Normally IFP will guarantee 90% conversion of all H_2S and SO_2 in the tail gas when the Claus plant is operated properly. This will yield a tail gas with 1,000-2,000-ppmv SO_2 to be vented to the atmosphere. If local regulations permit these levels of SO_2 emissions, the IFP process may be the best choice.

If higher emissions are permitted, there is a significant investment incentive to design the IFP unit for less than 90% conversion of the H_2S and SO_2 in the tail gas. Assuming a case of a 100-long-tons-per-day (lt/d) Claus plant with two catalytic stages and overall recovery of 95%, Table 20 illustrates the savings permitted in the IFP unit cost, provided the additional SO_2 emissions can be tolerated.[17]

Of all Claus tail-gas-cleanup processes which are able to reduce the SO_2 emissions down to 1,000-2,000 ppmv, the IFP process is probably the simplest in design and operation, and requires the lowest capital investment. As far as is known, there is no lower limit of size for Claus plants to which the IFP process can be economically adapted. However, if SO_2 emissions levels of 200-500 ppmv are required, some other process must be selected in most instances.

Stauffer Aquaclaus process. Stauffer Chemical Co. has developed the new

"Aquaclaus" process. This wet absorption system is said to be capable of producing a treated gas which contains less than 100 ppmv of SO_2.[19]

In this process, the Claus tail gas is first incinerated to convert all sulfur-bearing compounds (such as H_2S, COS, CS_2, etc.) to SO_2. Then the stream is cooled in a waste-heat boiler and/or a direct-contact cooler and fed to an absorption tower. The SO_2 is absorbed by the Aquaclaus solution, aqueous sodium phosphate.

The rich solvent from the absorber is contacted with fresh H_2S feed, from the front of the Claus plant, in a vessel to form elemental sulfur by the classic Claus reaction occurring in an aqueous phase. The solution is heated and liquid sulfur is withdrawn. The Aquaclaus solution is cooled and recirculated to the absorber after sulfur is separated.[19,20]

Several undesirable side reactions that can occur in the absorber and reactor are the formation of compounds such as polythionates, thiosulfates, and sulfates. Since some of the thiosulfates are destroyed in the sulfur melting step, the thiosulfate level remains relatively constant.[19]

The sodium phosphate solution serves as a buffer to shift the reactions toward formation of bisulfites instead of the undesirable compounds. A purge stream is still required.

The process has been thoroughly tested in pilot-plant operations but no commercial plants have been reported to date. Stauffer estimates that an Aquaclaus process plant to recover sulfur (20-25 lt/d) from the tail gas of a 400-lt/d Claus plant would require a capital expenditure of about $4 million (1973 basis). This represents about 125-135% of the cost of the parent Claus plant. Operational cost for such a plant (20-lt/d tail-gas sulfur recovered) would be about $1,000,000/year.[19,20]

Townsend process. The Townsend process is similar to the IFP process in that it uses an organic solvent (such as triethylene glycol) to allow H_2S and SO_2 to react (Claus reaction) to form elemental sulfur.[21,22] The reactor operates at a temperature above the melting point of sulfur, and liquid sulfur is produced from the bottom.

This process may be applied directly to treatment of Claus-plant tail gas without any preconditioning of the gas. As far as is known, it does not remove COS or CS_2 from the gas and, therefore, has some of the same drawbacks (for attaining very low emissions) as the IFP process.

The Townsend process was tested in pilot plants in Canada, but no commercial plants have been built to date.

ASR Sulfoxide process. In 1972, Alberta Sulfur Research Ltd. (ASR) announced its ASR Sulfoxide process for removing sulfur compounds from gas streams at better than 99.9% efficiency.[23] ASR is a sulfur-industry-supported research group in Calgary University's chemistry department.

This new process appears to be most applicable to treatment of tail-gas streams from Claus plants. It is capable of producing a treated gas stream containing appreciably less than 1,000 ppmv of sulfur compounds (more typically less than 500 ppmv) when receiving a feed gas containing in excess of 30,000 ppmv of sulfur compounds.[23] So far, the process has been operated only at the laboratory-bench-scale level.

The ASR Sulfoxide process uses an organic sulfoxide as a liquid-catalyst reaction medium for the well-known Claus reaction. The process chemistry involves the initial formation of an adduct between the sulfoxide and the H_2S, which in turn forms a complex with the other sulfur compounds.

The oxidation-reduction reactions occur in this complex to yield H_2O, CO_2, and sulfur. Low concentrations of H_2S and SO_2 in tail-gas streams can be reacted virtually to completion.

A most-important factor in the process is its ability to convert COS and CS_2 to sulfur. The process can convert better than 70% of the COS and CS_2 present to sulfur. As mentioned earlier on the IFP process, conversion of COS and CS_2 to sulfur has been a serious problem in reaching effluent sulfur levels below 1,000 ppmv.

The process operates at ambient temperatures and produces crystalline sulfur in solution. The recovered sulfur has the same high purity and color as the sulfur produced from the parent Claus plant.

The ASR Sulfoxide process appears to have good potential; but, commercialization of the process remains to be done.

Dry-bed processes

The first set of dry-bed processes is an extension of Claus reaction on solid bed:

SNPA/Lurgi Sulfreen process. In 1968, SNPA and Lurgi (Frankfurt, Germany) combined their efforts and developed the Sulfreen process for tail-gas treatment. In 1970, a Sulfreen plant capable of handling the tail gas from a 1,000-lt/d Claus unit was put on stream at Lacq, France. The cost of the new plant was about $3.6 million (1969 basis)[24] on a battery-limit basis.

It has a capacity of handling 110 MMscfd of Claus-unit tail gas. The sulfur produced from the Sulfreen unit is 99.9% pure. The original Sulfreen unit is operating at a 75% SO_2 removal efficiency on the original charge of catalyst, and the Claus unit is operating at a 91% recovery efficiency. An overall recovery efficiency of 97.8% is being realized.[24]

Following the initial Sulfreen unit in France, 16 units are now in operation or under construction (in France, Canada, and W. Germany). Two very large Sulfreen units (2,200-lt/d Claus-unit capacity each) are in operation at Aquitaine's Ram River plant (No. 1 and No. 2).

Although the original Sulfreen unit and the Ram River No. 1 unit employ activated carbon as the Sulfreen catalyst, subsequent units employ activated alumina as the Sulfreen catalyst. The use of alumina catalyst offers lower catalyst cost, lower energy cost due to lower regeneration temperatures, lower capital plant cost due to construction of plant from carbon steel permitted by lower regeneration temperatures, and long catalyst life.

The change to alumina catalyst, along with some other process developments, has substantially reduced the investment for Sulfreen units.[25]

The Sulfreen process converts H_2S and SO_2 contained in the tail gas to sulfur at low temperatures (260°-300° F.) by extension of the classic Claus reaction:

$$2\ H_2S + SO_2 \xrightarrow{\text{Catalyst}} 3\ S + 2H_2O$$

A special activated alumina is used as adsorbent and catalyst in the reactors. This material was selected because of its high adsorption capacity and ease of desorption of sulfur deposited on the surface.[25]

At the initial Lacq unit, the tail-gas stream is passed through four of the six parallel reactors while one of the six is on a "heating" cycle and the other is on a "cooling" cycle. The cycle of switching gas flow from reactor to reactor is set by the sulfur-holding capacity of the catalyst bed in each.[24]

In regeneration of a bed, inert heating gas circulated by a blower vaporizes the sulfur and is cooled in a standard sulfur condenser (50-psig steam production) where liquid sulfur is condensed and removed. When desorption is completed, the bed is cooled before going back on stream by a portion of the cool inert gas taken from the blower discharge and flowed through the bed.[24]

Table 21

Commercial Sulfreen process units

Plant operator	Location	Claus unit capacity, lt/d	Tail gas volume, MMscfh	No. of reactors	Type catalyst in Sulfreen reactors	Start-up Year	Capital cost, $ million*
SNPA	Lacq I (France)	1,200	4.6	6	Carbon	1970	3.6
Aquitaine Canada	Ram River I (Canada)	2,200	7.9	6	Carbon—conv. to Alumina	1972	6.5
Aquitaine Canada	Ram River II (Canada)	2,200	7.9	3	Alumina	1973	3.2
SNPA	Lacq II (France)	1,800	7.5	3	Alumina	1974	3.3
SNPA	Lacq III (France)	1,600	7.0	3	Alumina	1974	3.3
Westcoast Trans.	Fort Nelson (Canada)	400	7.9	3	Alumina	1975	
Petrogas	Balzac (Canada)	1,220	5.0	3	Alumina	1975	
Elf	Feyzin (France)	90	0.32	2	Alumina	1976	0.42
Hudson Bay	Fox Creek Kaybob (Canada)	1,100	3.7	4	Alumina	1976	
Texas Gulf	Whitecourt (Canada)	1,000	4.0	4	Alumina	1976	
Esso	Karlsruhe (W. Germany)	47	0.17	2	Alumina	1976	
Esso	Ingolstadt (W. Germany)	50	0.19	2	Alumina	1976	
U.S.S.R.	Orenbourg	1,060	4.4	3	Alumina	1978	
U.S.S.R.	Orenbourg	1,060	4.4	3	Alumina	1978	
Irving Oil-Chevron	St. John (Canada)	200	0.84	2	Alumina	1977	
Elf	Spire (W. Germany)	40	0.17	2	Alumina	1977	
Mobil/Veba	Neustadt (W. Germany)	40	0.17	2	Alumina	1977	

*Capital costs are based on value in the year the plant was built. For the initial Sulfreen unit, its capital cost was equal to about 50% of the parent Claus plant. In current larger sizes, the relationship decreases to a cost of 30-40% of the parent Claus plant.

Sulfur adsorption and desorption result in Sulfreen being a batch or semi-continuous process. A new, improved flow scheme and design is now being offered by SNPA.[33]

SNPA claims overall sulfur recoveries (Claus+Sulfreen) of up to 99%. The percent recovery attainable by the Sulfreen unit depends upon the concentration of H_2S and SO_2 in the tail gas fed to the unit. At feed-gas values of 15,000 ppmv (H_2S+SO_2), Sulfreen conversion of 84% can be expected.

At values of 8,000 ppmv (H_2S+SO_2), Sulfreen conversion of 70% is expected. For a typical Claus/Sulfreen system, an overall theoretical conversion of 99% is attainable. Also, a concentration of 2,500 ppmv of H_2S+SO_2 is expected in the treated tail gas, and a level of 1,500 ppmv of SO_2 (or less) after incineration is expected.[25]

Table 21 shows estimated capital costs and sizes of Sulfreen units.[33]

The Sulfreen process is a viable, commercially proven process. It is quite attractive for large Claus plants, due mainly to the possibility of eliminating a third or fourth Claus converter and reducing the size and height of the tail-gas disposal stack. Also, it may be attractive for smaller plants where an SO_2 emission level of 1,500-2,000 ppmv is permitted.

Amoco CBA process. Amoco Production Co. has developed their cold bed absorption (CBA) process through the commercial stage.[26] This process is similar to the Sulfreen process, except CBA uses a process stream indigenous to the Claus plant to accomplish regeneration of the sulfur-fouled catalyst beds in the CBA reactors. The Sulfreen process uses an external stream, circulated through a gas heater, for this purpose. As with Sulfreen, the CBA process is basically an extension of the Claus reaction over a cool bed (260°-300° F.) of conventional Claus catalyst.[26]

Amoco claims overall recoveries (Claus+CBA) of 98-99.5%. They state that the cost of adding a two-bed CBA process to an existing Claus plant is about 50% of the cost of the parent Claus plant. For a 100-lt/d Claus plant, the estimated CBA process cost is $500,000. For a 500-lt/d Claus plant, the estimated CBA process cost is $1,300,-000 (1974 basis).[26]

The first commercial CBA unit was commissioned in late 1976 at Amoco's East Crossfield plant (Alberta), where it treats the tail gas from an 890-lt/d Claus sulfur plant. No other CBA operating units are installed at this time.

The second set of dry-bed processes involves oxidation of all sulfurous compounds to SO_2 and subsequent adsorption or reaction.

Shell SFGD process. Shell Oil Co. developed their Shell flue-gas-desulfurization (SFGD) process, and announced it to industry in 1970. Although directed toward SO_2 recovery from stacks of large power-generation boilers, it can be applied to cleanup of Claus tail-gas streams.

The SFGD process uses a fixed bed of copper oxide-on-alumina to adsorb SO_2 from hot flue gases at an optimum temperature of about 750° F. Two or more beds are employed and a swing-bed scheme is used to adsorb, regenerate, adsorb, etc.

The SO_2 is desorbed from the adsorbent, at about 750° F., by addition of a hot reducing gas such as H_2 or H_2/CO mixture. The SO_2 may be used to produce sulfur, sulfuric acid, or other by-products. SO_2 recovery of the process is reportedly about 90%.[27]

In the version of the SFGD process which may be applied to Claus plants, the tail gas is first incinerated to oxidize all sulfur compounds to SO_2. The gases are cooled somewhat (to about 750° F.) and are passed to the CuO acceptor bed where SO_2 is removed.

The SO_2 produced upon regeneration is recycled back to the front of the Claus plant.[28] The treated tail gas from the CuO acceptor bed may contain from 1,500-3,000 ppmv of SO_2.

The first commercial SFGD process unit was installed at the Showa-Yokkaichi Oil Co. in Tokyo, Japan, to remove SO_2 from 120,000 nm^3/hr (4,500 Mscf/hr) of flue gas from power boilers and a Claus plant. The plant was started up in 1973, and appears to be operating satisfactorily.[27] The Claus unit recovers about 55 lt/d of sulfur.

The estimated cost of an SFGD unit to recover about 16 lt/d of sulfur (SO_2 equivalent) from a 100-lt/d Claus-plant tail-gas stream is about $3.2 million (1973 basis).[27] This equals 250% of the cost of the parent Claus plant.

The main potential application of the SFGD process would be in a large refinery, where the operator is forced to install an SO_2 removal process for cleanup of all boiler, furnace, and Claus-plant stack gases. The incremental cost in the SFGD process unit for the Claus-plant portion alone would be small compared with the overall project cost.

The SFGD process is licensed by Universal Oil Products (UOP).

Westvaco process. The Westvaco Corp. of Charleston, S.C., has developed a new activated-carbon adsorption process for SO_2 removal from stack gases.[28 29 30] The process is in development and is reportedly a few years away from commercialization.[30]

In the version of the process that can be applied to Claus plants, the Claus tail gas is first incinerated at 1,000° F. and diluted with air to bring the oxygen level to about 3.5 vol %. The gas is cooled in three steps to 200° F. and then flows to a three-stage SO_2 adsorber. This is a continuous, countercurrent, multistage, fluidized-bed adsorber, with carbon particles flowing down and tail gas flowing up.

The SO_2 is adsorbed from the gas as sulfuric acid by the activated carbon. The treated gas leaves the adsorber containing less than 200-ppmv SO_2. The SO_2 is released from the H_2SO_4 carbon in the regenerator and is recycled back to the front of the Claus plant.[30]

Westvaco has estimated the cost of its SO_2-recovery process to treat tail gas from a 400-lt/d Claus plant at $760,000 (1972 basis).[30] It is the author's opinion that this estimated cost is quite low and based on premature information, or either is a typographical error in the literature.[30]

SNPA/Topsoe catalytic oxidation process. Soc. Nationale des Pétroles d'Aquitaine (SNPA) of Lacq, France, and Haldor Topsoe of Vedbaeck, Denmark, have developed a wet-contact catalytic-oxidation process for treating Claus-unit tail gases. Such an SNPA process plant was built at Lacq, France, in 1965, for treating tail gases from a 1,000-lt/d Claus unit.

The SNPA unit produces about 200 lt/d of 94 wt % H_2SO_4 recovered from the sulfur compounds in the tail gas, which represent 130 lt/d of SO_2. It should be noted that the acid is produced without formation of sulfuric-acid "mist" which is usually formed in the wet-contact acid process.[31]

In the SNPA/Topsoe process, the Claus tail gases are first incinerated to transform all sulfur to SO_2. The gases are cooled in a waste-heat boiler to 790° F. They are then passed through a converter containing a vanadium oxide-base catalyst where SO_2 is oxidized to SO_3 with a 95% yield.

The converted effluent gases are cooled in a boiler feedwater economizer to 570° F., and then go through an acid concentrator and on to the absorber in which SO_3 is absorbed to form 80 wt % H_2SO_4.

This "weak" acid is then sent to the concentrator in which heat from the incoming gases evaporates part of the H_2O to produce a 94 wt % H_2SO_4. The product acid is cooled and sent to storage. The clean tail gas from the absorber may be reheated or sent to the stack directly.

The cost of such a plant, as built at Lacq, France, for treating 3.5 MMscfh of tail gas is in excess of $2 million (1970 basis).[31] An SO_2 removal efficiency of 90% is reported.[28]

As far as can be determined, the one plant built at Lacq, France, is the only commercial installation of this process. Therefore, the SNPA/Topsoe process is not considered to be of wide commercialization to date.

Lurgi Lucas process. In 1976, Lurgi Mineraloeltechnik G.m.b.H. of West Germany announced its Lucas process for cleanup of Claus tail-gas streams and lean-H_2S streams. One semicommercial plant was built and tested at Wintershall AG's Lingen refinery in West Germany. As far as can be determined, this is the only plant built to date.[34, 35]

The Lucas process consists of three main steps: (1) incineration of the Claus tail-gas stream, (2) slight cooling of the incinerated tail gas (from 580° C. to 380° C.) and treatment of the gas with hot coke, and (3) further cooling of the gas and absorption of SO_2 from the cooled gas stream with a solution of aqueous alkali phosphate.[34]

The coke treatment step converts certain undesired impurities to harmless compounds such as SO_2, CO_2, N_2, and H_2O. Various grades of coke may be used for the oxidation step; but "green" petroleum coke appears to be the best.[34]

The SO_2 absorbed by the solution is stripped from the liquid in a regenerator tower and recycled to the front of the Claus unit. Lurgi claims an overall sulfur-recovery efficiency for a Claus/Lucas facility of 99.9%.

The vent gas (offgas) from the Lucas absorber is reported to be practically free of H_2S and to contain less than 200 ppmv of SO_2 and less than 150 ppmv of COS and CS_2.[34]

Lurgi states that the cost of a Lucas unit for Claus tail-gas cleanup will equal about 70-80% of the cost of a parent two-stage Claus unit.[34]

Acknowledgment

This presentation is an updated paper, "Today's Claus Tail Gas Clean-up Processes," presented at the 1975 Gas Conditioning Conference, Norman, Okla.

References

1. U.S. Environmental Protection Agency, "Standards of Performance for New Stationary Sources, Petroleum Refinery Sulfur Recovery Plants," Federal Register, 40 CFR Part 60, 41, No. 193, Oct. 4, 1976, pp. 43,866-43,874.
2. "Cleaning Up Claus Offgas," Chemical Engineering, Sept. 30, 1974, pp. 56-57.
3. Naber, J.E., et al, "The Shell Claus Off-Gas Treating Process," paper No. 10F presented at 74th National Meeting of AIChE., New Orleans, Mar. 11-15, 1973.
4. Naber, J.E., et al, "Reduce Sulphur Emission With Scot Process," Energy Processing/Canada, September-October 1973, pp. 32-38.
5. Beavon, D.K., "Add-On Process Slashes Claus Tail Gas Pollution," Chemical Engineering, Dec. 13, 1971, pp. 71-73.
6. "The Beavon Sulfur Removal Process For The Purification of Sulfur Plant Tail Gas," Publication of Ralph M. Parsons Co., Los Angeles.
7. Beavon, D.K., "Prevention of Air Pollution by Sulfur Plants," Proceedings of 1971 Gas Conditioning Conference, University of Oklahoma, Norman, Okla., pp. 1-1 through 1-8.
8. Goar, B.G., "Today's Gas-Treating Processes-2," OGJ, July 19, 1971, pp. 84-86.
9. "Sulfur Stack Emissions Reduced," OGJ, July 27, 1970, p. 131.
10. "Cleanair," Publication of J.F. Pritchard Co., Kansas City, Mo.
11. Crow, J.H., and Bauman, J.C., "Versatile Process Uses Selective Absorption," Hydrocarbon Processing, October 1974, pp. 131-132.
12. "The Wellman-Lord Sulfur Dioxide Recovery Process," Publication of Davy Powergas Inc., Lakeland, Fla.
13. Potter, B.H., and Craig, T.L., "The Wellman-Lord SO_2 Recovery Process Experience in Commercial Installations," paper No. 52c presented at 71st national meeting of AIChE, Dallas, Feb. 22, 1972.
14. Morrison, J., "Japan Tries Three Stack-Gas Desulfurization Routes," OGJ, July 24, 1972, pp. 36-41.
15. Beychok, M.R., "Coping With SO_2," Chemical Engineering, Oct. 21, 1974, pp. 79-85.
16. Rosenbaum, J.B., et al, "Sulfur Dioxide Emission Control by Hydrogen Sulfide Reaction in Aqueous Solution—The Citrate System," Report of Investigation No. 7774, United States Department of Interior, Bureau of Mines, Salt Lake City, Utah, 1973.
17. Barthel, Y., et al, "The Technology of Claus Unit Tail Gas Clean-Up with the IFP Process," paper presented at 53rd annual meeting of GPA, Denver, Mar. 25-27, 1974.
18. Frankowiak, S., et al., "Partial and Total Sulfur Recovery Using IFP Process in Oil Refineries," paper No. 25d presented at 71st national meeting of AIChE, Dallas, Feb. 20-23, 1972.
19. Hayford, J.S., "Process Cleans Tail Gases," Hydrocarbon Processing, October 1973, pp. 95-96.
20. Hayford, J.S., et al, "Stauffer's Aquaclaus Process," Chemical Engineering, 69, No. 12, December 1973, pp. 54-55.
21. Townsend, F.M., and Reid, L.S., "Newest Sulfur Recovery Process," Petroleum Refiner, 37, No. 11, November 1958, p. 263.
22. Goar, B.G., "Today's Sulfur Recovery Processes," Hydrocarbon Processing, September 1968, pp. 248-252.
23. "New Process Removes Sulfur From Gas," OGJ, July 3, 1972, p. 47.
24. Martin, J.E., and Guynot, G., "The Sulfreen Process," Gas Processing/Canada, November-December 1971, pp. 26-30.
25. Cameron, L.C., and Grancher, P., "Sulfreen Process: Status of Development," paper presented at 53rd annual meeting of GPA, Denver, Mar. 25-27, 1974.
26. Goddin, C.S., et al., "CBA Process Ups Claus Recovery," Hydrocarbon Processing, October 1974, pp. 122-124.
27. Conser, R.E., and Anderson, R.F., "New Tool Combats SO_2 Emissions," OGJ, Oct. 29, 1973, pp. 81-86.
28. Barry, C.B., "Reduce Claus Sulfur Emission," Hydrocarbon Processing, April 1972, pp. 102-106.
29. Hyne, Dr. J.B., "Methods For Desulfurization of Effluent Gas Streams," OGJ, Aug. 28, 1972, pp 64-78.
30. Ball, F.J., et al., "Take Sulfur Out of Waste Gases," Hydrocarbon Processing, October 1972, pp. 125-127.
31. Bailleul, M.R., et al., "Pollution and Sulfur Plants," paper presented at 49th annual meeting of NGPA, Denver, Mar. 17-19, 1970.
32. Kouzel, B., et al., "Treat Low Sulfur Gases with Beavon Sulfur Removal Process and the Improved Stretford Process," Proceedings of 1977 Gas Conditioning Conference, University of Oklahoma, Norman, Okla. 1977.
33. Morin, M.M., and Philardeau, Y.M., "Sulfreen Process Experience at Ram River," Energy Processing/Canada, May-June 1977, pp. 30-40.
34. Doergas, A., et al., "Lucas Process Cleans Lean H_2S Streams," Hydrocarbon Processing, October 1976, pp. 110-111.
35. Anon., "Coke Key to Cleaning Claus Unit Tail Gas," OGJ, Nov. 22, 1976, pp. 142-143.

Methods for desulfurization of effluent gas streams

DR. J. B. HYNE
Alberta Sulphur Research Ltd.
Calgary, Alta.

COMMERCIAL METHODS for desulfurization of tail gas from sulfur plants, stack gases, and other effluent gas streams amount to some 20 techniques which are representative of more than 50 identifiable processes.

These techniques may be categorized by phase—whether in the gaseous state, or in solution. Which of the processes one selects for a particular application depends on such factors as:

- Allowable sulfur concentration in the final gas effluent.
- Condition of stream to be treated.
- Presence of COS and CS_2.
- End use for product.

Here is a roundup of what's now known about currently available commercial processes for effluent-gas-stream desulfurization.

Editor's note

CRITICAL attention frequently focuses on the effluent gas streams from processing plants and field installations, particularly on the sulfur concentrations in these streams. In response, many processes for sulfur reduction have been developed. This article lists 20 techniques which are representative of the more than 50 commercial processes available.

In the short time since this material was presented as a paper at the 1972 New Orleans NPGA meeting, three or four other processes have been announced. The Journal has described two — Alberta Sulphur Research's ASR Sulfoxide, July 3, p. 47; and Monsanto's Calsox, June 19, p. 78. A description of the Cataban process has appeared elsewhere.

Economics vs. concentration

There are two essential elements in the effluent-gas-stream desulfurization problem — economics and concentration. It is a well known chemical principle that pushing a reaction "to completion" becomes more difficult (and expensive) as the definition of "completion" approaches 100%. Traditionally any yield in excess of 90% total conversion has been considered excellent in industrial chemical processes.

Hardly ever has the economics of attempting to recover the last 10% been considered attractive.

The 10% "loss" from 2,000 tons/day sulfur production in a modern sour natural gas processing plant, however, clearly means 400 tons/day of sulfur dioxide vented to atmosphere. Environmental quality control—not direct economic advantage—is then the driving force to achieve conversions closer to 100%.

So the inclusion of environmental quality maintenance in the cost/benefit equation dictates that costs will have to be borne for the benefit of environmental protection.

Since the added product yield will not likely bear the added costs (particularly at $8 to $10/ton sulfur) it is essential to obtain maximum conversion enhancement for minimum cost, thus keeping the necessary product-price increase to the consumer as low as possible (see Fig. 1). The quality of the environment will be maintained only if the price is kept at an acceptable level.

The major influence of these economic considerations of tail-gas clean-up is reflected in the ease with which new processes can be incorporated into existing systems. In the long term, radical new technologies may lead to entirely different process systems which can be designed from the "ground up." The current capital investment in sour-gas processing equipment is such, however, that for some years the "add-on" improvement is likely to be the principal method for achieving the required conversion enhancement in the 90 to 100% region.

Sources of problems. The problem of tail-gas component concentrations is, in large part, the result of chemical procedures that have become traditional in the handling of sulfur-containing process gas streams. Where hydrogen sulfide is a major component of the process stream the choice has been to convert it (eventually) to sulfur and water. Since this process involves oxidation, the cheapest oxidant available, air, has been employed.

This results in massive dilution of the process stream at the tail of the plant. The water produced is carried through as steam, and the air carries in four parts of diluent nitrogen for every part of oxidizing oxygen. As a result, the unconverted 10% of the original gas stream may appear in the tail gas at concentrations of 2% (20,000 ppm) or less and at the stack exit at even lower concentrations of 0.2% (2,000 ppm). These dilution factors

Fig. 1

The economic problem

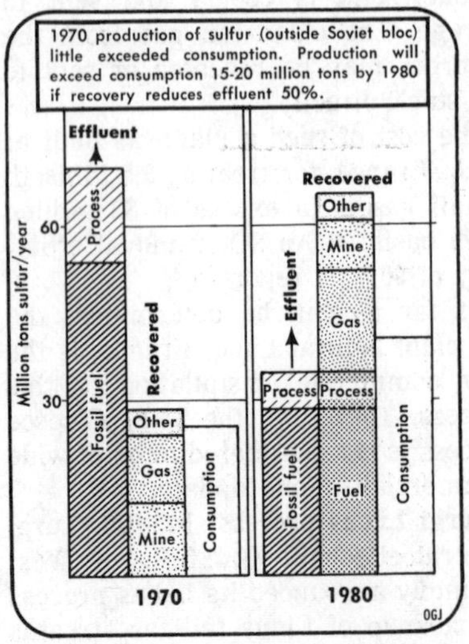

are of critical importance in considering the problems of tail gas and stack gas cleanup (Fig. 2).

The huge volumes of largely inert material that must be handled in order to react a few percent of the original components have prompted some workers to attack the problem by attempting to maintain higher concentrations. Enriched oxygen feeds at the front-end of the plant have been proposed (TRW -S100 Process[1]), thus minimizing nitrogen dilution.

Molecular sieves which selectively concentrate one of the major tail-gas components (e.g., EFCO process[2]) have been developed and tested (BHP Australia), and Lurgi Purisol[3] and GAF M-Pyrol[4] use of n-alkyl-pyrollidones represents a chemical method of attacking the concentration problem. Selective absorption under pressure and regeneration by depressurization, e.g., Selexol[3, 5], has also been examined as a possible method of overcoming high dilution problems.

While many of these procedures have attractive possibilities, virtually all represent additional steps in the overall process without actually achieving conversion to final product.

Add-on processes. Right now the problem of gas-stream desulfurization is one of finding an "add-on" type process which minimizes the added cost burden that must be carried by the existing product and which is capable of effective performance on a feed stream of very large volume and relatively low concentration of reactant species.

However, these are not the only considerations. Many other secondary factors such as corrosion, reagent losses, multiplicity of stages, secondary waste products, etc., must also be evaluated.

Basic chemistry of the desulfurization process

Sulfur can exist in a large number of oxidation states. Virtually every tail gas or stack-gas desulfurization process involves chemical reactions up and down this oxidation-state scale, as illustrated in Fig. 3. The most common oxidation states of sulfur are those typified by H_2S (-2), elemental sulfur (zero), sulfur dioxide (+4) and sulfur trioxide (+6).

Virtually every other oxidation state is represented by some chemical species or another, but these compounds are either unstable transient species found in combustion-type reactions, or are organosulfur compounds not likely to be found in commercial desulfurization processes.

The redox (reduction-oxidation) interconversion of sulfur between the oxidation states -2, 0, +4, and +6, however, is an essential feature of many current and proposed desulfurization processes. A detailed examination of the redox scale, therefore, aids in understanding and classifying the various desulfurization processes to be discussed here.

By and large the sulfur in typical tail-gas streams is mainly in the —2 and +4 oxidation states. Small amounts of SO_3 (+6) are often formed in Claus-type systems, especially when the reaction furnace at the front-end is run on the oxygen-rich side of stoichiometric ratios. But this inevitably leads to catalyst sulfation and is usually avoided. Entrained elemental sulfur mists are also common

Fig. 2

The concentration problem

Diluent effect of nitrogen in air used as oxidant in the classical Claus process is very evident. Use of pure oxygen or oxygen-enriched air feed would improve the concentration problem. Incineration of tail gas to convert all sulfur to SO_2 also introduces more dilution.

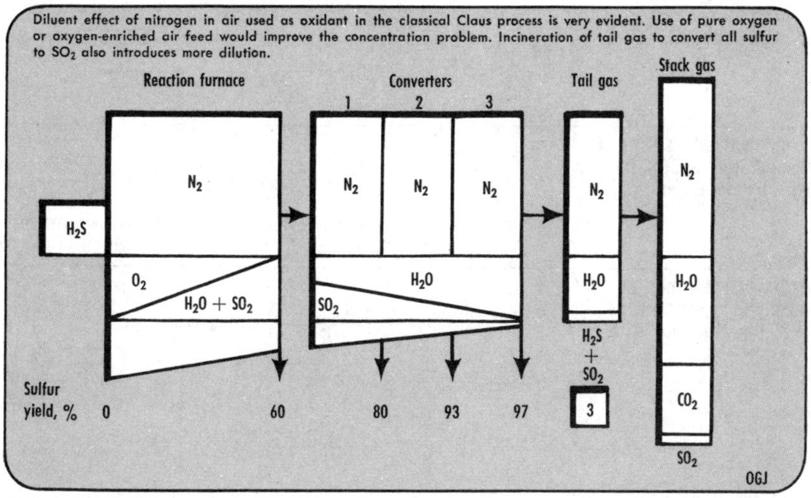

Fig. 3

Redox scale for sulfur

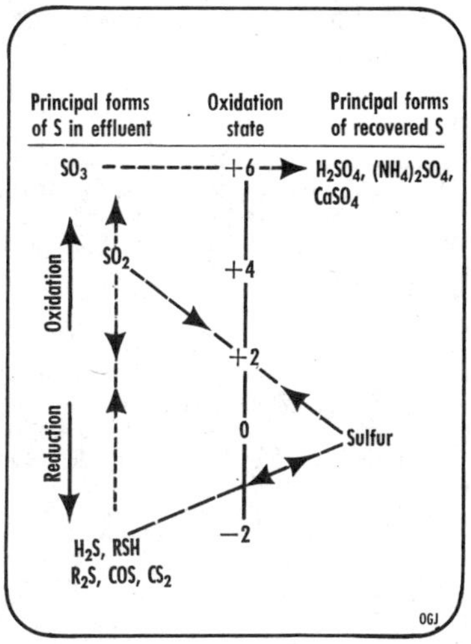

Table 1

Process for Desulfurization of Effluent Gas Streams

PROCESSES PRINCIPALLY IN THE GAS PHASE

No.	Process (Developer) (Reference)	Comments	Process Chemistry
YIELDING SULFURIC ACID OR SULFATE (S=+6)			
1	Dry limestone (TVA) (30)	Simultaneous reaction of SO_2 with lime and air oxidation of resulting sulfite to sulfate. End product slag requires suitable disposal. Swedish Bahco Process uses hydrated lime slurry (31). Lignite ash also used as absorber (Carl Still, W. Germany).	$CaCO_3 \longrightarrow CaO \xrightarrow{SO_2,\ SO_3,\ AIR} CaSO_4$ $+\ CO_2$ (TO ATMOSPHERE)
2	Manganese dioxide (Mitsubishi) (25)	Initial concentration and oxidation of SO_2 to metal sulfate with air regeneration of MnO_2 oxygen carrier and ammonium sulfate (fertilizer) production.	$SO_2 \xrightarrow[MnO_2]{} MnSO_4 \xrightarrow[AIR]{NH_4OH} (NH_4)_2SO_4 + H_2O$ REGENERATION
3	Active magnesia (Showa Hatsuden) (Chemico) (32)	Essentially a concentration process using MgO as a "collector" followed by regeneration of concentrated SO_2 stream for sulfuric acid plant feed.	$SO_2 \xrightarrow[MgO]{200°-300°F.} MgSO_3 \xrightarrow{1400°F.} 15\%\ SO_2$ GAS REGENERATION CONTACT PROCESS $\downarrow$ H_2SO_4
4	Modified contact (Monsanto-Penelec) (Tokyo Tech.) (26)	Essential Contact Process yield acid (Monsanto) or ammonium sulfate (Tokyo) but accepts hot dilute SO_2 gas stream rather than high SO_2 acid plant feed. (see also Topsoe Process)	$AIR + SO_2 \xrightarrow[V_2O_5]{900°F.} SO_3 \xrightarrow[H_2O]{} H_2SO_4$ $SO_3 \xrightarrow[2NH_4OH]{} (NH_4)_2SO_4 + H_2O$
5	Modified chamber (Tyco Labs. Mass.) (24)	Updated version of Chamber Process for sulfuric acid, modified to accept dilute SO_2 stream. NO/NO_2 couple as oxygen carrier.	$SO_2 + H_2O + NO_2 \longrightarrow H_2SO_4 + NO$ $NO \xrightarrow{1/2\ O_2} NO_2$
6	Activated carbon (Sulfacid-Lurgi) (Hitachi, Tokyo) (Reinluft, W. Germany) (Westvaco Corp. U.S.) (33, 34, 35, 36)	All method depend on absorptive powers of various forms active carbon to first concentrate and then catalyze oxidation SO_2 to SO_3 for acid or sulfate production. Fluidized, fixed and plugged flow beds variously employed.	$SO_2 \xrightarrow[ACTIVE\ CARBON]{AIR,\ H_2O} H_2SO_4$
YIELDING SULFUR (S=ZERO)			
7	Sulfreen (SNPA-Lurgi) (37)	Catalytic use of active carbon for high efficiency Claus redox reaction to yield sulfur rather than oxidation of SO_2 to $SO_4^{=}$. Requires both H_2S and SO_2 in stream.	$H_2S + SO_2 \xrightarrow[ACTIVE\ CARBON]{} H_2O + S$
8	Catalytic redox (Princeton Chem. Res.) (47, 48)	Essentially modified Claus redox with unspecified but claimed high efficiency catalyst. Where H_2S not present in feed can be generated from product S and natural gas.	$SO_2 + H_2S \xrightarrow[CATALYST]{250-350°F.} H_2O + S \longrightarrow$ $CO_2 \xleftarrow[CATALYST,\ HIGH\ TEMP.]{} CH_4$
9	CO/SO_2 redox (Chevron Research) (Univ. Mass. (29, 39)	Being examined with and without NO_x involved and at various temperatures; COS formation problem at lower temperatures.	$SO_2 + CO \longrightarrow 2CO_2 + S$ 1000°F, Cu ON Al_2O_3 NO PRESENT CONVERTED TO N_2
10	Alkalized alumina (U.S. Bur. Mines (dis-cont.) (U.K. Cent. Elect. Board) (27)	Concentration of dilute SO_2 stream on alk/alumina and in situ catalytic reduction to H_2S using reformer H_2, H_2S then to Claus with regenerated SO_2.	$SO_2 \xrightarrow[56\%\ \ 37\%]{Al_2O_3/Na_2O}$ ABSORBED $SO_2 \longleftarrow H_2 + CO$ (REFORMER) REGENERATION $S \xleftarrow[CLAUS]{SO_2} H_2S + CO_2$

PROCESSES WITH AT LEAST ONE PRINCIPAL SOLUTION STAGE

No.	Process (Developer) (Reference)	Comments	Process Chemistry
YIELDING SULFATE (S=+6)			
11	Ammoniacal solution (Cominco) (Showa-Denko) (TVA) (Fulham Simon-Carves) (21, 22, 23)	Absorption and concentration of SO_2 and air in ammonia solution yields bisulfite and thiosulfate which subsequent undergo solution redox to yield (+6) and sulfur (zero). One of few solution processes yielding sulfate.	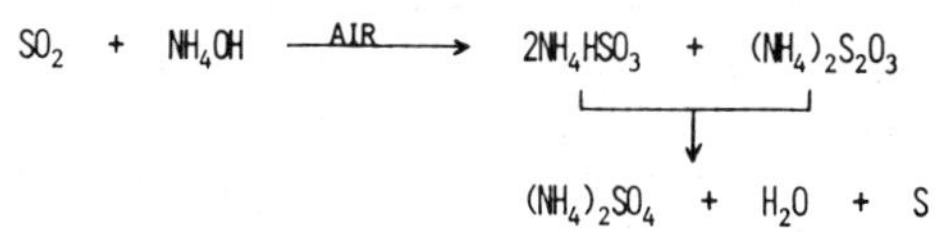
YIELDING SULFUR (S=ZERO)			
12	Molten salt (Atomics Internatl.) (Garrett, Res. & Dev.) (28, 40)	Dilute SO_2 concentrated by absorption in molten salt as sulfite and reduced by H_2 (atomics) or coke roasting (garrett) to sulfide and hence H_2S. Both processes feed Claus.	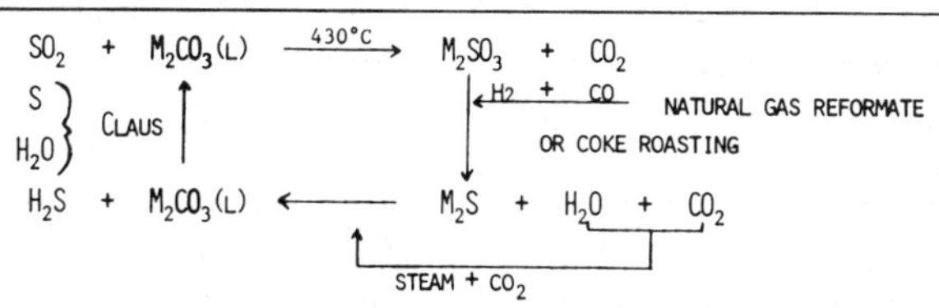
13	Solution Claus IFP (Inst. Francais du Petrole) (9) see also Townsend (7) Shell (Deal) (8)	Essentially Claus redox in solution with or without added catalyst. High boiling solvents preferred to accept hot gases without extensive cooling.	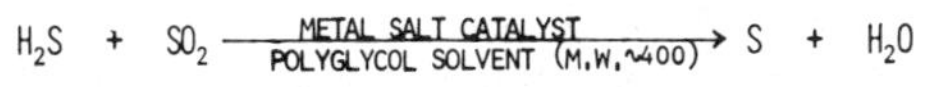
14	Bumines Citrate (U.S. Bureau of Mines) (43)	Claus solution redox here catalysed by forming intermediate citrate complex of SO_2 for reaction with H_2S. Catalytic H_2S formation from natural gas and sulfur if no H_2S in gas stream.	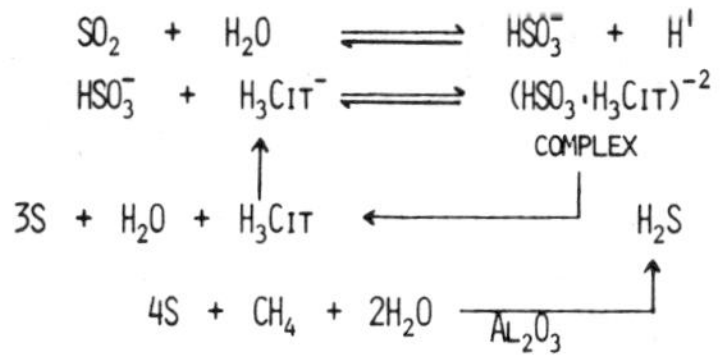
15	Giammarco-Vetrocoke (same) (16, 17) see also Thylox (7)	Solution oxidation of H_2S absorbed as thioarsenite with arsenate/arsenite air regenerateable redox couple as oxygen carrier. Several similar systems involving inorganic redox couples (thylox, manchester, lacykeller (17, 18) exist.	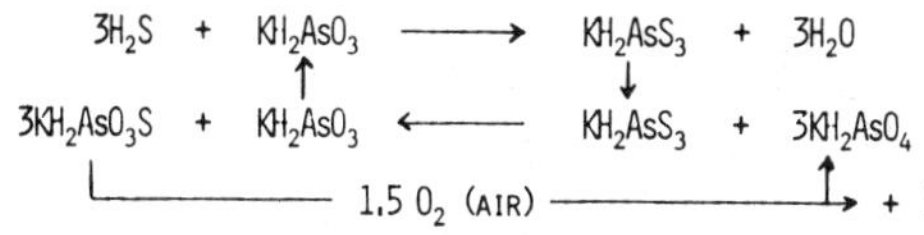
16	Stretford (U.K. North W. Gas Board) (10, 12) see also Takahax (11)	Solution oxidation of H_2S absorbed as Bisulfide by two stage redox couple involving vanadate and anthraquinone disulfonic acid as oxygen carriers. Takahax uses napthaquinone.	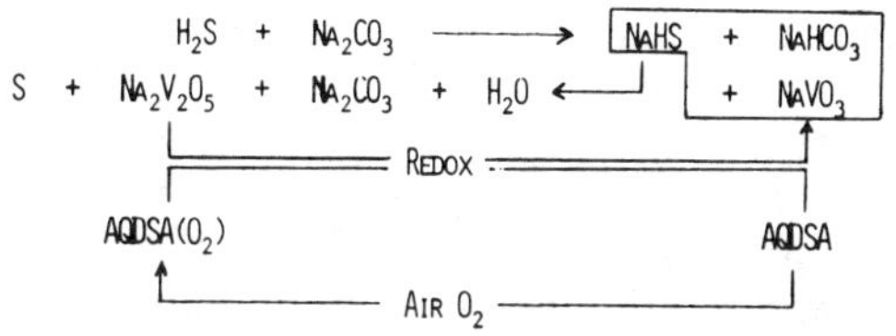
17	Wellman-Lord (same) (44)	A solution method for concentrating dilute SO_2 stream to rich feed for Claus by bisulfite formation, crystallization and thermal regeneration. No reduction or oxidation in solution step.	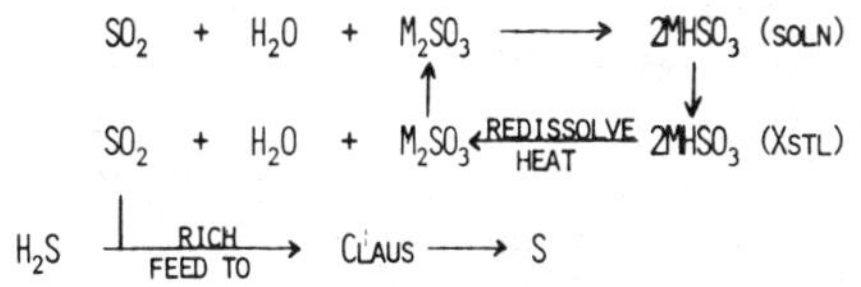
18	Alkazid (Shell) (45)	Preliminary gas phase catalytic hydrogenation all sulfur compounds to H_2S followed by concentration in amino acid salt solution to yield rich Claus feed.	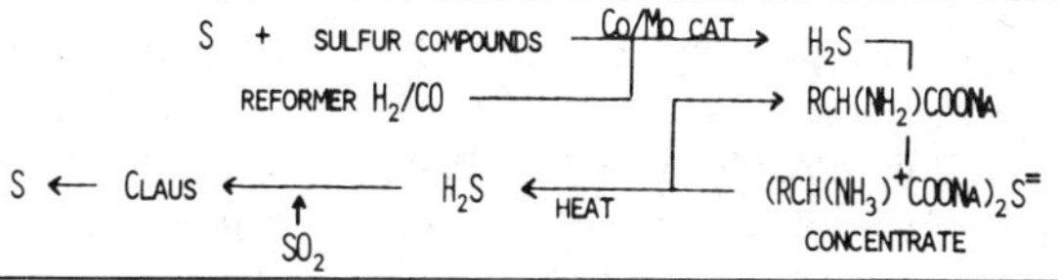
19	Beavon (Parsons Co.) (12, 13)	Preliminary gas phase catalytic hydrogenation all sulfur compounds to H_2S for feed to Stretford. In particular COS and CS_2 reduced.	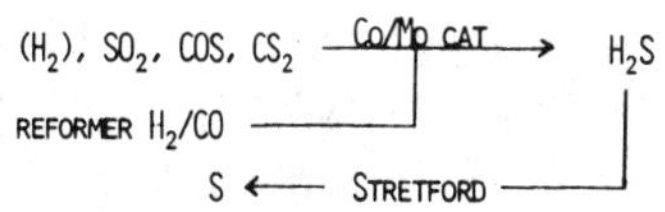
20	Cleanair (Pritchard) (TGS) (14)	H_2S rich tail gas water cooled to continue Claus and hydrolyze COS and CS_2; final H_2S to Stretford.	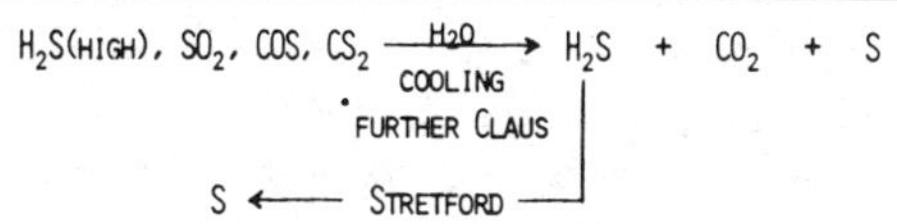

components of tail gases, but removal of this zero-oxidation-state sulfur is more a mechanical problem than a chemical one. With these exceptions, the input sulfur can then be regarded as in the —2 and +4 oxidation states.

The oxidation state of the product sulfur removed in the desulfurization is normally either zero (elemental sulfur), or +6 for sulfuric acid (e.g., Monsanto-Penelec, Topsoe), sulfates such as ammonium sulfate (e.g., Mitsubishi), or calcium sulfate (e.g., TVA).

The process interconversion of —2 or +4 oxidation states to zero or +6 might seem to be a simple matter of providing a suitable oxidant or reducer to carry the system in the desired direction along the oxidation scale. As will be seen from a cursory examination of the multitude of processes described in Table 1, this is hardly the case. The overall result of a process may well be the conversion of +4 sulfur to a zero oxidation state, but the route may involve taking the system through an indirect pathway, e.g., +4 to —2 to zero in the Beavon process.

The nature of the oxidizing and reducing agents used to effect the interconversions on the redox scale also varies. H_2S is itself a reducing agent (—2) and SO_2 (+4) is an effective oxidizing agent. The Claus reaction between these two reagents is itself a typical redox process and can be carried out under a wide variety of conditions yielding elemental sulfur (zero).

The variation of these conditions, from gas-phase surface catalysis (e.g., classical bauxite cat beds, Haines[6] zeolite catalysis) to solution phase in aqueous organic solvent media (e.g., Townsend[7], Shell Deal[8], IFP[9]), is the characteristic feature of many of the desulfurization processes.

In some instances sulfur in the —2 oxidation state only is available, and aerobic oxygen is provided for the Claus-type reaction by burning the sulfur compounds (—2) to yield SO_2 (+4). Sulfur dioxide can, in these cases, be viewed as the oxygen carrier. This, of course, is the primary method of treating the high volume H_2S from sour natural gas processing.

Oxygen carriers. In a number of processes specifically designed to handle low concentrations of sulfur compounds, the oxygen carrier is an organic compound chosen for its ease of transferring oxygen to the species being oxidized and for ease of reoxidation in a recycling process. This is the principal characteristic of the anthraquinone and naphthaquinones used in such processes as the Stretford[10] and Takahax[11] methods and other overall systems incorporating these techniques, e.g., Beavon[12, 13], Cleanair[14].

Inorganic arsenical oxygen carriers are used in the Thylox[15] and Giammarco-Vetrocoke[16, 17] methods; iron oxide plays a similar role in the Ferrox, Manchester, and old iron sponge techniques[3, 18, 19]. The Nalco process[20], currently used for Kraft mill effluent treatment, also uses an oxygen carrier which can be regenerated by aeration.

Although less obviously an oxygen carrier, $(NH_4)_2S_2O_3$ functions in this manner in the ammoniacal solution method [21, 22, 23] developed by Cominco, TVA, Simon-Carves, and others. Finally in this group, the various modifications to the old chamber process using the NO/NO_2 oxygen carrying system is a feature of the Tyco[24] process for direct conversion of SO_2 to sulfuric acid.

Direct oxidation. The direct oxidation of SO_2 (+4) to SO_3 (+6), and hence sulfuric acid and sulfates, is generally achieved catalytically. This step is the essential one in the contact process for sulfuric-acid manufacture which is where the bulk of recovered sulfur is used, whatever may have been its previous journeyings on the redox scale. Manganese dioxide (Mitsubishi[25]) and vanadium pentoxide (contact and Monsanto-Penelec[26]) are both well established catalysts for this one stage direct oxidation.

The Topsoe process used by SNPA at its Lacq, France, facility is also a direct SO_2-to-SO_3 conversion that operates on low sulfur dioxide concentrations typical of tail-gas composition. Since these cleanup processes require sulfur in the gas stream to be in SO_2, they are commonly employed where the sulfur in the fuel has been fully burned to SO_2.

By and large H_2S is not considered an acceptable end product for desulfurization processes. Yet in a number of methods the primary step is to reduce the entire sulfur content of the process stream to H_2S. Among these processes are some that are receiving very active consideration at this time. Here we have a prime example of the indirect route to the end point on the redox scale.

In virtually all cases hydrogen, usually from reformed natural gas, is the reducing agent and in effect converts all sulfur to the —2 oxidation state (H_2S), usually from the +4 state (SO_2). The Beavon process uses such catalytic hydrogenation as a preparatory step to provide an H_2S feedstock for the Stretford. The alkalized alumina method[27] of BuMines and the U.K. Electricity Board uses hydrogen to reduce SO_2 absorbed on the catalyst to H_2S for subsequent Claus treatment.

The Atomics International molten-carbonate process[28] reduces +4 sulfur in a metal sulfite to —2 sulfur, in the corresponding metal sulfide. In all of these cases subsequent oxidation of H_2S (—2) to sulfur (zero) takes place using some form of oxygen carrier. One of the few direct SO_2 (+4) to sulfur (zero) reductions is in the Chevron Research CO/SO_2 method[29]. This particular process has a very high temperature rating making it an attractive possibility for hot effluent-gas-stream treatment.

Organization of processes

Perhaps the most difficult task in reviewing and comparing the multitude of processes that have been and are being proposed for desulfurization of effluent gas streams is that of finding a suitable framework upon which to organize the data. Without such a structure the very volume of information is simply unmanageable. An attempt has been made at such a presentation in Table 1.

The first necessity is to reduce the number of individual processes considered by selecting typical examples of various types. Many of the more than 50 known processes are modifications of a general type of system, and considerable simplification can be achieved by selecting illustrative examples. Table 1 identifies and references related processes that are not discussed in detail. This selection process permits limitation of detailed consideration to some 20 processes listed in Table 1.

A major subdivision of processes is possible on the basis of the principal phase in which the reactions take place. This clearly cannot be a clear cut division since many processes have a number of steps which involve, in turn, the gas phase, the liquid or solution phase and the solid process is associated with either the gas phase or solution and on this basis a significant subdivision can be made.

Gas phase. Processes occurring principally in the gas phase are considered first in Table 1. These processes can be generally of the oxidation type leading to sulfuric acid or sulfates, with sulfur in the +6 oxidation state; or they can be overall reduction processes yielding elemental sulfur in the zero oxidation state. These characteristics permit a further binary classification of the gas-phase processes.

The first two processes involve absorption of SO_2 on an inorganic phase (CaO, MnO_2) and oxidation of the +4 sulfur of SO_2 to +6 in sulfate while in that phase. Both processes yield solid sulfate products. In the case of $CaSO_4$ the question arises as to whether a gaseous environmental contaminant has been converted into a solid one. Ammonium sulfate does, of course, have an agricultural use as a nitrogen and sulfur source.

The next four processes[3-6] all have the common factor of sulfuric acid as an end product. The active magnesia process is really a gas-phase equivalent of the solution phase Wellman-Lord process; both serve primarily to concentrate the dilute SO_2 gas stream to provide an acceptable feedstock for an established conversion process. In the active magnesia case 2% or less SO_2 is concentrated to a 15% feedstock for a Contact sulfuric acid process by absorption as the thermally regenerable $MgO/MgSO_3$ system.

One of the important factors in effluent-gas-cleanup systems is the ability of the process to accept relatively high temperature gas streams without the need for extensive cooling. In this regard, the MgO absorption is among the lower temperature processes.

One-step oxidations. Processes 4, 5, and 6 are all essentially one-step oxidations of SO_2 to SO_3, and hence sulfuric acid. The first pair are modified versions of the traditional contact and chamber processes for sulfuric-acid manufacture, with the principal change being in their ability to accept dilute SO_2 gas streams as the feedstock. The use of activated carbon as an air oxidation catalyst has clearly received international attention, with success or failure depending to a large extent on subtle modifications in catalyst preparation and catalyst presentation to the reactant gases. As indicated in Table 1, virtually every type of catalyst-bed configuration has been explored.

Generally speaking, Processes 1-6 are considered to be stack-gas (i.e., combustion effluent) desulfurization techniques. It is common practice, however, to incinerate all plant tail-gas streams so that sulfur content is almost exclusively in the SO_2 form, the required feed for all six processes. Any one of these processes would, therefore, meet the "add-on" criterion cited earlier with virtually no plant modification required. The question of disposal or marketability of the sulfate/sulfuric acid end product would be a factor in economic considerations.

Processes 7-10 have a common end product—elemental sulfur. They are, therefore, overall gas-phase-reduction processes. The Sulfreen, Princeton Cat-Redox, and CO/SO_2 systems are all direct conversions to elemental sulfur. The alkalized alumina process, however, involves concentration and catalytic hydrogenation of the sulfur components to hydrogen sulfide for subsequent oxidation to elemental sulfur in a Claus process. It is therefore related to such processes as the Alkazid and Beavon which involve preliminary hydrogen reduction to H_2S before either Claus or Stretford treatment.

Carbon catalysts. The Sulfreen process is perhaps unique in this group, in that it was developed specifically for sulfur plant tail-gas cleanup. This process will be examined in detail later, but the use of active carbon as a Claus redox catalyst should be compared with the role of carbon as an oxidation catalyst in Process 6. It is often forgotten that the primary role of a heterogeneous catalyst is to provide a maximum surface area with activating sites for adsorption of the species to be related. When SO_2 is adsorbed on carbon it is activated to react with either H_2S (reduction) or oxygen (oxidation).

Like the Sulfreen process the catalytic redox system provides a catalyst (unspecified) for efficient Claus conversion of a gas stream containing both H_2S and SO_2. In the event that H_2S is not a significant component of the effluent stream, e.g., incinerator stack gas, the process requires synthesis of the required H_2S from methane and product sulfur. This procedure is incorporated into a number of processes that require **both** H_2S and SO_2 in the feed gas; it is, however, somewhat less convenient to synthesize missing H_2S than it is to provide needed SO_2 by combustion of sulfur.

Hydrogen sulfide is replaced by CO as the reducing agent in the conversion of SO_2 to sulfur by a process being considered by Chevron Research and the University of Massachusetts. Details are limited, but higher temperature and copper activated alumina are involved in one modification. Nitrogen oxides can be removed by conversion to nitrogen, but the production of COS, especially at lower working temperatures may be a difficult problem to overcome.

Solution processes. Processes 11-12 are characterized by at least one principal stage taking place in solution. In all but one of the cases elemental sulfur is the end product and in the single instance of the ammoniacal solution process, sulfur is a coproduct with ammonium sulfate.

This process has been extensively examined and developed in a number of countries and is chemically interesting because of the unusual redox reaction that is suspected to take place between the air-oxidation products of SO_2 absorbed in ammonia solution. Both products, sulfur and ammonium sulfate, are normally salable commodities.

The use of molten salts in process chemistry is a relatively recent development. The high temperatures require special handling techniques but, with molten carbonates and thiocyanates, the temperatures involved are wholly compatible with the temperatures characteristic of the effluent gas streams to be desulfurized.

Both the simple molten carbonate (atomics) and carbonate/thiocyanate (Garrett) processes involve reduction of a metal sulfite to the corresponding metal sulfide, but the reducing conditions differ through use of reformed natural gas in one case and coke roasting in the other.

In both cases the resulting H_2S is fed to a Claus plant for reoxidation to elemental sulfur. Reduction to sulfur is not direct, but via the —2 oxidation state of H_2S and subsequent reoxidation to elemental sulfur.

Claus reactions in solution. Claus processes are typically thought of as gas-phase hetrogeneous catalysis reactions occurring at temperatures in the 400°-500° F. region. A number of solution-phase Claus-type reactions have been known for some time, however, and the Townsend and Shell Deal (Sulfinol) process have been patented.

More recently Conoco[41] has added to the list a patented process involving use of liquid SO_2 as the reaction medium. A Claus-type reaction is also claimed to occur in liquid sulfur.[42] Primary interest, however, appears to have been focused on the IFP process where a relatively high molecular weight solvent is charged with a metal salt catalyst to give a remarkably high conversion of H_2S/SO_2 to elemental sulfur.

Although at first sight, the citrate process may not appear to be in any way related to the traditional Claus, it is in fact an H_2S/SO_2 redox reaction in solution with the activating bauxite, carbon, or metal-salt-type catalyst replaced by a citrate complex with SO_2. The citrate process is yet another reduction process that may require the ancillary generation of H_2S from natural gas and product sulfur if the effluent gas stream contains sulfur only as SO_2.

Solution redox reactions. The Giammarco-Vetrocoke (G-V) and Stretford processes are classic examples of multiple coupled redox reactions in solution. Yet, despite their widespread application in Europe, they are little known in North America. One plant in Texas uses the G-V process and only recently has the Stretford been incorporated into more comprehensive desulfurization procedures such as the Beavon and Cleanair techniques.

The Stretford and G-V have as common features at least one oxygen carrier to bring aerobic oxygen into a chemical reaction for conversion of H_2S to elemental sulfur. The G-V process uses the arsenate/arsenite couple, and the Stretford the corresponding vanadium redox system. Additionally, the Stretford couples the vanadium carrier with an anthraquinone disulfonic acid (AQDSA) oxygen carrier. It thus bears a remarkable resemblance (in an inorganic sense) to nature's elegant series of coupled redox reactions involving the cytochromes, the very essence of cellular respiration.

Both processes are typical of several in their general class. The Takahax process is similar to the Stretford, but uses naphthaquinone rather than anthraquinone and has been widely tested in Japan. Companions to the G-V Process such as Thylox have met with interest in Europe. An inorganic oxygen carrier is also involved in related processes such as Manchester, Perox, Lacy Keller, and Nalco.

Process 17 is not in itself a conversion method, but rather a solution-phase technique for concentrating a dilute SO_2 effluent stream to provide a suitably rich feed for Claus redox conversion. The Wellman-Lord, when coupled with the Claus process, constitutes an overall desulfurization system which involves all three phases, gas, liquid solution, and solid crystalline. The heart of the Wellman-Lord concentration step involves the separation of SO_2 as the crystalline metal bisulfite and subsequent thermal decomposition of the redissolved material.

Combinations. The final three processes, Alkazid, Beavon, and Cleanair, are largely combinations of basic techniques. The importance of the processes as desulfurization methods lies in the particular way in which the various steps are combined and the high efficiencies that can be obtained by the particular modifications employed.

Both the Alkazid and Beavon processes require initial reductive hydrogenation of the sulfur values in the effluent gas to H_2S. In the Alkazid the product H_2S stream from the reduction is enriched to levels acceptable to a Claus process by treatment with an amino acid salt solution. A modification of this concentration step using an alkanolamine rather than the amino acid salt is known as the Adip process, and has been considered by Japan Gasoline[46].

In the Beavon process the H_2S stream from catalytic reduction is fed directly to a Stretford unit. In the Cleanair system the Stretford is also used as the final oxidation step to produce sulfur. But the upstream unit is essentially a Claus process in which the tail-gas-cooling step is used simultaneously to achieve a further Claus redox reaction between the H_2S and SO_2 content of the tail gas. The H_2S content of the tail gas is kept above stoichiometry with respect to SO_2, so that virtually all SO_2 is removed before the excess H_2S passes to the Stretford.

Basis of choice. The 20 process types illustrated in Table 1 cover virtually all of the "add-on" methods that are presently being considered as techniques for effluent gas stream desulfurization. The choice will clearly depend on the precise nature of the gas stream being treated and a variety of engineering and economic factors peculiar to each situation. Effluent streams from stationary hydrocarbon combustion sources such as power generating stations will, like smelter stack gases, be rich in SO_2 with little or no H_2S content due to prior oxidative incineration.

Such conditions would tend to encourage the use of processes that handle SO_2 directly and alone. The same situation will likely exist with sulfuric acid and fertilizer plants, and in some petrochemical effluent streams. Where hydrogen sulfide is either present in large quantities in the plant feedstock (e.g., sour natural gas processing), or is generated in the plant in some major reduction process (e.g., hydrodesulfurization, H-Coal SNG), those effluent-stream desulfurization processes designed to handle H_2S alone or H_2S and SO_2 combined will be preferred.

Even in these cases incineration of the effluent gas stream would convert the sulfur values to SO_2, but such incineration leads to further dilution of the effluent and compounds the problem of low concentration.

The temperature factor

Incineration or fuel-combustion-stack gas emission temperatures are characteristically near 1,000° F., while tail gases from multistage Claus sulfur plants are more commonly in the 300° F. region. The temperature of the effluent gas stream being desulfurized must be matched with the acceptable inlet temperature of the desulfurization process. A cooling stage may thus be necessary between the plant exit and the desulfurization unit. This temperature matching can be a particularly important feature in many of the finishing solution processes that have relatively low operating temperatures.

Processes primarily designed to handle 1,000° F. stack-gas desulfurization tend to have higher operating temperatures. The molten carbonate at 800° F. and the CO/SO_2 reduction system at 750° F. are easily compatible with stack gas emission temperatures. The Sulfreen and IFP processes both operate in the 260°-300° F. range, requiring little if any cooling of typical Claus tail-gas effluent.

The IFP process, though it is a classical solution procedure, can accept tail gas with little cooling requirement. Virtually all other solution processes require some form of effluent gas cooling before entering the desulfurization process. The G-V or

Vetrocoke process is rated at 70°-300° F., but few if any systems operate with high efficiency at the upper end of this range.

A typical sour-gas-treating Vetrocoke system in the U.K. operates at 85°-90° F., although temperatures as high as 135° F. occur in some sections. The Stretford, another highly regarded finishing process, has equally low temperature requirements. Above 100 F., undesirable side reactions are encountered (see section on reagent losses). Since the Stretford is used in both the Beavon and the Cleanair processes, it is important to note the need for gas cooling in both these systems.

The upstream unit in the Beavon (and in the Alkazid process) is a 600° F.+ catalytic hydrogenation, requiring cooling before the H_2S-rich gas is admitted to the Stretford. In the Cleanair process, the 280° F. Claus tail gas must also be cooled before being admitted to the Stretford, but in this process further sulfur conversion is obtained from a continuing Claus-type redox reaction during the cooling process.

An inlet gas cooler between the Claus plant tail gas exit and the sulfite solution reactor is also required in the Wellman-Lord solution-concentration method. Being an aqueous process, temperatures are limited to below 200° F., but corrosion and losses through oxidation to sulfate are both problems that may be amplified at higher temperatures.

The BuMines Citrate process requires initial cooling of the tail gas to 120° F., comparable to both the Stretford and the Vetrocoke conditions. In the citrate process, however, the sulfur product is recovered by reheating to 130° C. at 35 psi to melt out the sulfur from the aqueous phase. This additional thermal requirement may, in some instances, be a disadvantage.

Reagents and products

Virtually every industrial chemical process has its advantages and disadvantages and tail-gas desulfurization techniques are not exceptions. As in other instances, the operator finally must make a choice based on the best tradeoff of advantages against disadvantages for the particular situation being handled.

All desulfurization processes yield an end product. Normally this would be considered an advantage if this product were a salable commodity. Sulfur, the end product in the majority of favored tail-gas cleanup processes, is undoubtedly a widely employed base chemical, but its marketability fluctuates.

The effect of these fluctuations on price renders it virtually impossible to build in a meaningful cost-recovery component in desulfurization economics, based on sales value of the sulfur end product. Most operators tend to use a very low dollar-per-ton recovery from sulfur sales, recognizing that increased sulfur recovery from all sources in response to environmental protection regulations will not help the market situation in the immediate future.

Sulfur as the end product of desulfurization has at least the advantage of being relatively easily stored. There are environmental contamination problems associated with its storage and handling, but these can and are being overcome through such techniques as forming (slates, pellets, etc.) and liquid shipping.

Storage is not such a simple matter when the desulfurization end product is sulfuric acid. An obvious factor in choosing a desulfurization process will, therefore, be the ease of disposal of the end product. With sulfuric acid product, a conveniently proximate consumer is a necessity, since large tonnage storage or long distance haulage are not attractive possibilities.

Ammonium sulfate, the end product in processes such as the ammoniacal solution and manganese dioxide techniques, has an immediate use as a fertilizer. It is, however, water soluble, and this precludes its cheap surface storage outdoors in many climatic regions.

It is therefore essential that the continuous movement of the product be assured in order to avoid excessive capital investment in covered storage facilities.

The calcium sulfate end product of the dry-limestone process presents a problem in disposal. Suggestions for use of large tonnage of this "sulfur trap" have ranged from land fill to construction material such as aggregates and road beds. The need for careful consideration of the disposal problem in such an instance, however, is apparent.

Losses. Reagent-loss problems are primarily associated with the highly efficient low-temperature-solution methods of desulfurization. The gas-phase heterogeneous catalysis processes are not without their own loss problems, but these are generally associated with catalyst bed operations and are well understood and easily handled. The Stretford process used in both the Beavon and Cleanair systems has relatively complex chemistry to match its high efficiency.

Losses through side reactions and physical processing of foams, slurries, and solutions must, therefore, be expected.

Oxidation to thiosulfate is a major source of loss in the Stretford and is particularly temperature sensitive. At the recommended operating temperature of about 100° F., thiosulfate formation can be held to a modest 0.1-0.2%, and loadings of up to 20% can be tolerated in the working solution before cleanup and recycling is necessary However, at 160° F. thiosulfate yield jumps to 3%. The importance of close control of operating temperature (and inlet gas cooling) is obvious.

The product from the Stretford is a finely divided sulfur froth which is subsequently filtered. The efficiency of the filtering process determines not only the quality of the sulfur end product, but the extent of solution component losses including the easily water soluble sodium salt of the anthraquinone disulfonic acid. The filter cake washing process also generates large volumes of contaminated water which must be repurified before recycling or discharging as effluent.

Water waste is, of course, a problem with all large-volume aqueous solution processes.

Arsenical solutions. The use of arsenical solutions in the Vetrocoke process has long been considered, particularly in North America, to be one of its major disadvantages. There can be no question about the toxic hazards resulting from leakage of the working arsenical oxidizing solution. One wonders, however, about the equally serious hazards from chlorine, cyanide, phosphorus, and a myriad of other industrial chemicals in common use today. The importance of the end-product sulfur being kept free of arsenic contamination is of considerable importance also, especially in view of the likely end use of the sulfur as feed for a sulfuric plant and subsequently fertilizer production.

Oxidation of the sulfite solution to sulfate is a principal loss factor in the Wellman-Lord process. Makeup of

Fig. 4

The COS, CS_2 problem

Formation		Decomposition
$CO_2 + H_2S$	$\rightleftarrows$	$H_2O + COS$
$COS + H_2S$	$\rightleftarrows$	$H_2O + CS_2$
$CH_4 + S$	$\rightarrow$	$H_2S + CS_2$
$CH_4 + H_2S + (H_2O)$	$\leftarrow$	$H_2 + CS_2\ (COS)$

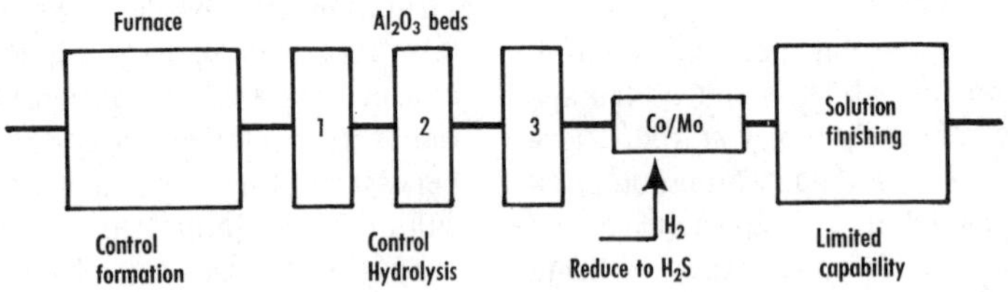

COS and CS_2 formed and hydrolyzed in equilibrium steps. Control may be possible in formation (front-end furnace), in hydrolysis (catalytic beds), in special total reduction to H_2S step ahead of finishing process, or in new developments in finishing processes themselves.

OGJ

caustic soda of as much as 1 ton/day from a unit handling 100,000 scfm of a 5,000-ppm CO_2 effluent gas stream might be expected. The nonregenerable sodium sulfate by-product is not difficult to handle, nor is its volume particularly great on a year-round basis. There are, however, only limited uses for sodium sulfate, and the present market would not be difficult to saturate.

Relatively few operating details are available on the BuMines Citrate process, but it is known that traces of sulfuric acid in the effluent gas stream being treated must be scrubbed out before the citrate solution is contacted. Sulfuric acid being a much stronger acid than the citric or sulfurous acids involved in the critical complex formation, pH control in the solution is clearly a matter of importance. The presence of nitrogen oxides in the gas stream is reported to have no adverse effects on the Citrate process.

Carbonyl sulfide and carbon disulfide

COS and CS_2 have proven to be the two most intractable components of Claus tail-gas streams in desulfurization (Fig. 4). They are not generally a problem in combustion flue-gas treatment, since the incineration step converts both components to CO_2 and SO_2. Depending upon the exact nature of the sour-gas feed stream and the operating conditions in the upstream Claus reaction furnace and catalyst beds, COS and CS_2 levels as high as 5,000 ppm have been encountered in the tail gas. Generally speaking values in the range 600-1,500 ppm are more common.

Equilibrium thermodynamics under the Claus conditions would indicate much lower levels should be encountered, suggesting that the higher levels actually observed are the result of kinetic considerations rather than thermodynamic factors.

None of the solution processes listed in Table 1, and highly rated for their efficient cleanup of SO_2 and/or H_2S, have any significant capability for handling COS and CS_2. The claim has been made for the Vetrocoke process that COS, CS_2 and mercaptans are partially removed, but the mechanism of the process is not clear. The potential for COS, CS_2 absorption and conversion may be present in the molten-salt system, but no specific claim has been made to this effect. One of the disadvantages of the CO/SO_2 redox technique is that COS is **generated** in the reaction and is a major hurdle to be overcome in devising a practical system based on this reaction.

The Beavon, Alkazid, Cleanair, and IFP processes all claim a certain capability in handling COS and CS_2, but in an indirect manner, not in the final solution-conversion step. Hydrolysis of both components is the preferred method of handling their removal, and it is presumably the slowness of this hydrolysis, coupled with the short residence times in the reaction zones, that lead to nonequilibrium kinetically controlled high-concentration levels of COS and CS_2 in tail gases.

In both the Beavon and Alkazid processes, cobalt/molybdenum catalytic reduction of all sulfur values to H_2S is effected prior to the solution (Stretford or amino acid salt solution) stage. The Co/Mo catalyst has been found to be effective also in catalyzing the hydrolysis of COS and CS_2 to CO_2 and H_2O. Accordingly, COS and CS_2 not catalytically reduced to H_2S are effectively hydrolyzed in the same step by water vapor in the gas system.

Hydrolysis reactions. The Cleanair process has a third stage addition which may be introduced ahead of the gas cooling and Stretford sections to effect hydrolysis of COS and CS_2 and H_2S. No details of this unit are available, but it is claimed to be necessary for achieving 250-ppm-SO_2 equivalent-effluent values. No information is available on the ability of the Sulfreen process to handle COS and CS_2, but the low operating temperature of the gas phase catalytic reaction (260°-300° F.) suggests that hydrolysis would not likely take place to any great extent. Carbon catalyst is not known to cause generation of COS or CS_2 at these low temperatures. The IFP recommendation is that the first Claus convertor be operated at a higher temperature than normal, with an improved catalyst resulting in reduced COS and CS_2 levels in the tail gas. The loss of overall Claus conversion efficiency resulting from the hot operation of the first stage can be compensated for by the improved conversion in the add-on IFP stage.

The presence of carbonyl sulfide and carbon disulfide will continue to be a particularly difficult problem to han-

Fig. 5

How clean for what cost?

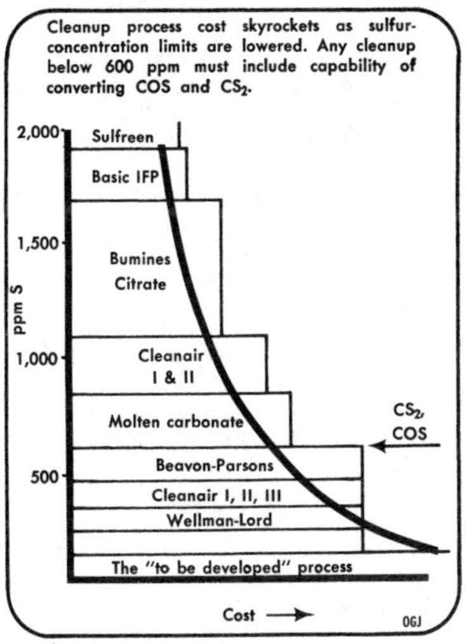

dle in the desulfurization of tail gas until a process becomes available which can handle them in the finishing step. As maximum permitted sulfur levels in effluent gas streams edge down toward the 250-100 ppm level the importance of COS and CS_2 removal will become critical.

Tail-gas-desulfurization process efficiencies

Data on the efficiencies of the various processes likely to be considered as suitable for desulfurizing tail gas are presented in the literature in a variety of forms. Fig. 5 is a qualitative summary. Because of the very considerable difference in the operating details of many of the processes it is more appropriate to discuss each of the several techniques individually.

The Sulfreen process is recommended as an add-on unit behind a two-stage Claus converter system. This configuration is claimed to yield 97.8% overall conversion which would indicate the need for a finishing unit to obtain sulfur emission levels at or below 2,000 ppm.

The IFP add-on unit again is recommended behind a two-stage Claus, but claims 99.3% overall conversion efficiency yielding an effluent stream with sulfur values in the range 1,500-2,000 ppm. Reduction to the 250-ppm range would require further treatment with one of the low-temperature solution processes.

The molten carbonate process claims a 75% working efficiency on a 0.2% (2,000 ppm) SO_2 effluent stream which would result in a 500-ppm sulfur content for final effluent.

The BuMines Citrate process, designed initially to handle smelter off-gases, operates on these streams with 99% efficiency. With lower concentrations of SO_2 in the 2% (20,000 ppm) range the claimed 90-95% efficiency would give 1,000-2,000-ppm sulfur levels in the final effluent.

250-ppm levels. Only three, all low-temperature solution processes listed in Table 1, are known to be capable of residual sulfur levels at or below 250 ppm. The Stretford and Vetrocoke both remove sulfur compounds and convert them to elemental sulfur. The Wellman-Lord removes the sulfur as sulfite and concentrates the regenerated SO_2 for feed back to the Claus plant.

Both the Vetrocoke and Stretford are capable of removing H_2S to levels of below 1 ppm. The 250-ppm level, however, is somewhat more realistic since a 1-ppm level is conditional on all the sulfur content being converted to H_2S, i.e., a 100% efficient upstream reduction process. The Vetrocoke process does have an upper limit value for H_2S content of 1.5% which might be a problem with tail-gas streams having in excess of 15,000-ppm sulfur content.

A number of examples of Wellman-Lord process operating efficiencies are available, including a sulfuric acid plant off-gas stream with imput values near 6,000 ppm and effluent value of 500-ppm SO_2. A 200-ppm effluent concentration of SO_2 is claimed for a Wellman-Lord process operating on a Japanese Claus tail gas with 13,000-ppm SO_2 content.

The costs

It is generally agreed that desulfurization of industrial effluent gas streams involves additional expenditures which generate benefits measured in environmental quality maintenance, not in improved financial return on investment. The fact that the cost/benefit relationship is not of the traditional kind, however, is no reason for not using established methods for maximizing the environmental benefit from the additional cost incurred.

How does one cost out a desulfurization process with this goal in mind? Carrying out the exercise for a particular process is not an unusually difficult task to accomplish and, indeed, for those processes which have reached a reasonably advanced stage of development data are available either in the open literature or from the developers. The complexity of the problem increases severalfold, however, when the question of comparison of processes is superimposed on the individual cost analysis. The reasons for the complexity are not difficult to identify.

Virtually every desulfurization process has its own chemical, engineering, economic, and operating peculiarities. Some processes are particularly well suited to meet one set of plant effluent conditions, while another process may be best articulated to an entirely different set of plant parameters. Since there are probably as many different plant effluent-gas-stream types as there are plants emitting gases, the permutations and combinations of effluent stream and desulfurization techniques are almost limitless.

So, quantitative examination of the comparative cost factors involved in the various processes is not possible until the precise nature of the gas stream to be treated is known. We can, however, examine and illustrate in a qualitative manner some of the factors which must be considered.

Plant size. An obvious first consideration is the size of the plant. Processes that appear most attractive for handling the desulfurization of an effluent stream from a 3,000-ton/day sulfur plant may not retain their attractiveness at the 50-ton/day level. The Cleanair process operating as an add-on unit to a 95% efficient Claus system at 1,000 ton/day is estimated to require only three times the capital investment needed to handle the effluent from a 95% efficient Claus system producing only 50 ton/day of sulfur—a 20-fold difference in size, but only a three-fold difference in cost.

The sulfur content of tail gases varies widely depending on the overall efficiency of the installed upstream equipment. This variation results in equally varied demands on the desulfurization system. Some systems vary little in cost as a function of gas stream composition variation. Others vary markedly. The Beavon process, including the Co/Mo reduction reactor and Stretford section, is estimated to have a capital requirement approximately equal to the parent sulfur plant.

The major variable in the cost, however, is in the Stretford H_2S removal section, and the higher the percentage H_2S and load on the system (almost irrespective of the volume throughput) the higher the cost. The IFP process, however, claims that a unit to handle the tail gas from any of three 200 ton/day Claus plants operating at 96.8% (three stage), 94.4% (two stage) or 86.1% (one stage) to give overall conversions of better than 99.3% in each case, varies in cost by only 20%. The differing cost sensitivity of the various processes to the composition of the feed stream must therefore be carefully considered.

Very few if any of the currently available technologies can effectively eliminate COS and CS_2. The Beavon and Alkazid processes presumably have this capability built in through the use of the catalytic reduction step, but the third-stage hydrolysis step in the Cleanair is optional. As in the case of "off-the-shelf" instrumentation as opposed to "custom built," the question must always be asked whether the process is optimally designed to meet the needs of the gas stream to be treated.

Product and waste disposal have already been identified as factors in selecting a desulfurization process. These are clearly important components of the overall comparative costing exercise. Suggestions have been made that industry be charged for the extent to which they use the environment to dilute and dispose of their wastes. Should this materialize, processes with minimal secondary effluent streams would have a direct cost advantage over those with heavier fluid and solid residue disposal requirements. Growing concern with thermal contamination of the environment may eventually lead to a much closer examination of heat losses in otherwise competitive processes.

References

1. "The TRW S-100 Process," TRW Systems Group, 1 Space Park, Redondo Beach, Calif.
2. K. J. Gustavson, M. J. Healy, Proc. of 1968 Gas Condition Conference, Univ. of Oklahoma, Norman, Okla.
3. C. D. Swain, Jr., Hydrocarbon Processing, **49,** No. 3 (1970), p. 127.
4. F. C. Boston, Proc. of 1968 Gas Condition Conference, Univ. of Oklahoma, Norman.
5. A. M. Hegiver, R. A. Harris, Hydrocarbon Processing, **49,** April 1970.
6. H. W. Haines, G. A. Van Wielingan and G. H. Palmer, Pet. Ref., **40,** No. 4 (1961), p. 123.
7. F. M. Townsend, L. S. Reid, Pet. Ref., **37,** No. 11 (1958), p. 263.
8. U.S. Patent, 3,363,989, Jan. 16, 1968.
9. Y. Barthel, Y. Bistri, A. Deschamps, P. Renault, J-C Simandoux and R. Ditriau, 36th Midyear Meeting, American Pet. Inst., Div. of Refining, San Francisco, May 12, 1971.
10. D. K. Beavon and P. Ellwood, Chem. Eng., July 20, (1964), p. 128.
11. N. Hasebe, Tokyo Gas Co. Ltd., Report.
12. D. K. Beavon and R. P. Vaell, 8th Ann. Technical Meeting, S. California Section AIChE, Apr. 20, 1971.
13. D. K. Beavon, Chem. Eng., Dec. 13, (1971), p. 71.
14. G. L. Thomas, E. Plum, Paper at 71st. National AIChE Meeting, Dallas, Texas, Feb. 20-23, 1972.
15. B. G. Goar, Hydrocarbon Processing, **47,** No. 9 (1968), p. 72.
16. E. Jennett, Oil and Gas Journal, Apr. 30, 1962, p. 72.
17. R. N. Maddox, M. D. Burns, Oil and Gas Journal, June 3, 1968, p. 90.
18. A. L. Kohl, F. C. Riesenfeld, Chem. Eng., June 15, 1959, p. 127.
19. Oil and Gas Journal, Dec. 13, 1971, p. 39.
20. J. C. Knudson, Paper at Pacific Northwest International Section, Air Pollution Control Assoc., Spokane, Nov. 16-18, 1970.
21. R. A. King, Ind. Eng. Chem., Vol. 42 (1950), p. 2241.
22. Simon-Carves Ltd., British Patent, 525,883, Sept. 6, 1940.
23. Chem. and Eng. News, Sept. 8, 1969, p. 48.
24. Sauder, Wallitt, Rissman and Keilin, Tyco Labs., Proceedings of the Symposium on Air Pollution Control Technology, 65th National Meeting, AIChE, Cleveland, May 1969.
25. R. F. Bovier, Proc. American Power Conference, **26,** 138 (1964).
26. Stites, Horlacher, Bachoffer and Bartman, Proceedings of the American Soc. Mech. Engineers, Winter Meeting, N.Y., December 1968, 68-WA/APC-2.
27. D. Bienstock, J. H. Field and J. G. Myers, U.S. Bureau of Mines Report of Investigations 7021, Washington, D.C., 1967.
28. "Sulphur," No. 81, March-April 1969, The British Sulphur Corp. Ltd., London.
29. J. Air Pollution Control Assoc., **17** No. 12 (1967), p. 796.
30. Tennessee Valley Authority, Knoxville, Tenn., 1968.
31. Chem. and Eng. News, Sept. 6, 1971, p. 19.
32. Chem. and Eng. News, Aug. 30, 1971, p. 18.
33. Staub-Reinhalt Luft, **28,** 6 (1968).
34. Staub-Reinhalt Luft, **28,** 1 (1968).
35. Chem. and Eng. News, Sept. 8, 1969, p. 48.
36. Oilweek, July 5, 1971, p. 9.
37. G. Guyot, J. E. Martin, Paper at CNGPA Meeting, Edmonton, Alta., June 11, 1971.
38. Chem. and Eng. News, Sept. 2 and 9, 1968, p. 10 and 22 respectively.
39. Chem. and Eng. News, Dec. 14, 1970, p. 54.
40. Chem. and Eng. News, Apr. 12, 1971, p. 65.
41. R. L. Every, R. L. Grimsley, U.S. Patent, 3,445,181, May 1969.
42. T. K. Wiewiorowski, Chem. and Eng. News, Apr. 27, 1970, p. 68.
43. Chem. and Eng. News, June 14, 1971, p. 31.
44. B. H. Poeter, Paper at CNGPA Meeting, Calgary, Alta., Nov. 26, 1971.
45. R. Leuhddemann, G. Noddes, H. Schwarz, Oil and Gas Journal, **57,** No. 32, (1959), p. 100.
46. Oil and Gas Journal, Dec. 13, 1971, p. 39.

Guidelines aid control of SRU tail-gas composition

J. W. Palm, Neal M. Caruthers, Amoco Production Co., Tulsa

THE CLAUS process is widely used for converting hydrogen sulfide to elemental sulfur.[1 2 3 4] Analysis and control of the tail gas are quite important.

A close relationship exists between the air-feed rate to a sulfur-recovery unit (SRU) and the composition of the tail gas from the final condenser which normally goes to an incinerator or to a tail-gas cleanup process.

J. W. Palm

Neal M. Caruthers

The authors . . .

J. W. (Jack) Palm is a research associate with Amoco Production Co. which he joined in 1952. He is engaged in developing design methods for sulfur-recovery plants. Prior to his present firm, he was associated with Cities Service Research & Development Co. where he was assistant chief chemist in charge of the firm's research laboratory at Tallant, Okla. Palm holds BS and MS degrees in chemical engineering from Oklahoma State University and the University of Michigan and is a registered professional engineer in Oklahoma.

Neal M. Caruthers is a senior research engineer with Amoco Production Co. He joined the firm's research organization in 1937 and has been involved with several bench-scale and pilot-plant studies, as well as the development and improvement of analytical methods. Currently, he conducts bench-scale studies related to Amoco's Claus sulfur plants. Caruthers holds a BS degree in chemical engineering from the University of Tulsa.

Several methods are available for use in analyzing the tail gas. Some parameters are more useful than others in relating tail-gas composition to air-feed rate to the plant.

The H_2S/SO_2 ratio is not a satisfactory parameter for control of tail-gas composition. If Tutweiler analyses are used, a better parameter is $NaOH/I_2$ ratio.

If stain-tube analyses are used, a suitable parameter is SO_2 fraction, defined as SO_2 mol %/(SO_2 mol % + H_2S mol %). However, the preferred parameter is excess SO_2, defined as SO_2 mol % − ½ (H_2S mol %).

If an instrumental analyzer is used, the preferred parameters are of the following forms:

1. Excess $SO_2 = SO_2 - \frac{1}{2} H_2S$
2. Stoichiometric air % $= 100 + A(SO_2 - \frac{1}{2} H_2S)$

Claus process. A Claus sulfur-recovery unit usually contains a thermal reactor or furnace, a series of catalytic converters, and sulfur condensers. The principal components in the acid-gas feed normally are hydrogen sulfide, carbon dioxide, water vapor, and hydrocarbons.

Air is fed at a controlled rate to the burner in the thermal reactor to react with the acid gas. The principal compounds in the thermal-reactor effluent gas are sulfur vapor, sulfur dioxide, unreacted hydrogen sulfide, carbon oxides, water vapor, and nitrogen.

Fig. 1

Air-correction-factor plots

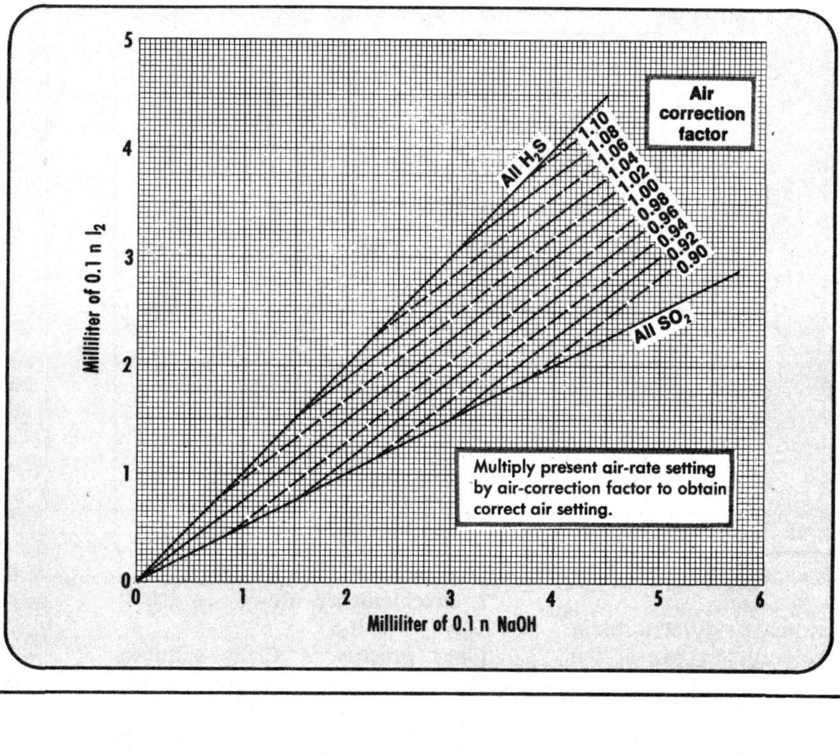

Figs. 2 and 3

Air-correction factors

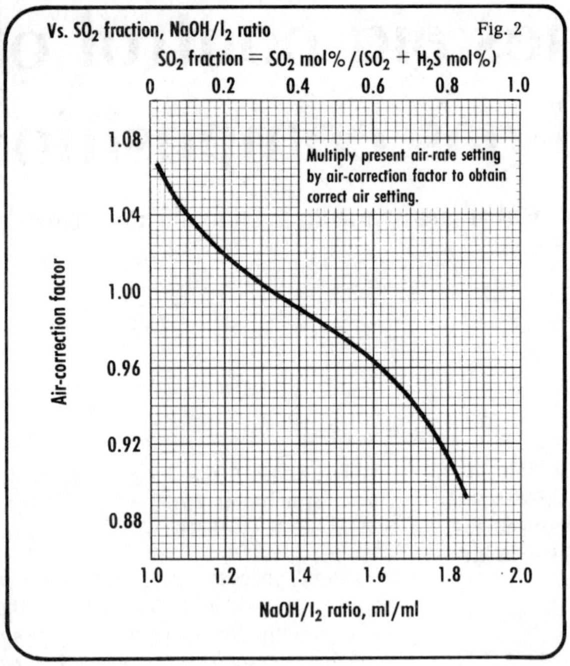

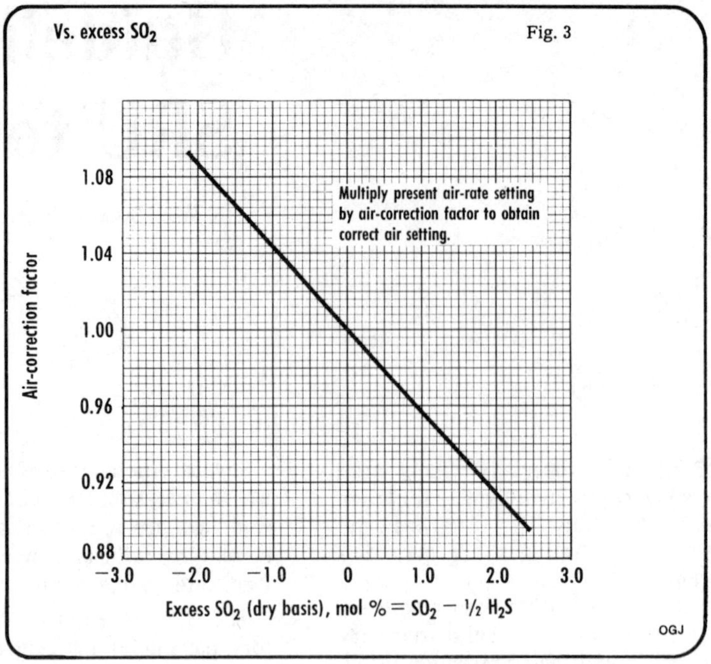

Fig. 4

Heated double window for light cell

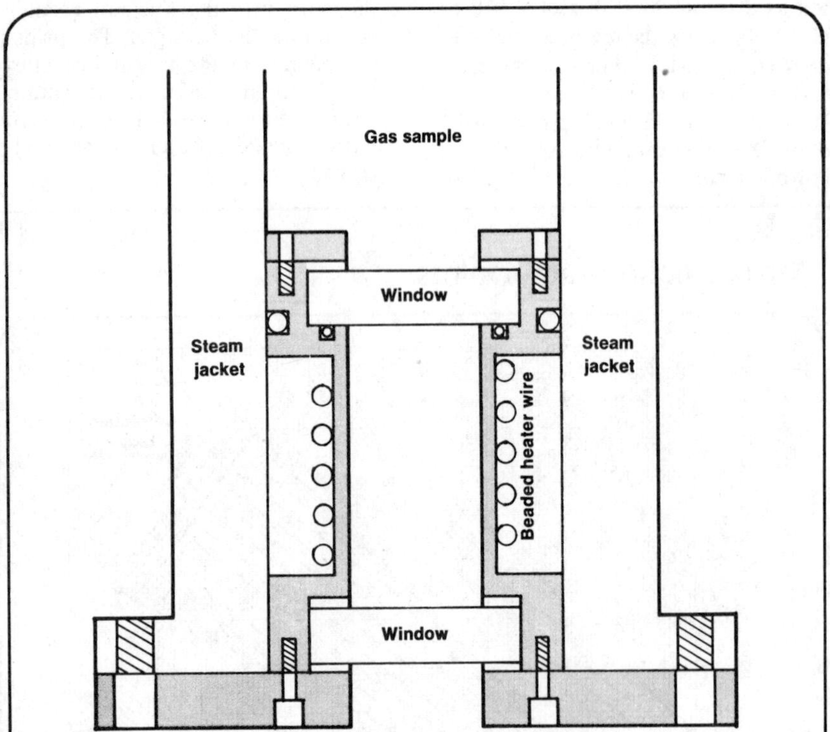

After condensation of sulfur from the thermal-reactor effluent, the gas is reheated and enters the first catalytic converter where the following reaction (usually referred to as the Claus reaction) occurs, producing additional sulfur product:

$$2\ H_2S + SO_2 = 3/e\ S_e + 2\ H_2O \quad (1)$$

S_e refers to the various molecular species of sulfur, such as S_2, S_6, and S_8.

After condensing the additional sulfur, the catalytic conversion process may be repeated. The effluent from the final catalytic converter and condenser are referred to as tail gas from the sulfur-recovery unit.

Stoichiometry and terminology. It is desired for the hydrogen sulfide and sulfur dioxide in the tail gas to be in the proper proportion for the Claus reaction to proceed to the maximum possible extent. A certain air rate is required at any instant to burn the hydrocarbons in the feed gas and also to result in the proper proportions of hydrogen sulfide and sulfur dioxide in the tail gas.

Too much air will result in an excess of sulfur dioxide, and a deficiency of air will result in an excess of hydrogen sulfide in the tail gas.

The historical concept has been that the molar ratio of hydrogen sulfide to sulfur dioxide in the tail gas should be 2/1. This concept was a direct result of considering the Claus reaction (Equation 1) where the coefficients on the left side of the equation are 2 and 1.

This was a poor choice of tail-gas control parameter, because the hydrogen sulfide to sulfur dioxide ratio does not have any simple mathematical relation to any process-control variable. Several other parameters are more useful.

In correlation and interpretation of plant operating and test data, the fraction of sulfur dioxide in the remaining

Table 1

Relation between air-correction factor and tail-gas composition

Air-correction factor	0.90	0.94	0.98	1.00	1.02	1.06	1.08
Air rate, % stoichiometric	111.1	106.4	102.04	100.00	98.04	94.3	92.6
Tail gas composition, mol % (dry)							
SO_2	2.60	1.78	1.02	0.673	0.41	0.11	0.04
H_2S	0.53	0.71	1.05	1.346	1.74	3.00	3.82
Excess SO_2, mol % (dry)	2.335	1.425	0.495	0	−0.46	−1.39	−1.87
Calculated Tutweiler readings							
I_2, ml	2.51	2.00	1.663	1.622	1.727	2.50	3.10
NaOH, ml	4.60	3.43	2.48	2.16	2.06	2.59	3.13
NaOH / I_2, ml / ml	1.83	1.715	1.49	1.33	1.19	1.036	1.010

Table 2

Acid-gas composition

(Used for calculations in Table 1)	Mol %
H_2S	86.46
CO_2	9.03
Methane	0.25
Ethane	0.25
Water	4.01
	100.00

hydrogen sulfide and sulfur dioxide is a useful parameter. The SO_2 fraction is defined as $(SO_2)/(SO_2 + H_2S)$, where each component is expressed on a mol basis.

This parameter has the advantage of having finite limits (it varies from zero to unity), whereas the hydrogen sulfide to sulfur dioxide ratio may vary from zero to infinity.

Another useful parameter is excess SO_2, which is defined as $(SO_2) - (\frac{1}{2} \times H_2S)$. Each component is on a mol basis, but this may be mol %, mols/hr, or other.

In this article, the units will normally be mol %. This would be on a dry-gas basis, if the analysis is by a chemical method, or on a wet-gas basis if the analysis is by an on-line instrumental method.

This parameter may be either positive or negative. Negative values do not cause any difficulty in applying the parameter, but merely indicate that the mixture has an excess of hydrogen sulfide and a deficiency of sulfur dioxide.

The excess SO_2 is related linearly to the excess of air which is fed to the plant.

Manual control using Tutweiler analyses. Most sulfur-recovery plants have a ratio-control system which maintains a specified ratio of the air rate to the acid-gas feed rate. Some method must be provided to analyze the tail gas to determine if the SO_2 fraction has the correct value. If not, then the control point setting of the ratio-control instrument is adjusted either manually or automatically.

This section applies to plants in which the tail gas is analyzed manually by the Tutweiler method, and the ratio-controller set point is adjusted manually.

The Tutweiler method was used universally until about 1968 and is still used in many plants. It uses titrations by two standardized chemical solutions (iodine solution and caustic solution) to determine the total concentration of reactants (H_2S and SO_2) as well as the proportion of the two reactants.

The textbook procedure is to use mathematical formulas to calculate the mol percent concentration of each component (H_2S and SO_2) and then calculate the H_2S/SO_2 ratio.

As indicated below, this detailed calculation is not necessary since simple short-cut methods can be used to relate the analytical titrations to SO_2 fraction and, more important, to the needed adjustment in air/acid-gas ratio.

In this article, I_2 refers to the volume of standardized iodine solution, ml, required in the Tutweiler analysis. Similarly, NaOH refers to the volume of standardized caustic (NaOH) solution, ml, used in the Tutweiler analysis.

Table 1 shows how the air-feed rate to a plant is related to (1) tail-gas composition and (2) the expected titration readings of the Tutweiler analysis. The tail-gas composition in this table can be calculated by a conventional design material and heat balance, which is normally done using a computer program but can be done by hand calculation.

The example in Table 1 was calculated in 1970 for a plant which had two catalytic converters with the acid-gas composition indicated in Table 2. In plants of more-recent design or containing three catalytic converters, the concentrations of hydrogen sulfide and sulfur dioxide in the tail gas would be lower but the air-correction-factor plot would be very similar to that in Fig. 1.

The principal reason for the calculation illustrated in Table 1 was to develop a guide for controlling the air rate to the plant. Therefore, the air-correction factor is shown, which is defined as the correct stoichiometric air rate divided by the existing air rate for a particular column in the table.

The graph in Fig. 1 was furnished for use in plant operation. To use the graph, the operator would run the Tutweiler analysis, then read the air-correction factor from the graph as a function of the I_2 and NaOH titrations which requires no mathematical calculations.

He then would use the air-correction factor to adjust the air-rate setting to the desired value. If the air-correction factor is very low (such as 0.90) or very high (such as 1.08), the operator normally would make only about half of the indicated correction, then run another analysis to confirm that further correction is needed.

Another method for generating the air-correction factor is based on $R = NaOH/I_2$, which is related to SO_2 fraction as follows:

$$SO_2 \text{ fraction} = SO_2 / (SO_2 + H_2S) = R - 1.0 \quad (2)$$

Therefore R, the $NaOH/I_2$ ratio, should be 1.333 to result in the correct SO_2 fraction.

The relation between R and air-correction factor obtained in Table 1 is plotted in Fig. 2. In this method, the operator would run the Tutweiler analysis, calculate $R = NaOH/I_2$, read the air-correction factor from Fig. 2, then make the indicated adjustment in the air-control setting.

Fig. 2 is not as accurate as Fig. 1 in some operating regions, but it is generally accurate enough for satisfactory use.

Manual control using stain-tube analyses. Many plants which previously used Tutweiler analyses are now using a method of analysis in which a gas sample of known volume is drawn through a glass tube packed with a chemical reagent. The reaction of the reagent with a specific component in the gas sample causes a color change of the tube packing.

The tube is calibrated so the length of the color-change zone indicates the concentration of the specific component in a gas sample. These tubes may be referred to as "stain tubes." They can be obtained commercially from several sources, including Drager, Gastec, and Matheson.

This method is generally considered to be easier to perform than a Tutweiler analysis and may be more accurate, particularly for an unskilled operator.

This method results directly in volume percent concentrations for hydrogen sulfide and sulfur dioxide. Various charts could be devised to convert these to air-correction factor.

The first inclination is to plot air-correction factor vs. SO_2 fraction as shown in Fig. 2. However, the preferred method is to plot air-correction factor vs. excess $SO_2 = SO_2 - \frac{1}{2} H_2S$, because it results in a straight line. This plot is shown in Fig. 3 using the data of Table 1.

Automatic control using an on-line analyzer. The principal methods which have been used for on-line analysis and control of sulfur-recovery-plant tail-gas composition are gas chromatography and ultraviolet (UV) spectrophotometer.

Gas chromatography has the advantage that it can be used to analyze for not only hydrogen sulfide and sulfur dioxide, but also various other compounds in the tail gas.

The UV spectrophotometer has been used on sulfur-plant tail gas to analyze for hydrogen sulfide and sulfur dioxide only. It furnishes a continuous reading for these two components whereas the gas-chromatograph readings are intermittent. However, either one can be used to periodically send a signal to the air-control system to optimize the air/acid gas ratio.

Amoco cooperated with Du Pont in developing the application of the UV analyzer to analysis of sulfur-plant tail gas. Data on the light absorption resulting from sulfur compounds and elemental sulfur in the gas were measured in the Amoco research laboratory, and certain features of the sampling system and instrument construction were suggested by Amoco.[5]

The UV spectrophotometer instrument and control system is also available from Western Research & Development.[6]

Most of the UV instruments installed to date furnish readouts for (1) mol percent H_2S, (2) mol percent SO_2, and (3) H_2S / SO_2 ratio. The latter output signal goes to the air-control system.

It is now recognized that H_2S / SO_2 ratio is not the preferred output signal from the analyzer. This parameter is difficult to apply because its value varies from zero to infinity, whereas the optimum value is 2.0.

Furthermore, the H_2S / SO_2 ratio is not linear with air rate to the plant, which is the function that must be controlled.

The preferred parameter to be used with UV or other instrumental control systems is one of the following forms:

a. $SO_2 - \frac{1}{2} H_2S = \text{excess } SO_2$

b. $2\ SO_2 - H_2S = \text{two times excess } SO_2$

c. $H_2S - 2\ SO_2 = \text{excess } H_2S$

These are different mathematical forms of the same parameter. The choice will depend on individual preference. Parameter a, which is related to air-correction factor as shown in Fig. 3, is convenient to use for correlation and interpretation of plant data because it is related linearly to plant air-feed rate.

Parameter c also is linear with air-correction factor and is the parameter used in one of the commercially available control instruments.[6]

At any instant, it is desired to operate the plant with stoichiometric air, which is the air rate that will result in an SO_2 fraction of one-third. Percent stoichiometric air is the relation between actual air rate and stoichiometric air, expressed as percent, which is related approximately to excess SO_2 as follows:

$$\text{Percent stoichiometric air} = 100 + A(SO_2 - \tfrac{1}{2} H_2S)$$

where:

SO_2 = mol percent SO_2 in the tail gas.

H_2S = mol percent H_2S in the tail gas.

A = (wet tail gas, mols/hr)/(O_2 to thermal reactor, mols/hr)

The value of A can be calculated using values from the plant design material balance or an updated material balance.

A more exact value could be determined by making calculations of the type illustrated in Table 1, then plotting percent stoichiometric air vs. excess SO_2 and determining the slope of the line. However, this refinement is seldom justified. The value of A will vary slightly with acid-gas composition.

In using these parameters for automatic control of air/acid gas ratio, the reading for excess SO_2 is sent from the instrumental analyzer to the air/acid gas ratio control system.

The control system is designed to accept the excess SO_2 signal and correspondingly modify the signal to the air-control valve to result in stoichiometric air.

Similarly, if desired, a signal for excess H_2S can be sent from the analyzer to accomplish the same final result.

This method was used in the UV analyzer and air-control system which was installed at the East Crossfield D-1 plant of Amoco Canada Petroleum Co. Ltd. in June 1976. A test in October 1976 indicated that the total loss in sulfur recovery caused by non-perfect control of the air rate was less than 0.1% of the feed sulfur to the train being tested.

Modification of UV analyzer cell. In some installations, air purge is used to prevent condensation of sulfur on the windows of the spectrophotometer cell. In cases where this is objectionable, various modifications can be made to eliminate the air purge.

Fig. 4 illustrates an arrangement used by Amoco in research applications of the analyzer in which double windows are provided at each end of the light-path cell, and an electric heater is installed between the double window.

The window heaters are powered to attain an interwindow temperature slightly higher than the temperature of the gas in the cell. This compensates for radiant-heat-loss cooling of the windows.

The steam jacket extends beyond the outer window. Sulfur condensation on the inner windows is thus prevented. For the laboratory application, where there is a limited quantity of gas available for flow through the cell, the heated-window arrangement eliminates the need for window purges and the consequent dilution of the sample.

Acknowledgment

The authors gratefully acknowledge the contribution of Elton B. Hunt, Jr., to development of these tail-gas control concepts.

References

1. Gamson, B. W., and Elkins, R. H., Chem. Eng. Prog., 49, 203-216, 1953.
2. Grekel, H., Kunkel, L. V., and McGalliard, R., Chem. Eng. Prog., 61, 70-73, September 1965.
3. Sudduth, L. F., Farmer, S. K., and Grekel, H., OGJ, 88-101, Dec. 14, 1970, and 8, Jan. 11, 1971.
4. Palm, J. W., Hydrocarbon Processing, 105-108, March 1972.
5. Saltzman, Robert S., and Hunt, Elton B., "A Photometric Analyzer System for Monitoring and Control of the H_2S / SO_2 Ratio in Sulfur Recovery Plants," ISA AID 72413, 55-60, 1972.
6. Beamish, Mike, Oilweek, Vol. 28, No. 49, 38-46, Jan. 16, 1978.

Chart helps estimate SO_2 emissions

V. GANAPATHY
Bharat Heavy Electricals Ltd.
Tiruchirapalli, India

THE accompanying nomogram or alignment chart gives stack sulfur dioxide (SO_2) emissions (A) in lb/MMBTU fired based on the percent sulfur in the fuel—either coal or oil—and on the higher heating value (HHV). It also gives the efficiency of the SO_2 removal system required (P) to limit SO_2 emissions to a particular value (C).

The chart is based on the equations,

$$A = 2 \times 10^4 \, S/HHV$$
$$P = 100\,(1 - C/A)$$

Example. When coal with 3% sulfur and 9,500 BTU/lb HHV is fired, what is the SO_2 emissions and the efficiency of the removal system required to limit emissions to 1.0 lb/MMBTU?

Solution. Connect 3 on the sulfur-content scale (S) with 9,500 on the HHV scale, extend to the stack emissions scale (A), and read SO_2 = 6.3 lb/MMBTU. Next, connect the 6.3 on the A scale with 1 on the SO_2 emissions limit (C) scale and read 84% on the SO_2 removal system efficiency (P) scale.

Note that the sulfur content (S), SO_2 emissions (A), and SO_2 emissions limit (C) scales are linear and may be extrapolated. END

Estimate SO_2 emissions

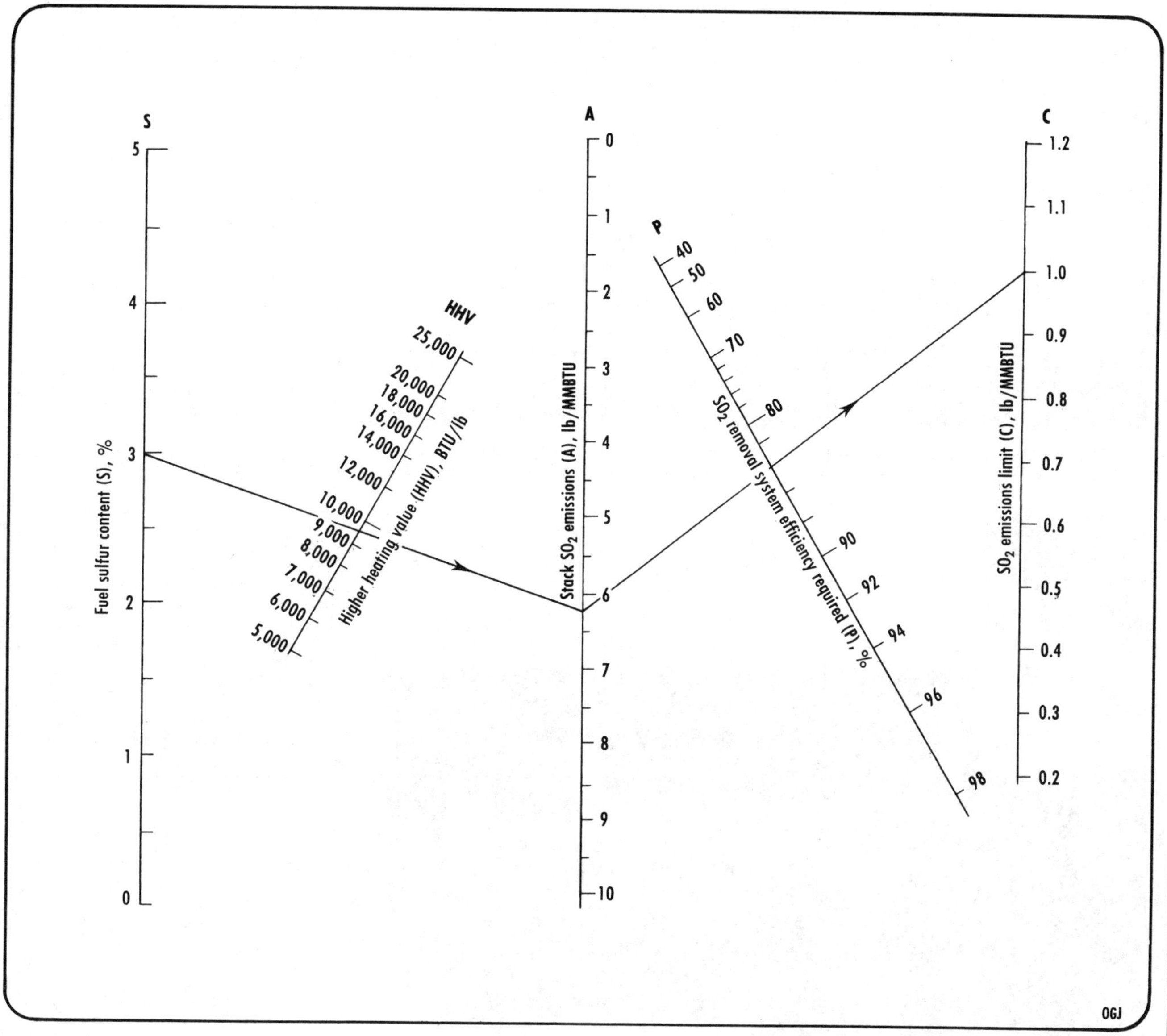

Aquitaine improves Sulfreen process for more H_2S recovery

L. C. CAMERON
Aquitaine Co. of Canada
Calgary, Alta.

INITIAL Ram River plant units (above) used activated-charcoal catalysts and required stainless-steel construction; later units using new alumina catalyst (below) permit carbon-steel material throughout.

THE Sulfreen process for reducing the sulfur in tail gas from Claus sulfur plants has been improved. With the improvements, including the use of a new alumina catalyst, the Sulfreen process now features:

- Overall recovery up to a minimum of 99% for dilute Claus gas streams based on performance in major industrial installations.
- Low capital, operating, and maintenance costs.
- Simple operation similar to Claus plant, using commonly available catalyst at normal operating temperatures and pressures.

This article is an abridged version of a paper presented at GPA Annual Meeting, Denver, Mar. 25-27.

Developers of the Sulfreen process were Soc. Nationale des Petroles d'Aquitaine (SNPA) with assistance of Lurgi Gesellschaft Fur Warme-Und Chemotechnik G.m.b.H. (Lurgi). The Sulfreen process is covered by several Lurgi patents for the use of the carbon catalysts and by SNPA patents for the use of alumina-base catalysts. Several pending patents cover the new development of the alumina process as described in this article.

SNPA and Lurgi have granted to Ralph M. Parsons Co. an exclusive right to design, engineer, and build Sulfreen plants in Canada, U.S.A., Puerto Rico, and Mexico.

• Conventional construction materials.

• No liquid or other waste-disposal problems.

Where improvement came. Improvements to the process came as the result of developmental studies in the laboratory, of bench-scale testing, and of application in the big SNPA plant at Lacq, France. Thrust of the work is in three areas:

• **Catalysts.** The best catalyst can now be selected for any Claus gas-effluent composition. The effect of particle size, residence time, and superficial velocity on the Sulfreen conversion was also investigated to determine optimum bed size and cycle time. The catalyst selected to replace the carbon catalyst originally used is an activated alumina available in North America at a cost of about 20c/lb.

• **Regeneration.** A technique was also developed which is used automatically during every regeneration to prevent sulfate formation. This technique effectively reduces catalyst aging and maintains the high activity and conversion efficiency. This improvement represented decisive step in Sulfreen process development. An overall sulfur conversion of 99% could be achieved with an activated alumina catalyst life of about 3 years.

• **Construction.** The alumina catalyst does not require a high temperature for regeneration. This fact permits using carbon steel through the plant. The reactors are designed like conventional Claus sulfur-plant converters with respect to general shape, support grating, and gas distribution.

Initial use of the new alumina catalyst in North America is at Aquitaine Co. of Canada's Ram River plant in Alberta. The first stage of this plant used the older activated carbon catalyst.[1] The second stage, with the alumina catalyst, uses three horizontal reactors which were completely shop-fabricated. The condenser is a standard horizontal sulfur condenser equipped with a stainless wire-mesh coalescer pad. The regeneration gas heater is an indirect-fired heater using a stainless-steel tube bundle.

These developments have substantially reduced the investment for Sulfreen units. As an illustration, investment for the Stage II Ram River Sulfreen (for alumina catalyst) is half of that for the Stage I carbon unit. Both units have exactly the same tail-gas capacity.

A side advantage of an alumina Sulfreen unit is the opportunity for improving Claus sulfur-plant operations. The Sulfreen regeneration loop can be used for regenerating and warming up the sulfur-plant converters during shutdown and start-up periods. This application would reduce times for shutdown and start-up. A longer life is also expected for the Claus catalyst which would not be exposed to overheating and sulfating.

In November 1973 the Stage II alumina Sulfreen unit of Ram River was put on stream with a minimum of problems. It processes 7.3 MMscfh of tail gas from two parallel two-stage 1,000-long ton/day sulfur plants. Since start-up, Stage II overall sulfur recovery has been in excess of 99% with a maximum at 99.6% during the first week of operation. Results of performance test are as follows:

Sulfur-plant conversion:	96.4%
Sulfreen conversion:	89.0%
Overall sulfur recovery:	99.6%

Figures for the conversions include H_2S, plus SO_2 plus COS, plus CS_2.

Additional alumina Sulfreen units are now under design or construction in both Canada and Europe (Table 1).

Process description. Basic reaction of the Sulfreen process is the Claus reaction:

$2H_2C + SO_2 \rightarrow 3S + 2H_2O$ which takes place over a solid catalyst at 260° - 300° F. and 1.4 to 1.7 psig.

Figs. 1, 2

Three-reactor Sulfreen units

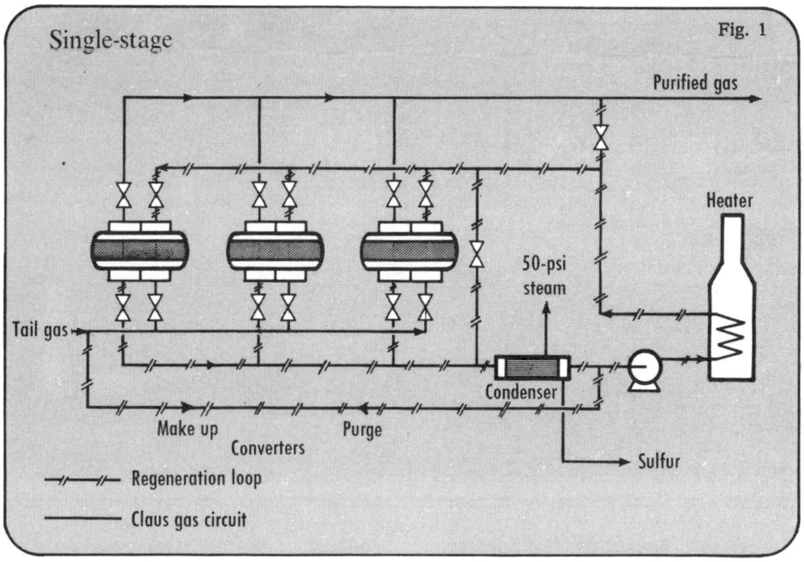

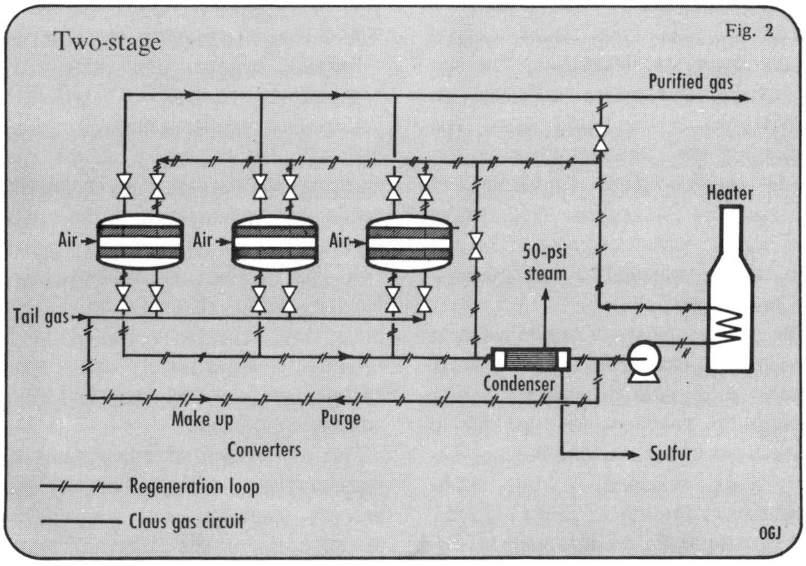

Figs. 3, 4

Typical combinations

One-stage Claus plus two-reactor Sulfreen

Fig. 3

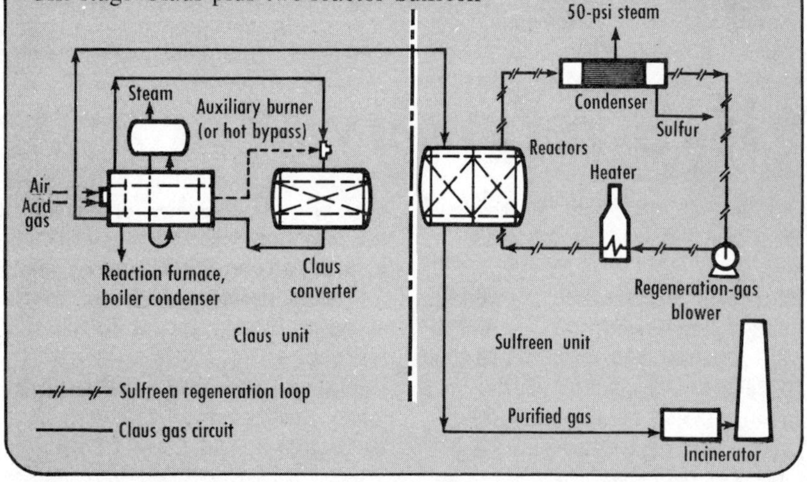

Two-stage Claus plus two-reactor Sulfreen

Fig. 4

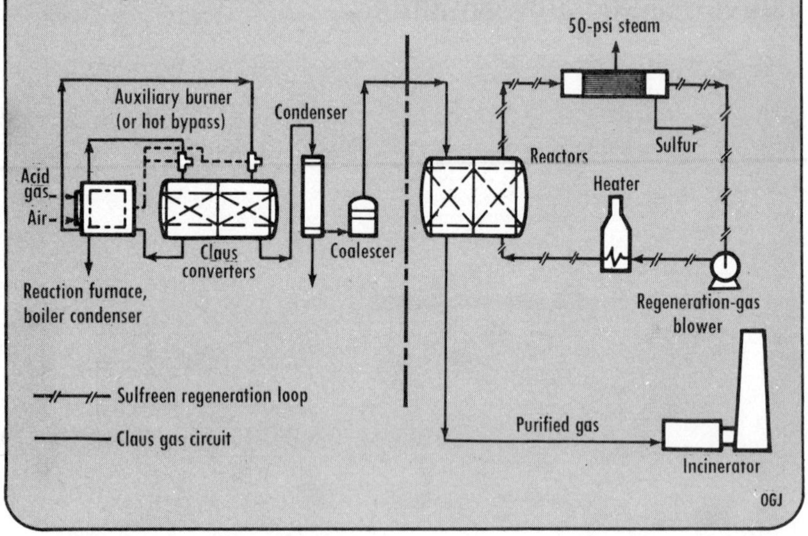

At this low operating temperature, the thermodynamic equilibrium is more favorable than under normal Claus conditions. Moreover, the sulfur produced remains adsorbed on the catalyst in the liquid form, thus displacing the Claus equilibrium towards the complete conversion of H_2S and SO_2 into sulfur. This results in a higher sulfur-recovery efficiency than can be achieved in a conventional Claus plant.

The plant consists of one or several parallel ractors filled with conventional Claus catalyst after passing through the reactors, the tail gas is incinerated in the normal manner before being released to the atmosphere from the stack. (Figs. 1, 2).

The reactor is maintained in adsorption until just before the sulfur content of the catalyst reaches a critical value beyond which the catalyst would lose its capacity of adsorption.

Periodic regeneration using Claus-type gas is required to vaporize the sulfur. The regeneration is in a closed loop with the gas being heated before entering the reactor and cooled in a condenser downstream of the reactor to recover the sulfur. The regeneration gas is then recirculated by a blower. When regeneration is complete, the reactor is cooled to the tail-gas temperature and then switched to the tail gas for a new adsorption period.

The alternation of adsorptions and regenerations results in a batch process operation, but the duration of the cycle makes Sulfreen comparable to a continuous process since switching occurs only once in about 48 hr. For this reason, the Sulfreen unit operates much like an additional Claus converter integrated to the sulfur unit. (Figs. 3, 4).

Performance. In the early stage of its development the Sulfreen process used a special activated carbon as the catalyst. Further development resulted in the use of the activated alumina. Since 1972, all Sulfreen units have been designed for alumina catalyst.

With the Sulfreen process in its present form, minimum guaranteed conversions of $H_2S + SO_2$ to sulfur depends on the total $H_2S + SO_2$ concentration in the tail gas. The following figures illustrate:

Tail gas $H_2S + SO_2$ % vol	Minimum Sulfreen conversion %
1.5	84
1.0	75
0.8	70

These conversion can result in an average overall sulfur conversion of 99%. Concentration of $H_2S + SO_2$ in the treated gas will be 2,500 ppm 2nd concentration of SO_2 will be 1,500 ppm in the stack after incineration under normal conditions (1,200° F., 5% excess oxygen). Higher efficiencies than shown above are obtained at the beginning of the run when the alumina catalyst is new (90% is achievable with a Claus tail gas containing 1.5% of $H_2S + SO_2$) The conversions given above are those obtained at the end of about 3 years of operation.

The Sulfreen process probably recovers all the free sulfur present in the form of vapor or entrained droplets, providing tail-gas temperature does not exceed 280° F.

COS and CS_2 present in the tail gas are unaffected and are not converted to sulfur. However, proven methods can minimize formation of these compounds in the Claus unit. One technique is to run the first Claus converter at a temperature hotter than normal. Another is to control the formation of COS and CS_2 which takes place in the reaction furnace.

Recent bench-scale studies by SNPA show that formation of COS and mainly CS_2 can be reduced significantly with a properly designed reaction furnace.

Pressure drop. Overall pressure drop through an alumina unit including the inlet valve manifold is 1.4 to 1.7 psig depending on pipe sizing.

This is low enough to allow a Sulfreen unit to be installed downstream of most Claus plants without modifying the existing plant.

With the proper quantity of catalyst to achieve an overall efficiency of 98-99%, pressure drop through the reactor only is 0.5 to 0.7 psi. This is the minimum pressure drop necessary to ensure an adequate gas distribution during the adsorption and regeneration steps.

The overall pressure drop of the regeneration loop is slightly higher than for the adsorption circuit.

Process development. The first two commercial units were designed to operate with an activated carbon catalyst specially developed by Lurgi. This catalyst has the characteristics of high adsorption capacity for sulfur compounds and a special pretreatment by impregnation to prevent aging.

The carbon must be regenerated at a high temperature (900° F.) and with an inert gas. This requires use of stainless steel throughout the plant.

The first unit was put on stream in the Lacq plant in France in early 1970 processing 4.6 MMschf of tail gas from a two-stage 1,150-long ton/day sulfur plant. This unit was built with six identical vertical reactors, four of which are in adsorption at any one time. The operation is essentially the same as previously described for alumina except for the higher regeneration temperature and inert gas.

The initial load of catalyst is still in the reactors in the Lacq unit. This demonstrates the excellent catalyst life, although a slight drop of activity has been observed.

Present operating efficiencies are:

Sulfur-plant recovery:	92%.
Sulfreen recovery:	68%.
Overall recovery:	97.4%.

The main process problem encountered in this unit was inadequate regeneration blower capacity. Results of this were a high residual sulfur content in the catalyst at the end of every regeneration step and a loss of adsorption capacity. This problem was solved by the replacement of the blower. Part of the first Lacq unit was also used as a pilot unit to test the behavior of the activated alumina in the Process.

The second unit designed to operate with the carbon catalyst is the Stage I Sulfreen unit of the Ram River Plant in Alberta. The Stage I unit is designed for an overall sulfur recovery of at least 98% to meet governmental requirements.

This unit, except for size is almost identical to the Lacq unit and uses the same catalyst and similar reactors.

This unit was put on stream in April 1972 to process 7.3 MMscfh of tail gas from two two-stage 1,000-long ton/day sulfur plants.

In June 1972 a performance test showed these conversions:

Sulfur plant:	96.25	96.2	96.2
Sulfreen:	58.7	61.7	60.2
Overall:	98.4	98.5	98.5

Even though these results meet overall sulfur-recovery requirements, they indicated that the Sulfreen unit was not performing properly. Cause of this deficiency was soon found to be poor design of the regeneration-gas inlet-nozzle baffles at the top of the reactors. Result had been displacement of the light carbon catalyst. This catalyst displacement occurred during the first cycles of re-

Table 1

Sulfreen installations

Location	Sulfur unit capacity long ton/day	Tail-gas volume MMcfh	Start-up date	Observations
In Operation				
Lacq 7 (France)	1,150	4.6	April '70	
Ram River II (Alberta)	2,000	7.3	April '72	Conversion to alumina planned for summer
Ram River II (Alberta)	2,100	7.3	November '73	
Under construction				
Lacq 5.6.9 (France)	1,800	7.0	April '74	
Lacq 8.10 (France)	1,600	6.7	July '74	
Fort Nelson (B.C.)	450	7.9	December '74	7.5% H_2S in acid gas
Balzac (Alberta)	1,200	5.0	January '74	
Feyzin (France)	90	0.32	1975	Refinery
Fox Creek (Alberta)	1,100	3.7	1975	

Table 2

Capital operating and maintenance costs

Plant operator	Catalyst	Tail-gas volume, MMscfh	Sulfur recovery long ton/year	Capital cost* $ million	Annual operating, maintenance cost
SNPA	Carbon 6 reactors	4.6	21,000	3.6	$190,000
ACC	Carbon 6 reactors	7.3	29,000	6.5	$175,000
ACC	Alumina 3 reactors	7.3	30,000	3.2	$140,000²
SNPA	Alumina 3 reactors	7.0	31,000	3.3†	$140,000²
SNPA	Alumina 3 reactors	6.7	29,000	3.3†	$135,000$/y²
Elf	Alumina 2 reactors	0.32	1,600	0.42†	

* Capital costs are expressed in present value for year of construction on battery limit basis.

† Estimated.

‡ Operation costs include utility consumption catalyst replacement, maintenance, materials, and labor.

Table 3

Utility consumption

Plant operator	Catalyst	Tail-gas-volume, MMscfh	Horsepower	Steam production or feedwater Maximum lb/hr	Fuel gas Mscfh
SNPA	Carbon	4.6	1,000	12,000	25
ACC	Carbon	7.3	1,700	24,000	40
ACC	Alumina	7.3	1,500 (design 1,900)	20,000	27
SNPA	Alumina	7.0	1,500	23,000	33
SNPA	Alumina	6.7	1,450	23,000	33
Elf	Alumina	0.32	40	850	1.25

generation and continued until the catalyst bed reached a stable shape causing poor gas distribution. Ballast and floating hold-down screens at the top of the catalyst bed corrected this problem.

An additional problem was poor design of the piping tied to the regeneration gas blower. The blower was prevented from operating at more than 3,000 rpm rather than the design 4,100 rpm because of vibration. As a result regeneration was incomplete and catalyst activity was reduced.

In spite of this, overall sulfur conversion remained quite constant at 98.3 to 98.4% while the sulfur plant efficiency was gradually falling off.

After repairs during turnaround, performance tests showed the following recoveries:

	At design rate	At 60% of design rate
Sulphur plant	95.5	95.3
Sulfreen	73.0	81.0
Overall	98.5	99.0

Since this turnaround, the Stage I Sulfreen Ram River unit has operated quite well with few mechanical problems.

Two-stage Sulfreen. A two-stage Sulfreen process for conversions beyond 99.5% has been tested in bench scale and will soon be available for industrial application. In this process Claus tail gas is desulfurized in two steps.

The first step consists of making the Claus reaction with an excess of H_2S. The inlet H_2S/SO_2 ratio is higher than 2 so that the SO_2 is completely eliminated. This reaction takes place under the same operating conditions as the one-stage Sulfreen process.

In the second step, the H_2S leaving the first catalytic step is oxidized to sulfur using controlled air injection by the following reaction:

$$H_2S + \rightarrow S + H_2O$$

The kinetics of this reaction are more favorable to the formation of sulfur than the Claus reaction. This reaction takes place on a solid catalyst at the temperature of the gas leaving the first stage (300°F.). The sulfur is adsorbed by the catalyst as the reaction proceeds.

Both catalytic stages are regenerated at the same time and under the same conditions and therefore only a single loop is required. The special catalyst used in the second stage is capable of withstanding the simultaneous presence of oxygen, H_2S, and sulfur.

The final aging tests are now being conducted in the Lacq pilot unit. The expected Sulfreen recovery based on $H_2S + SO_2$ is 95%, and the Lacq units presently under construction are designed for future second-stage addition. Start-up is scheduled for early 1975.

For refinery sulfur plants. Several process designs have been developed to fit the Sulfreen process for refineries Because of the small size, the most practicable scheme for existing plants may be a two-reactor Sulfreen unit with the reactors combined in a single vessel.

Where a new Claus plant is to be built, the most economical design consists of integrating the Sulfreen unit into the sulfur plant rather than adding them together. This reduces capital cost substantially eliminating an intermediate Claus catalytic stage. This design is being used in one Canadian plant.

SNPA has just been commissioned by the Elf Co. to design a Sulfreen unit to process tail gas from two parallel sulfur plants having a total sulfur production of 89 long tons/day. to fit the Sulfreen process for refineries. Because of the small size, the most practicable scheme for existing plants may be a two-reactor Sulfreen unit with the reactors combined in a single vessel.

Capital, operating, maintenance cost. Tables 2 and 3 summarize the capital cost of all the units which have been built or are sufficiently advanced in design to have accurate costs developed. The operating and maintenance costs available at present are related to the carbon Sulfreen process only.

The operating and maintenance costs for the alumina process are expected to be lower because of the lower price of the catalyst and also the lower temperature for regenerating. A substantial reduction of fuel-gas consumption is expected.

Work is continuing on the development of sulfur conversion beyond 99.5% using the two-stage Sulfreen process. The results to date obtained in the pilot scale indicate that the design of a commercial plant will be available in the near future.

Reference

1. McGhee, E: "Aquitaine cleans Claus tail gas with Sulfreen unit," OGJ June 26, 1972.

Aquitaine cleans Claus tail gas with Sulfreen unit

THE NEW Ram River, Alta., gas plant of Aquitaine Co. of Canada Ltd. is now operating the first commercial-scale cleanup unit for Claus tail gas in North America.

The unit recovers sulfur as product from tail gases off two 1,000-ton/day Claus sulfur units.

Aquitaine says that the addition of the Sulfreen cleanup made it possible to construct only a two-stage Claus plant. Otherwise, a four-stage plant would be necessary for the throughput. A further benefit was that the waste-gas stack height could be practically reduced from a probable 500 ft to 300 ft.

Aquitaine's contractor, Ralph M. Parsons, has built the new plant in two stages. The first—with feed capacity of roughly 200 MMcfd—came on stream in early February. The second stage—almost a mirror image of the first—is expected on stream in September or October.

A second Sulfreen unit will be built to accommodate the second-phase throughput. However, Aquitaine will use a different catalyst and different design which will reduce construction costs.

Ram River plant is notable for other reasons than the tail-gas cleanup:

- A separate, individual optimizer will be used on each of the four 1,000-ton/day sulfur plants.
- A supervisory-control system keeps close watch on certain processing variables so that most upsets can be quickly detected and rectified.
- The same supervisory system brings into the plant control room the surveillance and control of all the wells feeding the plant.
- Total sulfur output when both phases are on stream will be 4,000 ton/day, among the greatest of any plant in the world.
- Extensive use is made of air coolers in order to avoid possible water pollution and to avoid drawing on the limited supplies of water locally.

$65-million investment. The Ram

SIX Sulfreen reactors are first to be installed in North America. Use of a different catalyst and different design will allow use of fewer reactors in next construction stage of plant. (Top photo). At full capacity, Ram River's Claus units will be producing 4,000 tons/day of elemental sulfur. Tail-gas cleanup units allow two-stage Claus plants instead of four-stage usually required. (Center photo). Central control room in plant oversees not only plant operations but also the fields from which gas is drawn. Any upset in the field is detected immediately and offsetting changes can be made in plant operation. (Bottom photo).

DIGITAL computer in central-control room makes possible an immediate calculation of plant balance. Operators see availability of this balance as one of their best possible tools in optimizing desulfurization.

ACTUAL well control is performed in the enclosure on the gas heater at the site. Both temperature of the existing gas and flow rate can be controlled remotely.

DEA regenerators have unusual reduced-diameter section at top for direct water-contact cooling. This eliminates need for outside reflux condensers.

River plant draws gas from wells in the Strachan and Ricinus West fields. Sulfur content of the formation gas varies from as little as about 3% to as much as 34%. Average in-plant feed exceeds 20%.

In designing the installation, Aquitaine was guided by the most recent environmental regulations for the Province of Alberta. Any plant is given limits on the amount of sulfur which it can emit over any 3-month period, over any 24 hr within that, and over any 30 min in that. The general rule is that by 1975 all plants must remove 98% of the sulfur entering as feed.

Achieving 98% sulfur removal continuously with standard Claus units is difficult. Usually a minimum of a four-stage Claus process with frequent catalyst changes is required. Else, the units must be operated at a restricted throughput. Aquitaine figures that its two-stage units remove 95% of the sulfur. The Sulfreen units bring that figure to 98% and more.

The investment in optimizers and supervisory control will, according to Aquitaine, ensure that the plant does the very best job possible for the greatest percent of the time possible. That is, the plant should only very seldom be required to restrict throughput in order to avoid infringing environmental rules on stack-gas emission.

Already tested at Lacq. Ram River's Sulfreen unit is the second ever installed. The first has been operating for 2 years in the Lacq, France, plant of Aquitaine's parent company, SNPA. That unit serves a 1,000-ton/day sulfur unit, half the size of the one at Ram River.

Fig. 1

Individual sulfur-plant optimizer

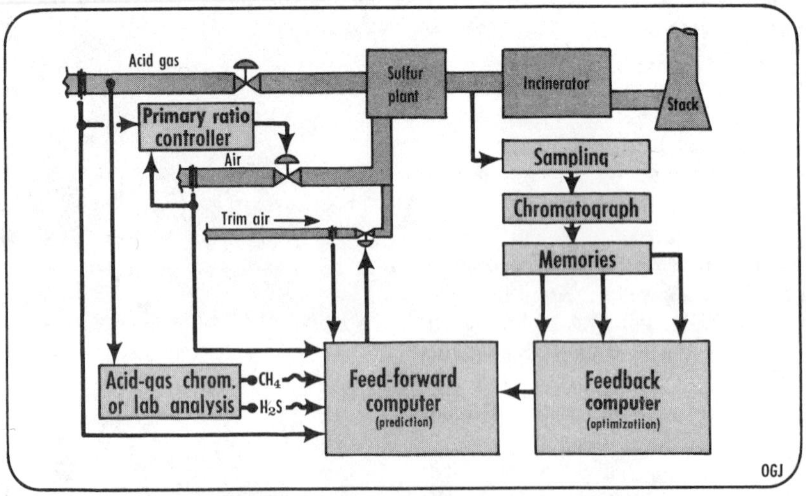

The process converts H_2S and SO_2 in the tail gas to sulfur at low temperatures, 260° to 300° F., using the Claus reaction:

$$2H_2S + SO_2 \rightarrow 3S + 2H_20 + 35 \text{ Kcal}$$

In the first two installations, a special activated charcoal has been used as adsorbent and catalyst. It features high adsorption and easy desorption of the elemental sulfur. An alumina-based catalyst is being considered for the next Ram River unit; it may make it possible to reduce to three the number of reactors. The presently operating units have six reactors.

The inlet gases to the unit pass through four of the six parallel reactors which operate on a regenerative cycle. A reactor remains on line until it approaches the sulfur-adsorbing capacity of the catalyst. Then, it is regenerated with hot, inert (atmospheric nitrogen) gas. Standard sulfur condensers cool the regeneration gas and its load of sulfur vapor. The condensing sulfur generates 50-psi steam which is used elsewhere in the plant.

The reactors are cooled by unheated inert gas before being returned to

ELECTRICALLY heated flow line is being tested to carry molten sulfur to pour area. If successful, such a line may be used to carry the sulfur to a railhead.

WELL site in the field includes heater, glycol pumps, and inhibitor pumps in building at left. These functions, as well as flowing pressures and temperatures can be monitored from the Ram River plant.

service.

The Sulfreen Process was a joint development of SNPA and Lurgi Gesellschaft. Recently, the firms arranged a licensing agreement with Ralph M. Parsons Co., who along with Lurgi, will make the process available to other firms in North America.

Wells integrated with plant. Wellsite control is centralized in the plant control room. This permits regulation of gas flow right from the wellhead on through to the sales-gas meters downstream from the plant and to the sulfur-pour towers in the stockpile area. This total integration allows the operators in the plant's control room to take immediate steps when a problem arises. No upset, no matter how minor, need be ignored for lack of corrective procedures.

The supervisory-control system has installed features which will facilitate an even greater degree of sophistication with little increase in cost.

Each well has its own on-site heater, fueled by sweet gas piped back through a 3-in. line from the plant. It also has its own flare with a constant pilot flame. Upon exiting the wellhead, the high-pressure gas passes through a remotely controlled wellhead-shutdown valve. It then makes an immediate pass through the heater and takes its initial pressure drop through an automatic (not remotely controlled) pressure regulator. After a second pass through the heater, the gas passes through the main flow-control valve which is remotely controlled. Following a third pass through the heater, the gas enters the flow line to the plant at about 1,200 psi.

At the plant control room, the operators can control the well-shutdown valve, gas-flow rate, heater temperature, methanol injection, and inhibitor injection. They receive alarms in the case of high lease-exit pressure, low fuel-gas pressure, low level in methanol or inhibitor tanks, power failure, control-relay failure, fire detection, gas detection, well-shutdown valve in abnormal position, and inhibitor or methonal pumps off. They can read the position of choke valve, flow rates, heater temperatures, wellhead temperature and pressure, and lease-outlet temperature and pressure.

Quick material balance. Operators in the average gas-treated plant can expect to receive a complete material balance only once per day. Even then, the data is several hours old when received and represents the operations of the previous day. Ram River operators can get immediate material balances at any time they choose. They expect that this will enable them to keep the plant operating at peak efficiency as nearly 100% of the time as modern technology permits.

The PDP8/E minicomputer which makes this possible automatically calculates and prints out the plant balance every 4 hr in any case. It also performs these calculations either automatically or on demand:

1. Flow rates and daily volumes for all wells and all plant meters.
2. Minimum, maximum, and average stack-emission temperature.
3. Sales-gas rate and volume.
4. Daily volume of sulfur emission up the sulfur stack.
5. Plant sulfur-recovery efficiency.
6. Field-proration factors, as a check on measurement accuracies.

Having this information at hand, the operators' task of keeping plant operation at best possible level comes easier. Hence, no supervisory control, need be provided inside the plant.

Optimizers. With stable plant conditions thus assured as nearly as possible, it remains only to optimize the Claus sulfur plants themselves. This is done with individual SNPA-developed optimizers which supplement the conventional air/acid-gas flow-ratio controller by maintaining the H_2S/SO_2 ratio at optimum, Fig. 1.

Input to the analog computer includes:

- H_2S and SO_2 concentration in the tail gas as measured by chromatographs.
- Differential pressure, static pressure, and temperature from both the air and the acid-gas feed streams.

The computer performs mass-flow and ratio calculations for air and acid gas. It relates these to the tail-gas H_2S/SO_2 ratio and adjusts a 20% trim-air flow to the reaction furnace.

This system has maintained the Claus units at a maximum efficiency regardless of variations in operation conditions such as changes in the content of H_2S, water vapor, and hydrocarbons in the acid-gas stream.

The SNPA optimizer uses both feed forward and feedback controls.

Even in the short time the plant has been on stream, it has already demonstrated the value of its supervisory-control and optimizing systems. During a late-winter storm, electric power was lost to a wide area, including Ram River. Aquitaine's plant manager says the plant came back on stream very quickly when power was restored. He considers that the previous throughput level was reached very quickly compared to what conventionally controlled plants could expect.

Gas flow through plant. On the gas side of the plant, processing is fairly standard. Three inlet separators (one for each gas source) handle flow for the stage-one plant and will also accommodate flow for the second stage when it comes on stream. Due to the hilly nature of the local terrain, the separators must be capable of taking a fairly large slug of liquid which might accumulate in the two-phase flow lines.

Acid gases are scrubbed out of the stream in a DEA treater.

The DEA regenerator units do not have a conventional outside reflux condenser. Rather, direct-contact water cooling is used in the uppermost section of the towers, with cool makeup flowing into a saddle-pack section where it cools and condenses any rising water vapor.

Dry desiccant absorbers downstream of the DEA treaters control sales-gas dew point and remove any hydrocarbon liquids not already separated out. These absorbers are of the packaged, regenerative type.

Pollution control. Aquitaine calculates its investment in pollution control at Ram River, including the Sulfreen units, to be about 10% of the total plant investment. In addition to the typical sour-gas-treating plant's tall stack, the plant has invested in:

- Large-scale air coolers for the processes.
- An injection well through which formation water and drainage from process streams containing amine is put back into the gas-producing reservoir.
- Extensive treating and storage capacity for boiler-blowdown and like water so that it can be returned to the nearby Baseline Creek.

A principal part of the water-treatment system is a 10-unit aeration system which has a capacity of blowing 400 cfm at 7.5 psi into the water pond. The aeration system is kept operative even in extreme winter weather by running steam lines into the ponds.

A 150-ft flare stack has been installed to accommodate the two emergency-flare systems. The stack can, if the need should ever arise, flare all of the plant's possible 440 MMcfd feed.

Electrically heated line. An interesting and promising adjunct to the Ram River plant equipment is its electrically heated sulfur flow line. The line keeps the sulfur hot enough to flow several hundred feet from the storage pits to the stockpile area.

JUST beginning are sulfur blocks which, when the plant is complete, will be receiving 4,000 tons/day.

LONG blocks of air coolers are used, not only to avoid any possible water-pollution problems but also to conserve local water supplies.

The electric system is a patented development of Chisso Engineering Corp., Tokyo, and adds heat to the line by induction.

Sulfur must be kept above 250° F. in order to flow. The polyurethane insulation, which Aquitaine engineers believe to be the most practical for such an application, has sufficient temperature tolerance but is being watched closely for any signs of deterioration with time.

If the polyurethane isulation proves long-lived at molten-sulfur temperatures, Aquitaine thinks that the electrically heated line system might be practical for fairly long distances. It might make possible pipelining Ram River sulfur a dozen miles or more to an existing rail-loading point. In fact, Aquitaine engineers believe the idea might eventually be worthy of investigation for moving sulfur from the Alberta foothills to tidewater in British Columbia.

References

1. Guyot, G., and Martin, J. E., "The Sulfreen Process," presented to Canadian Natural Gas Processors Association meeting, Edmonton, Alta., June 1971.

Japan tries three stack-gas desulfurization routes

JIM MORRISON
European Engineering Editor

INTEREST in flue-gas desulfurization is renewing everywhere.

Rising demand for low-sulfur fuel oil is making it technically and economically difficult for industry to rely solely on heavy-oil desulfurization. While new desulfurization routes are being studied intently, some form of stack-gas treatment will be imperative in many installations.

This article presents three new tail-gas cleanup processes developed in Japan. One of these is now being installed in three commercial installations. A second, following pilot testing, will be tried out in a 60,000-scfm prototype plant. The third route to cleaner gas will get extensive testing in two commercial plants due on stream late in 1972.

Industry in Japan has taken a lead in investigating flue-gas cleanup. The Japanese national, prefecture, and city governments have imposed stringent limits on SO_2 emissions. Maximum allowable concentrations, now limited to 0.38 ppm in highly populated areas, will be progressively reduced to 0.27 ppm in 1973 and to 0.2 ppm in 1978. These criteria can be changed downwards by local authorities. Already industry in Yokkaichi has to meet an ambient SO_2 limit at the property line of 0.009 ppm as an hourly average.

Chiyoda Thoroughbred 101

A 600-scfm pilot plant has already tested Chiyoda's Thoroughbred 101 route to clean gas. It will now be

THIS pilot plant of Chiyoda's stack-gas cleanup process operated continuously for more than 250 days on flue gas containing 1,500 to 3,000 ppm SO_2. Fig. 1.

tested in four commercial plants, two scheduled to come on stream in September and two others in December.

One unit will treat 20,000 scfm of tail gas from a Claus unit at Nippon Mining's Mizushima refinery and another 8,000 scfm tail gas from Tohoku Oil's Sendai refinery. A third will remove SO_2 from 95,000 scfm of stack gas from oil-fired boilers at Fuji Kosan's Kainan refinery. The fourth will treat 53,000 scfm stack gas from a power plant in Mitsubishi Rayon.

Following bench-scale laboratory testing of the process, Chiyoda built at its Kawasaki laboratory a pilot plant including an 18-ft high packed column, Fig. 1. It was operated continuously for 100 days on a slip stream of stack gas from vacuum-residue combustion containing 3,000 ppm SO_2 and another 150 days on a gas stream from bunker-oil combustion containing 1,500 ppm SO_2. The polluted gas was contacted countercurrently with a 3 wt% solution of sulfuric acid.

Results from pilot testing were completely satisfactory. SO_2 content of the gas vented to atmosphere was held lower than 50 ppm. The plant also successfully handled variations in feed composition and other upsets.

Six important features. Advantages claimed for the process are:

- Stable plant operation because no slurry is used in the absorption and oxidation sections.
- High desulfurization efficiency, said to exceed 97%.
- Process simplicity resulting in lower construction, operation, and desulfurization costs.
- Continuous solution recycle which eliminates all waste streams and guarantees no secondary pollution.
- Excellent turndown ratio provides maximum flexibility of operation.
- Improved pollution-control cost since end product is useful gypsum with characteristics equal to or better than natural gypsum.

How it works. The SO_2-rich stack gas is prescrubbed against water to remove particulate dust, Fig. 2. The cooled gas is passed to the bottom of an absorber where SO_2 and SO_3 contained in the gas are absorbed and removed by dilute sulfuric acid at 120°-160° F.

The absorbent, which is recycled from an oxidizer, is saturated with O_2 and the dissolved oxygen reacts with the SO_2 absorbed in the presence of a catalyst. The resulting oxidation reaction in the absorber column substantially reduces the volume of water required for complete SO_2 absorption.

Clean gas is taken off the top of the absorber and is reheated to a temperature that prevents steam-plume formation from the vent stack.

Sulfurous acid, dilute sulfuric acid, and accelerator catalyst from the bottom of the absorber are mixed with air in the bottom of the oxidizer. Preferably, this is a bubble-type column using air as oxidation agent.

The sulfurous acid (H_2SO_4) portion of the stream is oxidized completely

Table 1

Investment, utility costs, Chiyoda 'Thoroughbred 101' flue-gas desulfurization

Item	Description	Unit	Case 1	Case 2
Design conditions	Flue gas	scfm	441,400	1,413,000
	Power generating capacity	mw	250	800
	SO_2 content in flue gas	ppm	1,500	1,500
	Flue gas temperature	°F.	284	284
	Desulfurization rate	%	more than 90	more than 90
	Service factor of boiler	%	90	90
	Fuel-oil consumption in boiler	bbl/year	3,032,000	9,687,000
	Sulfur content in fuel oil	%	2.7	2.7
By-product	Gypsum ($CaSo_4 \cdot 2\ H_2O$)	ton/hr	7.8	25.1
Plant cost (A)		$	4,970,000	11,850,000
Fixed cost (B)	18% of (A)	$/year	894,600	2,133,000
Direct cost (C)	Limestone powder (0.25 ¢/lb)	$/year	202,400	642,400
	Power (0.7 ¢/kw/hr)	$/year	263,200	812,000
	Industrial water (8¢/1,000 gal)	$/year	8,000	22,400
	Reheat fuel oil ($3.18/bbl)	$/year	241,000	770,000
	Catalyst	$/year	6,000	19,000
	Labor ($12,000/yr/capital)	$/year	96,000	144,000
	Maintenance (2% of (A))	$/year	99,400	237,000
	Subtotal	$/year	916,000	2,646,800
Net operating cost (D)	(B) + (C)	$/year	1,810,600	4,779,800
Overhead (E)	12% of (C)	$/year	109,900	317,600
Operation cost (F)	(D) + (E)	$/year	1,920,500	5,097,400
By-product credit	(0.25 ¢/lb)	$/year	343,200	1,104,400
Cost of desulfurization	Without by-product credit	$/bbl	0.63	0.53
		$/mw-hr	0.96	0.80
	With by-product credit	$/bbl	0.52	0.41
		$/mw-hr	0.79	0.62
Space required	Including 7-day storage	1,000 sq ft	38.7	123.7

Fig. 2

Chiyoda 'Thoroughbred 101' flue-gas desulfurization process

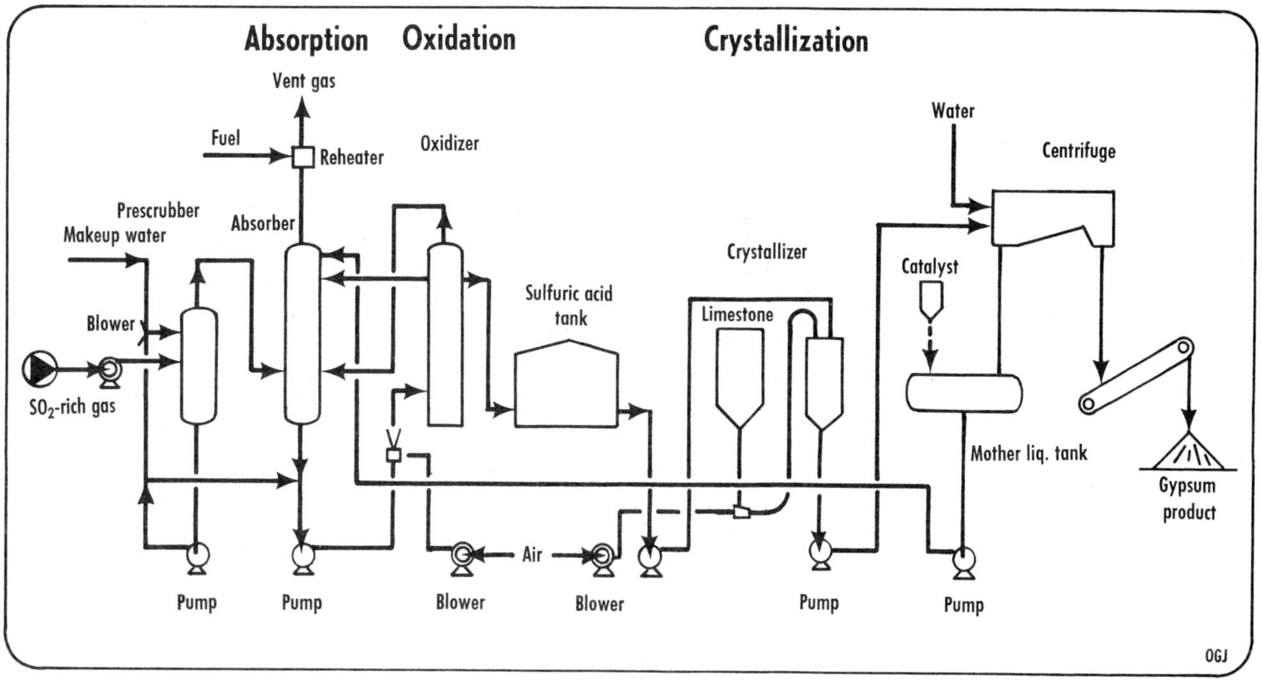

to sulfuric acid. Air consumption for this reaction is about five times the theoretical requirement.

In the oxidizer, the catalyst is regenerated and passes with bleed sulfuric acid to storage and crystallization. It is recycled to the absorption-oxidation system together with mother liquor centrifuged out from the gypsum crystals.

Since the by-product gypsum contains about 10% water, some catalyst will be lost with the water and will require making up. The lost catalyst is not harmful.

The accelerator catalyst used by Chiyoda for its process has not been disclosed but is said to be inexpensive and commercially available.

Sulfuric acid taken off as a side stream from the oxidation column is passed to a storage tank. Acid in a volume corresponding to the amounts of SO_2 and SO_4 recovered from the stack gas, is continuously withdrawn from the storage tank and pumped to the crystallizer.

Calcium compounds, preferably limestone, are passed to the crystallizer to neutralize the acid and produce gypsum by crystallization.

Gypsum crystals are continuously separated by centrifuging to produce the product gypsum.

Mother liquor is sent to a tank where makeup catalyst is added. This stream is pumped to the absorber column.

There is no effluent water drained out from any point in the whole system.

Economics. Chiyoda estimates a plant in Japan to treat 450,000 scfm of flue gas would require an investment of about \$5 million; for 1,500,000 scfm, it would be about \$12 million. In both cases, flue gas is assumed to contain 1,500 ppm SO_2 with desulfurization 90% or more efficient.

Net operating costs are given as \$1,800,000/year for the smaller plant and \$4,800,000 for the larger. Without credit for by-product gypsum, total cost of desulfurization would run \$0.63/bbl of fuel oil for the smaller and \$0.53/bbl for the larger. Crediting gypsum at a market price of \$0.25/lb would reduce these figures to \$0.52 and \$0.41/bbl respectively. Table 1 summarizes the economics.

Mitsubishi's lime-gypsum process

A semicommercial unit treating 60,000 scfm of flue gas is now on stream at Kansai Electric Power Co.'s Amagasaki-higashi power station in Japan. It uses Mitsubishi Heavy Industries' newly developed lime-gypsum desulfurization process. Aim is to remove 90% of the SO_2 from a flue gas containing 0.11 vol%, resulting in production of 19 tons of gypsum daily.

The wet system uses lime slurry and gypsum is recovered as by-product through air oxidation of calcium sulfite formed during an absorption process.

Pilot-plant tests were conducted at the power station, continuously treating 3,000 scfm of flue gas from an oil-fired boiler. A 37,000-scfm unit using essentially the same process has been operating successfully for 6 years treating waste gas from a sulfuric-acid plant at Nippon Kokan Co.'s

Fig. 3

Mitsubishi lime-gypsum flue-gas desulfurization process

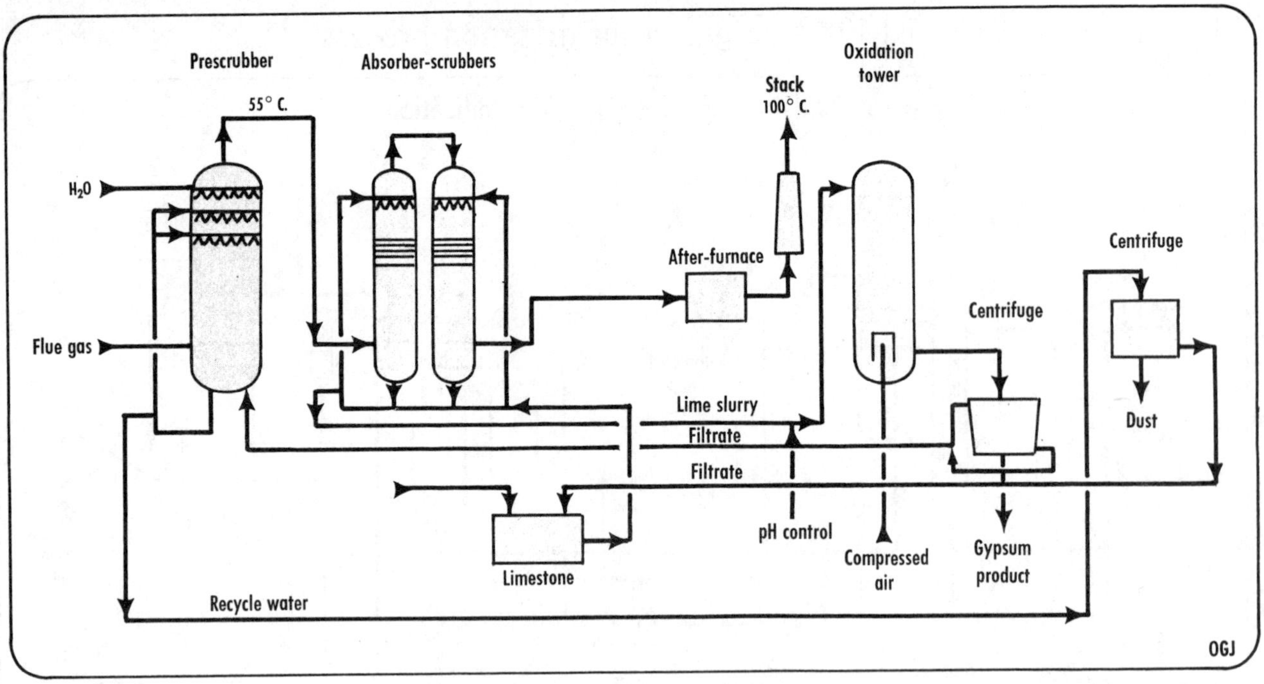

Koyasu works. A second pilot-plant study which continuously treated 1,000 scfm flue gas from an oil-fired boiler was used by Mitsubishi to study the absorption reaction and to develop techniques to prevent scale formation.

Main reaction of SO_2 absorption produces calcium sulfite, but some calcium sulfate is also produced depending upon SO_3 and O_2 contents of the flue gas and operating conditions of the absorber.

How it works. SO_2-rich flue gas is water scrubbed and cooled prior to being sent to the absorption section. Effluent water from the scrubbing tower is centrifuged to remove dust and impurities after neutralization with slaked lime. The clean water is recycled to the lime tanks for slaking, Fig. 3.

The absorption reaction in the absorber section is controlled mainly by the dissolving rate of lime.

Testing revealed that higher pH plus higher liquid-gas ratio increases absorption efficiency. On the other hand, lower pH improves the conversion ratio of lime to calcium sulfite or sulfate (the Ca reactivity). Mitsubishi accommodates this conflict by using two absorption scrubbers in series.

One of the scrubbers operates at low pH with the scrubbing liquor flowing countercurrent against the gas stream. In the second, high-pH scrubbing liquor flows cocurrently with the SO_2-rich gas.

Scrubbing liquor is firstly introduced into the second absorption scrubber and the effluent is then sent to the first absorption scrubber. The volume of each scrubber necessary for 90% SO_2 removal and 98% Ca reactivity is related with the pH value of the second absorption scrubber.

For operational economics, the optimum is obtained at pH 9. However, pH of the scrubbing liquor is held slightly lower to prevent formation of calcium carbonate in the scrubber when the flue gas has a high CO_2 content.

SO_2 contained in the gas is removed in absorption towers equipped with plastic grid packings. This fill can be operated at a higher superficial gas velocity and lower draft-pressure loss without scale formation. Between 90 and 98% SO_2 removal from a flue gas containing 0.1 vol% sulfur dioxide was attained at a superficial gas velocity of about 1.2 x 10^4 cu m/sq m/hr and a total draft-pressure loss of about 90 mm of water. Scrubbing liquor flow rate was 70 cu m/sq m/hr.

A strip down of the unit after 500 hr continuous operation revealed no signs of scale formation or plugging.

SO_2-free gas taken off the lower side of the second scrubbing tower is heated in an after-furnace before being vented to atmosphere.

Oxidation. Lime slurry (calcium sulfite), at 50° to 80° C. and containing dissolved sulfur oxides, is taken off the side of the first absorption-scrubber tower. Before entering the top of

Table 2

Typical operation data Showa-Denko SO_2-removal process

	Case 1	Case 2
Stack-gas flow rate, scfm	43,000	21,000
Absorbent circulation rate, g/hr	30,000	16,000
Absorption tower inlet, mmAg	300	180
Gas temperature at inlet, °F.	327	320
Gas temperature at outlet, °F.	131	128
Ambient temperature, °F.	40	38
SO_2 concentration at inlet, ppm	1,490	1,012
SO_2 concentration at outlet, ppm	7	3
Liquid temperature, °F.	137	131
Recovered liquid concentration, %	21.84	22.07

Fig. 4

Showa-Denko SO_2-removal and gypsum-recovery process

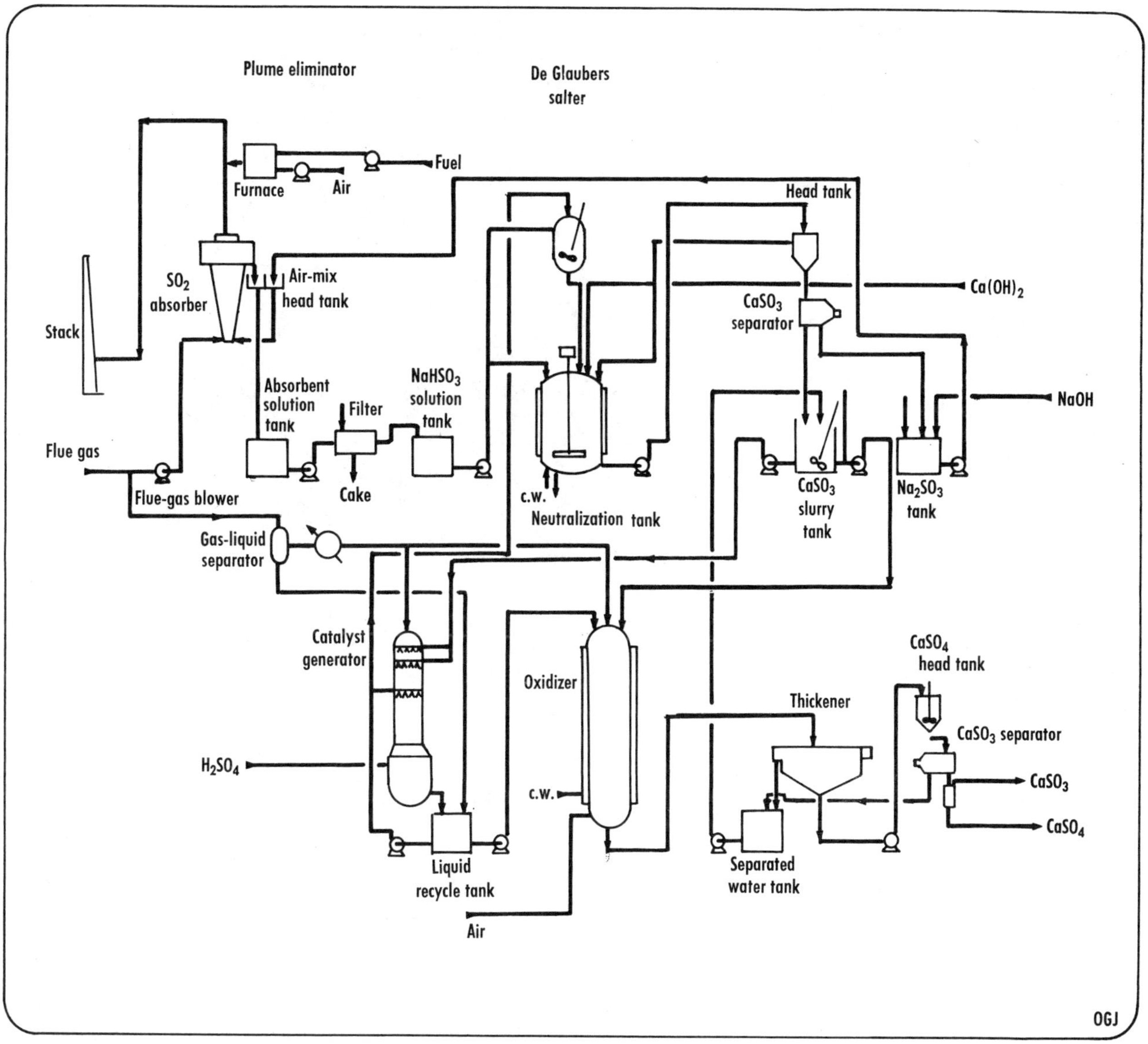

the oxidation tower, pH of the slurry is adjusted to between 4 and 4.5. Compressor air 4 - 5 atm for oxidation is introduced into the bottom of the tower through a rotating-cylinder-type air atomizer.

Seed crystals added to the scrubbing liquor promote the growth of gypsum crystals.

In a downstream filtering section gypsum and mother liquor are separated out of the slurry. Gypsum is sent to storage while the mother liquor is recycled to the precooling tower.

Showa-Denko sodium-sulfite system

Now successfully operating in Japan are three Showa-Denko process plants for gas desulfurization. Two others are under construction, one for 295,000 scfm and another for 160,000 scfm of SO_2-rich stack gas.

The process is a wet regenerative system in which SO_2-rich flue gas is contacted with a sodium sulfite solution. The resulting sodium bisulfite can be converted to sodium sulfite, purified, and crystallized to a high purity product. Or, it can be neutralized prior to catalytic oxidation to gypsum product.

SO_2-rich gas is blown into an absorption tower, Fig. 4, where it is scrubbed with a 20% solution of sodium sulfite. SO_2 and dust are absorbed from the gas as this passes upwards through the suspended layers of liquid particles.

A vertical upward-flow Venturi-type absorber was chosen for the absorption step. It has a high gas-liquid contact efficiency that is capable of absorbing more than 95% of the SO_2 contained in the flue gas. A gas-liquid separator is mounted on top of the absorption tower.

From 0.1 to 0.2 g/cu m of unburnt carbon and metallic fly ash may be present in the stack gases. These are separated to about 80% out of the gas with the SO_2 and pass with the sodium sulfite solution out of the separator back into the solution recycle tank.

Gas enters the bottom of the absorption tower at 320° and 350° F. Its velocity at the throat is about 115 ft/sec. Dependent on gas velocity in the throat, pressure losses in the absorber section range from 150 to 250-mm of water. Turndown range of the tower is given as 40 to 100%.

Since the SO_2-free stack gas from the top of the separator is saturated, it is passed through a preheater prior to venting to eliminate steam plumes.

The SO_2-rich sodium bisulfite solution recovered from the bottom of the separator is returned to the sodium-sulfite recycle tank where its pH value and density are controlled by the addition of water and a caustic makeup solution.

Part of the sodium bisulfite solution is withdrawn continuously from the recycle tank and sent to a purifier where the entrapped dust is recovered by sedimentation and filtration.

When sodium sulfite is to be recovered as product from the process, the 20% solution taken off the bottom of the purifier is passed to an evaporation-crystallizer. Water is evaporated from the solution, SO_2 is released, and sodium sulfite crystals precipitate out.

The SO_2 released during evaporation can be converted to elemental sulfur or sulfuric acid.

Valuable by-product gypsum can be obtained from the sodium bisulfite solution by adding $Ca(OH)_2$ and separating out sodium sulfite and calcium sulfite. The sodium sulfite is recycled to the SO_2 absorber while the calcium sulfite is catalytically oxidized to gypsum.

Process economics. For the SO_2-removal section alone, without downstream crystallization or gypsum conversion plant, Showa-Denko estimates a cost of $2 to $3 to reduce the sulfur content of 1 ton of heavy oil from 2.7 to 0.15%.

While credit for the sulfur recovered will be negligible, process costs depend on the prices of sodium sulfite and gypsum. Both of these are currently in high demand in Japan and prices are buoyant.

Since caustic soda consumption as makeup solution is relatively high, its price will also affect SO_2-removal costs.

Approximate utilities consumption for treating the stack gas of a 266 ton/day oil-fired boiler (delivering 90,000 scfm stack gas) are:

Water, 3,435 g/hr.

Cooling water, 2,642 g/hr.

Electric power, 350 kw.

Labor requirements are given as 0.5 men/shift. Sodium sulfite recovery is given as 2,244 lb/hr in a 20% solution. Caustic soda as makeup solution is consumed at a rate of 1,700 lb/hr in a 48% solution.

EACH of these two absorption towers treats 44,000 scfm of oil-fired-boiler flue gas, using Showa-Denko's new process. Single-tower designs are also available. Fig. 5.

Utilities consumption for a plant treating 176,000 scfm stack gas containing 1,785 ppm SO_2 and producing 9,000 lb/hr gypsum from the recovered sodium sulfite solution is given as:

Steam, 3,750 lb/hr.

Process water, 13,000 g/hr.

Cooling water, 190,000 g/hr.

Electric power, 1,430 kw.

Caustic consumption is 185 lb/hr. The plant will also require 780 lb/hr sulfuric acid plus 4,500 lb/hr lime at 93%.

Where its working. The three installations already operating with the Showa-Denko SO_2 removal process are:

1. Showa-Denko, Kawasaki: oil-fired-boiler stack gas, 88-Mscfm; inlet SO_2, 500-1,400 ppm; outlet SO_2, 50 ppm.

2. Nippon Rinsan, Chiba: sulfuric-acid tail gas, 47-Mscfm; inlet SO_2, 1,000 ppm; outlet SO_2, 50 ppm.

3. Ajinomoto, Kawasaki: oil-fired-boiler stack gas, 160-Mscfm; inlet SO_2, 1,250 ppm; outlet SO_2, 50 ppm. END